Transitioning to Autonomy

Whenever automation is introduced to control real-time activities or processes, the role of the human changes from being a *manual* controller to being a *supervisory* controller. Whether the activity is the control of vehicles industrial processes, or is in defence, healthcare, or elsewhere, the work performed by the people who are expected to monitor and supervise the automation places new demands on their attention, perception and cognition. Those demands can be significant and challenging and this book aims to address that.

Transitioning to Autonomy: The Psychology of Human Supervisory Control focuses on the transition period when automation is being introduced, and the human needs to learn and develop the competence to perform their new role effectively. The first Part extracts general lessons from the author's experience taking ownership of a new car which, under certain circumstances, was capable of driving autonomously. Part 2 explores the psychology behind the lessons extracted in Part 1 and proposes a comprehensive model of human supervisory control. The final Part focuses on six principal risks associated with human supervisory control and examines how the expectation that people will be proactive in monitoring for threats to the automation's performance is often relied on as a defence, or "Barrier", against serious adverse events. The core benefit for the reader is a deeper understanding of what it takes, cognitively, emotionally, and organisationally, to ensure safe and effective human oversight in the age of automation. It aims to give the reader the lowdown on delivering safer systems.

The book is for managers, engineers, safety professionals and those from other technical disciplines who have responsibility for the design, development and/or assurance of products that automate the control of real-time activities; it's for regulators and others responsible for setting policy and ensuring products automating real-time activities are safe; and it's for Human Factors and other professionals who need to understand and develop competence in aspects of the psychology associated with automated systems.

Transitioning to Autonomy

The Psychology of Human Supervisory Control

Ronald W. McLeod

CRC Press is an imprint of the
Taylor & Francis Group, an **informa** business

First edition published 2026
by CRC Press
2385 NW Executive Center Drive, Suite 320, Boca Raton FL 33431

and by CRC Press
4 Park Square, Milton Park, Abingdon, Oxon, OX14 4RN

CRC Press is an imprint of Taylor & Francis Group, LLC

© 2026 Ronald W. McLeod,

Reasonable efforts have been made to publish reliable data and information, but the author and publisher cannot assume responsibility for the validity of all materials or the consequences of their use. The authors and publishers have attempted to trace the copyright holders of all material reproduced in this publication and apologize to copyright holders if permission to publish in this form has not been obtained. If any copyright material has not been acknowledged please write and let us know so we may rectify in any future reprint.

Except as permitted under U.S. Copyright Law, no part of this book may be reprinted, reproduced, transmitted, or utilized in any form by any electronic, mechanical, or other means, now known or hereafter invented, including photocopying, microfilming, and recording, or in any information storage or retrieval system, without written permission from the publishers.

For permission to photocopy or use material electronically from this work, access www.copyright.com or contact the Copyright Clearance Center, Inc. (CCC), 222 Rosewood Drive, Danvers, MA 01923, 978-750-8400. For works that are not available on CCC please contact mpkbookspermissions@tandf.co.uk

For Product Safety Concerns and Information please contact our EU representative GPSR@taylorandfrancis.com. Taylor & Francis Verlag GmbH, Kaufingerstraße 24, 80331 München, Germany.

Trademark notice: Product or corporate names may be trademarks or registered trademarks and are used only for identification and explanation without intent to infringe.

ISBN: 978-1-041-15446-4 (hbk)
ISBN: 978-1-041-15005-3 (pbk)
ISBN: 978-1-003-67947-9 (ebk)

DOI: 10.1201/9781003679479

Typeset in Times
by codeMantra

I ain't dead yet, my bell still rings.

Bob Dylan, "Early Roman Kings"

For Kath, Fraser and Ross.

Contents

PART 3 Assuring the reliability of human supervisory control

Foreword

The widespread use of automation and artificial intelligence is a significant challenge facing pilots, drivers, soldiers, health care providers, and operators in many high-risk, high- consequence domains. While these systems make many routine tasks faster and easier to perform, they conversely can be brittle and error prone leading to new problems for safe effective performance.

A long history of people working to overcome these challenges shows that implementing automation is not as simple or straightforward as many might think. Automation changes the very nature of the job of the people who are called upon to oversee it, leaving them to become system monitors and supervisory controllers, with a predictable loss of the situation awareness that is so critical for performance.

McLeod provides an overview of the many challenges for human operators who are called upon to act as supervisors of automated systems, drawing upon his extensive professional experiences as well as the extant research base. This book is specifically aimed at engineers, managers and others who want to understand how people interact with automation and factors they need to consider to make their projects successful. He works to explore the expectations system developers often have of human operators and reveal their many limitations. He concludes with a practical tool for helping designers think through their expectations of what people overseeing automation can reasonably do.

With the increased emphasis on artificial intelligence and plans to implement it for so many different tasks in industries across the board, this is a timely volume that will be highly useful for many.

Mica Endsley

Acknowledgements

I owe a debt of gratitude to many people for helping me complete this book. In some cases, that help was provided explicitly, by reviewing or commenting on sections of the text as they developed, or by acting as a sounding board for some of my ideas. In others, the help was implicit, through things they have said, done or written at some point over the nearly five decades of my education and career: things that have stayed with me and influenced my interest in and thinking about human supervisory control.

In the first group, my old friend from our days as undergraduate Psychology students, Allan Mackenzie, despite having spent a career a long way from the world of Human Factors or Engineering Psychology, was heroic enough to act as a "lay reader" of the entire first draft. As well as making many suggestions to improve the clarity of my writing, his comments frequently made me laugh.

Three anonymous reviewers gave their time to provide comment on my book proposal, including answering questions which, while of value to a publisher, can have been of little interest to them. Without the willingness of dedicated professionals like them to give of their time in such thankless endeavours, our scientific and technological world would be greatly diminished.

Martin Anderson was kind enough to conduct a detailed review of two draft chapters greatly improving them. My friend and former colleague Johan Hendrikse reviewed a few chapters and made insightful comments. His golfing buddy, and vastly experienced instrumentation engineer, Greg Melrose reviewed my models of supervisory control and provided valuable insight, as well as a little history, from an instrumentation engineers perspective. And I am grateful to Niav Hughes Green, Jenny O'Donnell and Alistair Cowin for volunteering to review the final draft and providing such positive and constructive comment.

As well as kindly agreeing to provide a Foreword, Mica Endsley, someone whose work I have admired - and drawn on – for many years, was generous enough to provide background on some of the research in the area. She also made a number of suggestions of relevant publications from her vast catalogue I wasn't aware of. And, despite their heavy workload and responsibilities working at the front line of healthcare, Dr Wendy Russell and Dr Calum Nicholson from University Hospital, Crosshouse, were generous in helping me understand the roles and processes involved in brain scanning in Scotland.

The second group includes more people who have influenced my career and thinking about Human Factors in general, and human supervisory control in particular, than I could possibly identify. But I acknowledge, and appreciate, the debt I owe them all.

The book would never have been completed if it had not been for the enthusiastic response I received to some of the material from students on the combined Glasgow University and Glasgow School of Art Product Design Engineering programmes. In 2024, at a time when I had nearly convinced myself that there would be little interest in what I had to say, I included some of the early material in the annual lecture

I deliver as part of their course. The enthusiasm and interest shown by the students, and the response to the ideas I expressed, reinvigorated my motivation to return to the task and complete the work.

My family – my wife Kath, my sons, Fraser and Ross, and their partners (and dog) – provided the love and support that sustained the unexpectedly long gestation of this work. I thank them for never laughing when I told them, yet again, that it was "nearly finished". It is now.

Tayvallich

July 2025.

Prologue

I'm not someone who believes in fate or predetermination: though sometimes events do make you wonder. Three events happened to me in 2021 and 2022 – two of them on the same day - that seem somehow to have been more than simple coincidence.

By mid 2021, as the restrictions imposed by the COVID pandemic were slowly being lifted and having found that I had rather enjoyed a quiet life at home, I decided to draw a line under my career, take my pension, and retire. Though on the very day my first pension payment hit my bank account, I received a completely unexpected invitation from a company less than ten miles from my home (a branch of a global corporation) to work on a project part-time for six months. The aim was to update the user interface of a near obsolete system that needed to remain operational for another few years. I had worked on the original design more than twenty years before. In the intervening time I'd had no contact whatever with the company: indeed, I had hardly worked in Scotland, where I live, at all over that period. So this was an unusual and unexpected opportunity right on my doorstep.

Having agreed to work on the project, I decided that, once it was finished, and the fees were in, I would treat myself to an electric car. So it was that, on the morning of 10th April 2022, I signed an agreement to buy a new car: my first electric vehicle.[1] And in the afternoon, I travelled to the annual conference of the Chartered Institute of Ergonomics & Human Factors (CIEHF) to launch a new White Paper on the subject of 'Human Factors in Highly Automated Systems' [1]. It was more than six months later that the curious relationship between these events became clear.

The White Paper was the result of more than 18 months effort. I had led a team of specialist volunteers to prepare a summary of Human Factors challenges associated with highly automated systems. We wanted to raise awareness of key principles that could be used to guide thinking about Human Factors issues in the development of these kinds of systems.

I had embarked on the project with little knowledge of the Human Factors issues associated with autonomous vehicles. Though in the course of the work I learned a great deal from a range of individuals, academic researchers as well as those directly working for vehicle manufacturers. I was introduced to some of the work addressing those issues in the development of self-driving cars. I had also become aware of what had impressed me as a significant body of high-quality thinking and applied research around the issues. And I had taken a growing interest both in the increasing volume of marketing associated with these new products, as well as the growing body of media coverage, including reports of accidents and the response of government, regulators, and industry to them.

At no point had I actively considered myself as a future owner of a self-driving vehicle. When I paid the deposit on my new car, I did not realise I had committed to buying a vehicle that, under some circumstances, could drive itself. It was certainly not something I had intended. Rather, I had been focused on the electric feature. I have no doubt the marketing material, as well as the salesman's efforts to sell me the car, will have featured the driver assistance capabilities. But these were not my priorities; they had no impact on my purchasing decision. If anything, based on what I had learned in preparing the CIEHF White Paper, I was inclined towards scepticism that the world was ready for the widespread adoption of the technology.

"There should be no automation surprises" is one of the nine principles in the CIEHF white paper. Yet here I was, more than surprised when, in mid-November 2022, the salesman completed his short briefing on the features and abilities of my new car (being able to open the boot with my foot featured prominently), handed me the keys, and let me loose into the rush-hour traffic in central Glasgow. Having started out with full confidence that I was simply driving another car, within five minutes of leaving the showroom, I found myself on a busy motorway, confused and wondering if I'd made a big mistake. The three chapters in Part 1 of the book recount and reflect on my experiences transitioning to being the owner of a car that had the ability to drive itself.

MOTIVATION

My immediate motivation for writing the book lies in the sense of deep frustration I experienced during the transition I had to make to become familiar and confident in the capabilities of my new car. Frustration allied with a desire to make one more attempt, before riding off into the sunset of full retirement, to contribute to encourage those who invest in the development of advanced technological systems, to treat the psychological dimension of how human beings interact with the products and systems they develop as seriously as they take the hardware and software engineering elements.

That sense of frustration arose from my academic training in Psychology and Ergonomics and a career of more than 40 years working as a Human Factors professional. It was clear to me that the design of my new car, as well as the way I had been allowed to take control of it, had ignored many of the basic tenets of psychology and human performance that I had been taught and understood from the 1970s, and that had been the cornerstone of my professional life since.

My experience did not, of course, occur in a vacuum. Most significantly it followed closely on the heels of the publicity surrounding the emerging findings of the investigations into the crashes of the Boeing 737 MAX aircraft into the Java Sea in Indonesia in 2018 and then again in Ethiopia in 2019.[2] Though worlds apart in their consequences and impact, I saw clear similarities between what I was learning about those incidents and my experience with my new car. It also followed a regular flow of media reports on incidents, up to and including fatalities associated with the use of self-driving vehicles.

To my mind those two air tragedies and other incidents derived from the same sources as the frustration I experienced taking ownership of my new car. Either;

1. A seeming inability among organisations responsible for the development and operation of highly automated and autonomous systems to recognise and take seriously how the new systems were going to change the demands placed on the humans they would continue to rely on to monitor, support and use their systems. Or;
2. An unwillingness to invest the effort and resource needed into ensuring the risks associated with those new demands were properly understood and addressed during the design, implementation, and operation of those systems.

These both reside in corporate culture, thinking and decision making, rather than in the minds of any individual engineer or manager.

Those initial motivations were reinforced by two further experiences. I am occasionally invited to act as a reviewer of articles submitted for publication in peer reviewed scientific journals. During the initial stages of writing the book I received two invitations to review papers, both of which reinforced the frustrations described above.[3]

The first was a review paper written by representatives of a National Regulator in the oil and gas industry. Based on a series of audit findings, the paper reviewed how companies active within their national borders were managing the Human Factors issues associated with the introduction of autonomous systems in their drilling and oil and gas production activities. The findings they reported were shocking: essentially, no special effort was being made to identify or address Human Factors risks associated with introducing the new technologies. Systems were being introduced based purely on economic considerations, with little or no assessment of Human Factors risk. Operators were usually given no training, and systems were being introduced against a background of continuing full production.

The second paper was a report on a literature review performed by a national technical institute to identify Human Factors principles that should be applied in the design of automation and remote systems for use in the petroleum sector.

Reviewing the two papers, it was clear that both sets of authors showed little recognition of how introducing automation can fundamentally change the role of the people involved. Neither showed any understanding of how the human's role changes from being manually in control of operations, to becoming supervisory controllers. They suggested little awareness that the kind of Human Factors analysis and system development activities regulators expect and that have become routine in developing many conventional systems would not be effective for highly automated systems.

For example, the authors showed no awareness that the kind of approach needed to demonstrate that the risk of human error had been reduced to a level that could be demonstrated to be As Low As Reasonably Practicable (ALARP)[4] for operators conducting manual drilling operations will not be effective when operators are tasked with supervising an autonomous drilling system.

Coming on the back of my experience taking ownership of my new car, this lack of awareness among authors who were clearly committed to trying to help industry get these systems right, contributed to my feeling that perhaps I needed to publish my own experiences and opinions in the form of this book.

UNREALISTIC EXPECTATIONS OF HUMAN CAPABILITIES

One other experience that had been niggling at me for some years contributed to my motivation to write this book. A few years ago, I had been invited to support an investigation into a major incident that had led to multiple workers losing their lives. Inevitably, the investigation was complex and fraught with legal challenges. The investigating authority were reluctant to accept what I felt then (and still believe today) was the most important and far-reaching Human Factors conclusion. The reluctance of the authority to accept my recommendation was because they felt my concerns relied too much on subjective judgement about what had motivated certain behaviours at the worksite in the period immediately before the incident. They felt we did not have the hard facts needed to stand up to the legal scrutiny the conclusion would inevitably have received. Consequently, the learning – which I continue to believe is not only important and far reaching but fully justified by the evidence – remains hidden in files.[5]

In numerous documents, the organisations involved in the incident (mainly specialist contracting companies) expressed great confidence not only in their commitment to safety, but in their safety records and the robustness of their safety management systems. What struck me in reviewing this material, was just how unrealistic the company's expectations were of the psychological abilities of their workforce. As one example, at the heart of the processes one company relied on to ensure the safety of their real-time operations, was something they referred to as "Dynamic Risk Assessment".

The company described Dynamic Risk Assessment as a continuous, intentional process of assessing hazards to identify and mitigate risk. They talked of "Cognitive Risk Assessment" as being the basic level of practical thinking and judgement that their employees use to assess and mitigate risk. Once a job was in process therefore, the company expected that their employees, however busy they were, and whatever else they were engaged in at the time, would be intentional in continually assessing the current risk to the safe outcome of each job. The safety management system relied on the workforce performing these "cognitive risk assessments" continuously, in real-time, throughout the performance of their work.

Human beings are not capable of consciously paying continuous attention to any details of their environment or task performance for more than relatively short periods of time: indeed, if such an expectation were to be realised, the demands on cognitive processes such as attention and working memory would be such that people would be incapable of performing even moderately complex tasks. The expectation that, on top of performing physically and cognitively demanding work, often for 12-hour shifts and in challenging conditions, workers can be "intentional" in "continuously assessing" risk is unrealistic and inconsistent with a large body of knowledge about how humans perceive the world around them, think, make judgements and assess risk.

Clearly, the ability of people to maintain attention and perform cognitive tasks over sustained periods varies widely. Some people – astronauts, test pilots, some surgeons, Chess Grand Masters and the like – seem to be at the higher end of these abilities. The ability of the astronaut Neil Armstrong to remain calm and focused while making the mentally demanding decision where to land Apollo 11's Lunar Module under perhaps the most extreme pressure any human being has ever faced, remains truly extraordinary by any measure.[6] Though whether even such superhumans as these would be genuinely capable of performing "dynamic risk assessment" over a busy 12-hour working shift seems unlikely. More to the point, the kind of companies I was reviewing don't come near to employing those kinds of people.

The concept of "Dynamic Risk Assessment" as it was defined and relied on by the company in the incident that cost multiple lives, is similar to what the automotive industry, as well as governments, insurers and others appear to be relying on to assure the safety of autonomous vehicles. That somehow the human in the driving seat will be not only capable but will continuously and dynamically assess the performance of the vehicles and the state of the world around them and make real-time judgements about whether there is a need to take control.

In their joint definition of terms related to driving automation J3016, the Society of Automotive Engineers and the International Standards Organisation describe this in terms of the "Receptivity [of the user]". This is defined as; "An aspect of consciousness characterized by a person's ability to reliably and appropriately focus his/her attention in response to a stimulus" [2, para 3.22]. In a note, the definition clarifies that; *"In Level 3 driving automation, a DDT[7] fallback-ready user is considered to be receptive to a request to intervene and/or to an evident vehicle system failure, whether or not the ADS[8] issues a request to intervene as a result of such a vehicle system failure."*

The clarifying "...whether or the not the ADS issues a request to intervene..." is the essence of the issue. It is simply not realistic, whatever the SAE and ISO believe, to expect human beings to be continuously "receptive" all the time, over extended periods. Especially if they have little mental engagement with the system, they are relied on to attend to.

WHO AM I WRITING FOR?

Over the second half of my career, I spent a great deal of time developing and delivering training on aspects of Human Factors to managers, engineers, operations people and other professionals. Most, though far from all, were involved in large capital projects in manufacturing industry. Many were in operational roles. I also worked and collaborated with a lot of people who had been appointed to the role of company Human Factors specialists. While many of these were qualified professionals, generally accredited to either the Chartered Institute of Ergonomics & Human Factors (CIEHF), the Human Factors and Ergonomics Association (HFEA), or the International Ergonomics Association (IEA), many were not.

Even among those qualified professionals, Human Factors is such a broad discipline that, for many, their training and expertise lay in areas such as biomechanics or physical and environmental ergonomics, rather the cognitive and applied psychology that I had been grounded in. Those in Human Factors roles who lacked professional qualifications were generally from a background in either engineering or operations and were in the process of transitioning to develop competence in Human Factors based on an enthusiasm and deep interest in the subject. While some of these clearly under-estimated the depth and complexity of Human Factors as a discipline, in the main, their thirst to learn, and most especially to learn about the psychological basis of human performance, regularly impressed me. Many of what I came, through experience, to consider as the most capable Human Factors practitioners have come from this latter group.

Based on that experience, I have a clear idea in my head of the kind of people I have been trying to "speak" to in writing this book. They fall into three general groups;

1. Managers, engineers, safety professionals and those from other technical disciplines who have responsibility for the design, development and/or assurance of products and systems that seek to automate the control of real-time processes. Readers in this group need to be aware of the impact introducing automation to control real-time processes can have on the people who need to work with and support it. They also need an understanding of the risks they need to mitigate through the way the products and systems they contribute to are designed, verified, implemented and supported.
2. Regulators and others responsible for setting policy and ensuring products and systems intended to automate real-time control activities, whether in controlled industrial or commercial settings or by the public, are safe and can be used without exposing their users or others to significant risk of harm or loss. Readers in this group need to understand the risks that can arise when activities previously controlled manually become automated. And they need to understand the kind of questions and challenges those responsible for developing and profiting from those products and systems should be able to answer to give confidence that they have acted responsibly in recognising and addressing issues of human supervisory control in the design, implementation and support of their products.
3. Human Factors professionals who don't come from a psychological background, but who, in filling their professional responsibilities, need to understand and develop competence in aspects of the psychology associated with complex automated control systems. That is a large and increasingly complex topic. This book will only serve to help develop that understanding in one part of the problem space – when people need to fill the role of a human supervisory controller of a highly automated, perhaps autonomous, product or system. A space that, currently, is filled mainly by publications intended for a scientific, and largely academic, audience.

Unless there was a clear and obvious need to acknowledge the work of others, I have not thought it necessary to fully reference my opinions. Though I hope I have not failed to acknowledge someone else's original work or ideas where I have drawn on them. If I have, I apologise – it was not intentional.

The book is not written for an audience of academics or other professional research scientists. The content and any opinions expressed are entirely my responsibility. Though I hope it might be useful to those, such as postgraduate students and research assistants, who may need to be aware of and understand sufficient about the topic of human supervisory control, to be able to integrate some of the ideas into their research. Wherever I have thought is useful to a reader who wants to learn more about a topic, I have tried to include references to background reading and sources. The purpose has been to help those who wish to go deeper into the topic, rather than trying to justify the content.

Finally, If I give the impression I know anything about how automation works, I apologise. I don't. I did spend a significant part of my early career working in teams on large software intensive systems that included the design and development of user interfaces to many automated functions. That experience, however, pre-dates the kind of sophisticated and increasingly Artificial Intelligence-based systems that are coming to dominate most marketplaces.[9] But I like to tell myself that I do know something about how and why such sophisticated automation can create problems for the humans who are expected to use, monitor and support it. I also understand the kind of questions that should be asked, and the analysis and verification processes that need to be carried out throughout the conception, development and introduction of those systems if those problems are to be avoided. I hope, if you manage to read at least part of the book, that you might agree.

NOTES

1 Actually, a plug-in hybrid-electric.
2 Learnings about human supervisory control from these crashes are discussed in Chapter 6.
3 At the time of writing, neither of these articles has yet been published.
4 Some countries and industries use the nearly identical concept of As Low As Reasonably Achievable (ALARA). See for example the US Centre for Disease Control and Prevention: https://www.cdc.gov/radiation-health/safety/alara.html.
5 The issue I was concerned about was where an individual who was a stakeholder and director of a private company, having responsibility for the financial performance of the company, simultaneously held the role of safety manager responsible for a highly hazardous operation. The potential for conflict of interest between the goals of financial performance of the company and the safety of the operation are clear. From my reading of the evidence, it seemed sufficiently likely that exactly such a conflict had influenced at least one individual's priorities and attention in the events preceding the incident that it needed to be identified in the investigation. There are similarities with the position Stockton Rush was in as both CEO of OceanGate and head of the technical team responsible for the safety of the design of the TITAN submersible before he was one of the five people who died when the craft imploded during an attempt to visit the wreck of the RMS Titanic on 18 June 2023.
6 For one of the most detailed descriptions of what was involved in this task, see Chapter 9 in David Mindell's excellent book 'Digital Apollo' [2].
7 DDT is the "Dynamic Driving Task. All the real-time operational and tactical functions required to operate a vehicle in on-road traffic..."

8 ADS is the Automated Driving System … the hardware and software that are collectively capable of performing the entire DDT on a sustained basis…"

9 I recall - it must have been around 1988 - working with John Harrison to understand the user interface principles behind the design of the Apple Mac computer to support the design of the interface for a new real-time control system we were working on. John is one of the most naturally gifted Human Factors engineers I ever worked with, possessing an insight and intuition into how users are likely to experience a user interface that seemed to come naturally to him.

REFERENCES

1. CIEHF. Human Factors in Highly Automated Systems. Chartered Institute of Ergonomics & Human Factors. 2012. Available from: https://ergonomics.org.uk/resource/human-factors-in-highly-automated-systems-white-paper.html
2. SAE. Surface Vehicle Recommended Practice: Taxonomy and Definitions for Terms Related to Driving Automation Systems for On-Road Motor Vehicles. J3016. SAE International. 2021.

About the author

Ronald W. McLeod holds an Honours degree in Psychology, an MSc in Ergonomics and a PhD in Engineering and Applied Science and has spent more than 40 years as a Human Factors Specialist working primarily in safety critical and high hazard industries.

Ron has been influential by helping industry learn, develop, and apply best practice in Human Factors in the management of major hazards. As well as his 2015 book, *Designing for Human reliability: Human Factors Engineering in the Oil, Gas and Process Industries,* Ron has published more than 30 papers in the technical and scientific press in his own name, as well as contributing to numerous industry guides and best practices produced by a variety of professional and industry bodies. He has taken an active, and often a leadership, role in organisations including the International Oil and Gas Producers Association (IOGP), the Society of Petroleum Engineers (SPE), the UK's Process Safety Advisory Committee (PSAC), the Energy Institute (EI), the Chartered Institute of Ergonomics and Human Factors (CIEHF) and the Centre for Chemical Process Safety (CCPS).

Ron holds a position as Honorary Professor of Engineering Psychology at Heriot-Watt University, Edinburgh, and has been Visiting Professor of Human Factors at Loughborough University, School of Design. He is a Fellow, and has been a Trustee and Member of Council, of the UK's Chartered Institute of Ergonomics & Human Factors (CIEHF) and has served as a member of a peer review panel for the US National Academy of Sciences, Engineering, and Medicine.

In 2020, Ron was awarded the CIEHF's Lifetime Achievement Award, and in 2022 he was made a Fellow of the International Ergonomics Association in recognition of his *"...outstanding and sustained performance in the field of ergonomics and human factors at an international level..."*. He retired from full-time working in 2022, but remains active lecturing and writing.

List of acronyms

A2CR	Authority, Ability, Control and Responsibility
ACC	Adaptive Cruise Control
ADS	Automated Driving System
AELTC	All England Lawn Tennis Club
AHARP	As High As Reasonably Practicable
AI	Artifical Intelligence
ALARA	As Low As Reasonably Achievable
ALARP	As Low As Reasonably Practicable
ANSI	American National Standards Institute
AOA	Angle-of-Attack
ATG	Automatic Tank Gauging
ATM	Automated Teller Machines
AV	Automated Vehicle
BOEM	Bureau of Ocean Energy Management
BSEE	Bureau of Safety and Environmental Enforcement
BTA	Bowtie Analysis
CA	Competent Authority
CAT	Computerized Axial Tomography
CCPS	Centre for Chemical Process Safety
CIEHF	Chartered Institute of Ergonomics & Human Factors
CO	Commanding Officer
COMAH	Control of Major Accident Hazards
CSB	US Chemical Safety Board
d'	("D prime") Indication of signal strength in TSD
DDT	Dynamic Driving Task
DOA	Degree of Automation
DSS	Driver Support System
EEG	Electroencephalography
EI	Energy Institute
ELC	Electronic Line Calling
FCOM	Flight Crew Operations Manual
fMRI	Functional Magnetic Resonance Imaging
FO	First Officer
GEMS	Generic Error Modelling System
GIGO	Garbage-In-Garbage-Out
HASO	Human-Autonomy Oversight Model
HFEA	Human Factors and Ergonomics Association
HF-IRADS	Human Factors in International Regulations for Automated Driving Systems
HIS	Human Interface System
HTA	Hierarchical Task Analysis
IAS	Indicated Airspeed

IDHEAS	Integrated Decision-Tree Human Error Analysis System
IEA	International Ergonomics Association
IHLS	Independent High Level Switch
INS	Inertial Navigation System
ISO	International Standards Organisation
ITCZ	Inter-Tropical Convergence Zone
KAL	Korean Airlines
KISS	Keep It Simple Stupid
LCA	Lane Change Assist
LOCA	Loss-of-Coolant-Accident
LTA	Lane Tracing Assist
MAIB	Marine Accident Investigation Branch
MCAS	Manoeuvring Characteristics Augmentation System
Moc	Management of Change
MRT	Multiple Resource Theory
NASA	National Aeronautics and Space Administration
NES	NHS Education for Scotland
NUREG	US Nuclear Regulatory Commission
ODD	Operational Design Domain
OOTL	Out-of-the-Loop
OOW	Officer of the Watch
PID	Proportional, Integral and Derivative
PMQsc	Proactive Operator Monitoring (Supervisory Control) Score
POM	Proactive Operator Monitoring
POMAT	Proactive Operator Monitoring Assessment Tool
PRA	Probabilistic Risk Assessment
PWR	Pressurised Water Reactor
QRH	Quick Reference Handbook
ROC	Receiver Operating Characteristics
RSA	Road Sign Assist
SA	Situation Awareness
SAE	Society of Automotive Engineers
SFAIRP	So Far As Is Reasonably Practicable
SPE	Society of Petroleum Engineers
TEM	Threat and Error Management
TIS	Task Interactive System
TLX	NASA's Task Load Index
TSD	Theory of Signal Detection
UNECE	United Nations Economic Commission for Europe
β	('Beta') Indication of subjective bias in TSD

1 Introduction

> ...we are neophytes in knowing how to design systems in which a person nominally directs and oversees processes controlled by a computer, an interaction known as supervisory control. We know little about what strategies are or might be used by the human operator or the computer, how to relate available resources to control demands, how best to allocate functions to people and computers, or how errors arise in the interaction. Nevertheless, the very availability of the new computer-based technology seems to have set its own imperative that both government and industry adopt it, whether or not the art and science of design are ready. [1]

The above quote is taken from a report on a workshop organised by the US National Research Council in 1984. The workshop brought together nine of the world's leading researchers[1] to address issues associated with two questions; (i) how to conduct research into the emerging role of humans as supervisory controllers of otherwise automated machines, and; (ii) how to communicate the results of that research with designers involved in developing and implementing those systems. The subsequent 40 years have seen exponential growth not only in the power, sophistication, and abilities of automation, but in the scope and diversity of its uses. The benefits and achievements delivered by the technologies are impressive indeed.

However, other than in a few special cases - most prominently automated cockpits of modern aircraft, though there are still many current issues even there - the core argument made by the NRC workshop remains largely true. Technologies supporting ever more sophisticated automated systems continue to be implemented at a seemingly ever-increasing pace. While at the same time industry seems to remains limited in it's understanding – or if not understanding, the practice - of how best to optimise the relationship between the technology and the human users that continue to be relied on to perform the role of supervisory controllers to ensure the combined socio-technical systems operate reliably, particularly in unexpected and abnormal situations.

A great deal of thinking and research has gone into understanding the role of people in highly automated systems, particularly in the domains of military systems, aviation, nuclear power and transportation. What is clear is that introducing autonomous capability – whether to the complete system or, more often, to some aspects of an overall systems performance – can significantly change the role of the people in those systems. The most significant change, and one that seems likely to continue to accelerate over the coming decades, is that the work performed by the people who are expected to monitor and control processes is increasingly perceptual and cognitive, rather than manual. Those cognitive demands can be significant and challenging.

I was first introduced to the concept of human supervisory control through the teaching of the late Professor Neville Moray in the late 1970s. I was fortunate to be

DOI: 10.1201/9781003679479-1

an undergraduate psychology student at the University of Stirling during the time Neville was Professor and Head of Department. Not only a world-renowned psychologist, Neville was an inspired teacher, engineer, philosopher and artist – a true rounded intellectual. Part of his brilliance and inspiration was being able to integrate those seemingly diverse disciplines and focus them on giving his students insight into how to think about and understand the psychology of human performance in complex engineered systems.

Neville taught an undergraduate class that included content on Human Factors in control systems. He introduced us to the use of Control Theory, including what, to psychology undergraduates often with little mathematical background, were the seemingly impenetrable complexities of Optimal Control Theory and Information Theory. As an undergraduate psychologist searching for a potential career, the idea of adopting engineering approaches to understanding human performance in advanced systems fascinated me and captured my imagination. Though I lacked the mathematical skillset needed to fully engage with Control Theory, I received a solid grounding in the psychology involved.

I went on to do a PhD studying the effects of vibration on pilots engaged in manually controlling high performance aircraft. Having to immerse myself in the literature associated with Human Factors in control systems, I became familiar with Human Operator Transfer Functions, Laplace Transforms and Fourier analysis as tools to study what was happening. Though it turned out that, over the long term, Neville's teachings about supervisory control and the psychology of how we, as humans learn to allocate our attention in an optimal way across the many competing information sources in the world around us would have a far bigger and more useful influence on my professional life.

THE SCOPE OF HUMAN SUPERVISORY CONTROL[2]

Human supervisory control has generally been considered in the context of automation controlling some real-time process or activity: i.e. where there is some continuous activity that, provided everything is stable and within the constraints of what the technology has been designed to be capable of, can be maintained within those constraints by a technological solution without human intervention. Constraints can take many forms: physical location in space either geographically or relative to surrounding objects; chemical or thermodynamic properties (temperature, pressure, volume, etc), or even economic (fuel efficiency, resources consumed). Obvious examples include controlling the movement of some sort of vehicle - a car, plane, or ship – the movement and performance of a robot or drone, or some industrial process, such as chemical manufacturing, refining, or drilling.

Thomas Sheridan, Emeritus Professor at Massachusetts Institute of Technology, is by some margin the most influential thinker and writer about human supervisory control. He has also made significant contributions to related topics including adaptive automation. His publications, both individually and in collaboration with many colleagues, running from the early 1970s through to the late 2010s, have been extremely influential. Sheridan describes direct control, in control engineering language, as being the situation where;

> ...the controlling agent[3] is in the same control loop, performing direct control in an inner loop on a controlled process such as a vehicle, robot or an industrial machine.

Supervisory control, by contrast;

> ...is executed in an outer loop, and boils down to changing parameters of elements of the inner loop, or executing simulations that inform decisions to change parameters. [2]

The role of providing supervisory control over a direct control system can be provided by either the human or a technological (hardware and/or software) process. It is now normal practice in complex real-time IT systems - whether or not they provide automated control over a real-time process or activity - to design-in functions that perform that role as an outer-loop supervisory controller. Functions that monitor the status and performance of the system, make adjustments and, if they have the capability, intervene to exert control.[4] In the world of aviation, automated systems designed to supervise an aircraft's control system are sometimes referred to as "envelope protection systems". We therefore need to distinguish between situations where the supervisory function is performed by the human or by technology. Throughout this book, I have used the term *"human supervisory controller"* to refer to the supervisory control role when it is filled by a human.

The advances in technology that occurred over the period of Thomas Sheridan's working life, reflected in the sophistication and complexity of the operational systems he and his colleagues studied, are staggering. Despite that, many of the core messages and learnings from his research about the need to design automated systems that properly support human supervisory control are as true today as they were when the original research was published. Many of the predictions and concerns about the Human Factors implications expressed so long ago have come to fruition, sometimes disastrously. It is well worth taking the time to become familiar with some of the key thinking and findings from that body of research, as well as lessons that can be learned from incidents when things go wrong with automated systems, and the implications for the design of those systems today. That is the purpose of much of this book, especially the discussion of models of human supervisory control (which includes Sheridan's own model) and the summary of lessons that can be learned from incident investigations in Part II.

The book is concerned with understanding the psychology of the role of people as supervisory controllers of an automated real-time activity. One of the difficulties in bringing together material for this book, has been to be clear about the kind of situations it is concerned with. To try to reach for that clarity, the book is concerned primarily with situations where;

1. Automation is being introduced to fully or partially control the performance of a real-time activity. That activity may previously have been performed mainly or entirely manually. It includes situations where automated features are being added to a system or product design that already exists. It also includes situations where automation is used to perform activities that were not previously possible, where there was no manual control task to be automated, such as real-time automated control of robots or drones in deep space, military or other challenging environments.

2. Because of the introduction of automation, someone is expected to fill the role, at least part of the time, of being a supervisory controller of the automation. Where the activity was previously under manual control, that means the user's role will change from being a manual to a supervisory controller.
3. The individual involved knows they are required to fill the role of a supervisory controller. Generally, the individual will have performed some action that passes control authority to the automation. The scope excludes situations where the users are not aware they are expected to fill the supervisory role: if someone does not know they are in the supervisor's role, they cannot be expected to be proactive in allocating attention to monitoring the automation - the task which is at the heart of human supervisory control. While I can give no examples of products or systems where that is currently the case, it could occur, for example, where one individual engages the automation in the belief or expectation that someone else will perform the supervisory role.
4. The human needs to transition through the process of developing the knowledge and skills needed to be effective in the role of a supervisory controller: they have not yet developed the ability to perform the supervisor's role at a skilled level.

Boundaries

The psychological demands change significantly when an individual crosses the boundary from being a manual controller and moves into the role of being a supervisory controller, or vice versa. It seems clear that the boundary from manual to supervisory roles is going to be crossed at the moment the human who is manually controlling a real-time activity passes the authority to control the activity over to the automation. Typically through some manual activity such as pressing a physical button or making a selection on a user interface.

Coming the other way is not quite as clear. It seems that the boundary will be crossed when the human, acting in the role of a supervisory controller, is forced, or decides to take some action to influence either the automation, or the real-time activity that has been subject to automated control. Though this boundary needs clarifying. There are (at least) four possible scenarios worth considering;

i. the automation is not able to continue to exert control and, whether explicitly or not, passes control to the human;
ii. the human has decided that the automation is not able to exert effective control, so makes the decision to step in and take over real-time control;
iii. the human has decided, without being prompted by the system, that they need to make some sort of discrete or temporary intervention, such as changing a system setting or making a temporary change to the automation's set limits, mode or state;
iv. the automation has done something to attract the human's attention that requires some sort of response. That would include situations where the automation needs to alert the user that they may need to intervene soon,

though without actively passing control to the human at that time. It would also include situations where the system provides the human with information but there is no implication of the system being unable to continue exerting control.

In the first two of these scenarios, the human is clearly no longer a supervisory controller: they have exerted their authority to control the activity and have reverted to manual control.

In the third scenario, it is difficult to be clear about whether there is a significant change in the role of the human controller without knowing details about the system, the activity and the context. Consider an example. I am in my car and have engaged the vehicle's autonomous driving function. I can, if I choose, take my hands off the steering wheel and allow the system to control the car's speed and position while maintaining a safe distance from surrounding vehicles. I decide to use the controls on the steering wheel to change the current set speed. Doing so does not require me to take over control from the autonomous driving function, but it does change the parameters the automation uses to exert control. In performing the task of changing the set speed, I have certainly been mentally and physically engaged with the automation, at least for the time it took me to decide to act and then ensure my action achieved the desired effect. Although I exerted my authority to override the automation and make an adjustment, control of the vehicle's movement and position on the road always remained under the control of the automation. At no point did I take over manual control: I remained a supervisory controller.

However, consider if, rather than using controls on the steering wheel to make the adjustment, I decide to reduce the current set speed by pressing the brake, an action which disengages the automation and allows the car to slow to the speed I want, at which point I re-engage the automation. In this case, as soon as I press the brake, I override the automation. For at least these few seconds, I am no longer a supervisory controller but am in manual control of (at least) the vehicle's speed.

Of course, the way in which the human interacts to adjust the settings or limits used by automated processes will vary in all kinds of ways between different kinds of systems and activities. However, for the purpose of this book, I am going to consider scenario (iii) in the list above as falling within the boundary of the role of a supervisory controller. If a rationale is needed, it is because this third scenario represents the human having a specific goal of making a relatively minor (which is not to say unimportant) adjustment, and one where there is no intention to take over control of the real-time activity for longer than is necessary to achieve the immediate goal. None of the fundamental functions the human needs to perform to be effective as a supervisory controller have changed.[5]

Note that the difference between scenarios (iii) and (iv) is the difference between the human being *proactive* as opposed to being *reactive*. In (iii) the human identifies the need to make an intervention without any prompting from the system: they are behaving *proactively* making the decision to act based on actively monitoring what is going on. By contrast, in (iv), the user is *reacting* to a prompt generated by the system based on the system having detected a situation or event the human needs to know about, and that the user is assumed not to have detected for themselves.

This raises another distinction that needs to be made. That is between the terms "human supervisory control" and "proactive operator monitoring". To be clear:

- the term "Human Supervisory Controller" refers to the ***role*** an individual adopts when they are put in the position of being expected to oversee an automated or autonomous process;
- Proactive Operator Monitoring (POM) is the key ***task*** that needs to be performed to be effective in the role of a human supervisory controller.

To summarise, the scope of the book is limited to understanding the demands and behaviour of the human as a supervisory controller. From the moment the individual intentionally steps in and takes control over the real-time activity, other than for the purpose of making a temporary adjustment to the automation's settings or limits, they are no longer a supervisor. Though they will revert to the role once they make the decision to hand authority for real-time control back to the automation. Issues around what is needed to ensure the human has a good chance of making the transition from supervisory to manual control successfully are beyond the scope of this book. As we know from much applied research, and learnings from many incidents, the Human Factors and Psychology associated with the ability of people to make this transition quickly and effectively can be exceedingly complex. They go deeply into issues of maintaining situation awareness, mental models, and including issues to do with competence and skill retention.

Real-time and non-real time systems

Although there are clearly differences, many of the key psychological challenges involved in human supervisory control of real-time activities also apply to situations that don't occur in real-time. That includes situations where automation (increasingly nowadays supported by some form of Artificial Intelligence, and machine-based learning) is used to make complex decisions or recommendations that requires interpretation of large amounts of complex or uncertain data. The role of these systems is to usually to analyse, reason with and interpret the data and offer the human a solution. Once that is done, the human either accepts and implements the action, allows the system (or initiates another system) to do so, or decides on some other course. There is not necessarily any real-time control involved in these systems. While there may be time pressure, the activity itself does not carry on until the human has made their own decision whether to accept what they are offered. An example is the use of auto-contouring in radiotherapy, where imaged and other data from patients is fed into an AI-based system that automatically identifies organs-at-risk (OAR) and delineates areas for radiotherapy planning.

Systems using image processing, machine learning and/or AI, that are purely concerned with interpreting, mining, and reasoning with data, but that do not take any real-world action on anything in the world, are outside the scope of the book.

Adaptive automation

Over virtually the whole of my working career of more than 45 years, there has been a great deal of interest, both in applied research as well as in the design of operational systems, in the idea of *adaptive automation.* Sheridan defines adaptive automation as referring to systems where;

> …the allocation of control function (to either human or computer) changes with time to accommodate changes in the conditions of either the physical environment or the human…This requires an allocation authority or allocation agent that decides whether the control is by human or automation, a function that is distinct from the direct controlling function (human or automation) itself. [2]

This book does not attempt to address Human Factors or psychological issues associated with adaptive automation as such. The interest is specifically on situations where real-time control is being provided by an automated component, and the role of the human is to monitor and supervise the performance of the automation – including, if necessary, adjusting its limits, parameters or state.

IMPLICATIONS OF THE TRANSITION TO AUTOMATION

As the CIEHF put it in the second of its nine Principles to guide the development and implementation of highly automated systems, one of the most fundamental tenets of Human Factors associated with automation is to;

> Recognise that automation nearly always changes, rather than removes, the role of people in a system. Those changes are often unintended and unanticipated. They can make the tasks people need to perform more difficult and can disrupt established relationships, lines of communication and the ability to exert authority. [3]

The focus of the book is on understanding and preparing for that change. It is on the awareness that organisations introducing those systems need to have, as well as the kind of activities they should be conducting throughout design, development, and deployment, to be confident the introduction of the automation will go smoothly.

That change process involves understanding the implications of transitioning away from systems where people are directly involved, physically as well as mentally in performing some task or operation using, or in collaboration with machines.[6] And transitioning to a role where, while they may retain responsibility for the performance of an activity, for significant periods of time they may have little or no hands-on physical engagement with it. And where their mental role has changed to one where, rather than being actively involved in real-time task-related decisions and judgements, they are now expected to pay attention, to monitor, interpret and make decisions about how well the system is performing. And only to engage directly if they have sufficiently strong grounds for believing that the automation is, or is about to be, unable to do what is expected of it within acceptable constraints. That is, they have changed from being a manual control of a real-time activity or process, to being a *human supervisory controller.*

Frequently, a significant part of this change is that the activity that has been automated has become part of the background to an individual's role: something the user is expected to look after and be responsible for, while getting on with other activities. Other activities that may be new to their role or that may now be seen as being more important and have a higher priority for their attention.

That change, from being an agent actively involved in task performance, to being a largely hands-off monitor of a largely reliable system, introduces cognitive demands on people that are profound, far reaching and which, for the main part, humans simply have not evolved to meet. These include;

- the demand for sustained attention over long periods of time;
- the ability to maintain an adequate mental model of how a complex system works and behaves, despite little directly available information and little direct experience of how it does what it does;
- being able to maintain awareness of what has happened, why, what is happening in real-time, and what is likely to happen in the immediate future, despite little or no active engagement in the controlled process;
- the ability to make good judgements, decisions and effective trade-offs about the risks and rewards of intervening, often with little warning and under time pressure, while exposed to sources of bias, emotion, and other influences;
- the ability to detect, interpret and understand signs that there is a need to intervene, to plan quickly and effectively to know when and how to exert control, and to know what effect control actions are likely to have.

And all of this while being expected to retain the skills and knowledge learned as a manual controller and to be able to take over control in real-time when that is what is needed despite increasingly rare opportunities to practice and hone those skills.

In addition to the cognitive demands, the introduction of automation impacts on the characteristics and abilities needed of the people relied on to fill the role of a supervisor. People vary. And they vary across innumerable psychological dimensions, many of which are critical to being able to perform the role of a supervisory controller. These include the abilities;

- to sustain attention, maintain situation awareness and to retain an accurate mental model of how a system works and behaves despite long periods of inactivity;
- to reason and make judgements about the likely future behaviour and states of objects and elements in the world in space and time;
- to be aware of risk, and have a tolerance for risk that strikes an effective balance between the reasons for using the system and the potential consequences if the activity gets out of control; this includes the ability to imagine potential consequences if unexpected events are allowed to continue unchecked;
- to prioritise and direct attention in real-time to signals and events that matter while avoiding being distracted by those are not important to current task demands; as well as dynamically updating those priorities as the state of the world and events change;
- to make good decisions confidently and to make accurate and effective trade-offs between often conflicting priorities and objectives in the presence of uncertainty;

- to make good judgements about risk under emotional as well as time pressure;
- being able quickly and effectively to recall and perform critical tasks that, while they may have been learned in a training or learning environment, may never have been performed under pressure in real-world conditions.

These all rely on mental abilities that demonstrate a large variability both within and across people. They vary across personality types, between the sexes, with experience, as we age and as we experience physical and mental health challenges. They change over the course of a day as we get tired and fatigued and as our personal circumstances change. And they change as our managers or peer groups create expectations or set targets that motivate certain behaviours over others, and that can bias our opinions, our judgements, and our appetite for risk.

Transitioning is temporary

Transition, by definition, is a temporary state. How long it will take for someone who previously acted as a manual controller to complete the transition to being effective as a supervisory one will depend on many things;

- most importantly – and the principal subject of this book – the quality of the work that went into designing and deploying the system to support the user in their new role as a supervisor;
- close behind, is the effectiveness of the effort taken to prepare the user for their new role and how they are supported during the transition phase;
- the individual's own characteristics, including their perceptual and cognitive abilities, prior experience, expectations and even personality and attitudes. Someone who is highly risk averse may have a very different threshold for intervening than someone with a higher tolerance of risk. That risk tolerance can influence behaviour towards or away from being inclined to intervene in the presence of signs of potential trouble. Similarly people with high self-esteem and self-confidence may make a different decision to intervene than someone with low self-esteem in the presence of the same signs of potential problems;
- the social and/or organizational culture in which the system is used. In particular the user's perception of the relative trade-offs between overriding the automation when there was no need (a "false alarm") and the costs of not intervening when, in reality, the automation needed assistance (i.e. "misses"[7]).

Part I of the book recounts my experiences in the days and months immediately after taking ownership of a new car with autonomous capability. During the transition period, whenever I engaged the fully self-driving features, I was conscious of a feeling of unease, and having to pay heightened attention to what the car was doing and what was happening around me. And I regularly discovered new features, abilities, and limits to the automation's capability. After more than two years, I have certainly now completed the transition[8]: On reflection, the transition probably took somewhere up to four months in my case. My experience in the role of "driver-not-in-control" (or "user-in-charge") is very different now from what it was in those first few months. Chapter 5 reflects on how, with this experience, I now view the autonomous capabilities of my car, and my role as its human supervisory controller. I now feel I understand the features and functions the car offers, I know where to look for information and what that information means, and I feel comfortable allowing the automation to take control.

Most crucially, I now have what I believe is a robust understanding both of the capabilities of the autonomous features of my car, as well as its limitations: I know when I can and cannot rely on it. And, having learned through experience the limits of the automation's design envelope, I have a developed a great deal of trust in the car's capabilities.[9] I refer to the state I now adopt when I use the automated features of my car as being "highly engaged supervision". But my experience now could hardly be more different from what it was when I first took ownership: my experiences at that time are best described as being surprised, nervous, uneasy and even, at times, startled. As we will see in Chapters 3 to 5.

A HYPOTHETICAL SYSTEM: THE AUTOMATED NEUROSCIENTIST

As a means of exploring the scope of this book, consider a hypothetical future system[10] whose domain is the management of brain scans gathered by technologies such as Electroencephalography (EEG), computerized axial tomography (CAT), neuroimaging, functional magnetic resonance imaging (fMRI) and other brain imaging technologies.

Both the manufacturers and the healthcare providers believe the system offers major benefits that justify the investment:

- the time and effort to set up, perform and interpret scans for each patient will be significantly reduced, increasing the number of patients that can be processed;
- the quality of images produced is expected to be higher, and the proportion of wastage, in the form of unclear or unusable images, is expected to be negligible;
- the process of diagnosing conditions and preparing recommendations is virtually instantaneous. Previously it may have taken days or weeks to complete;
- because of the vast amount of data the systems AI and Machine Learning components draw on, including access to the most recent published research and clinical feedback, the system can detect, recognise, and interpret rare conditions that most clinicians are unlikely to have experienced;
- the systems recommended treatment plans are expected to be optimal, as they can take account of the most up to-date clinical findings research and guidance, taking account of pre-existing condition the patients may have, and supported by relevant case studies.

Although this hypothetical system is located sometime in the future, society has not yet moved to the point where it is prepared fully to trust its health to technology. There remains a legal responsibility for hospital staff to monitor and provide overall assurance of both the performance of the system and the care of the patient.

The previous, manual, process

Before the new system was introduced, a clinician would submit a request for a brain scan for a patient. A radiologist would review the request along with details of the patient's clinical condition and suggest changes to best answer what the clinician has asked. Once the patient was prepared, a radiographer would review the proposed scan against local protocols, select the programme parameters, perform the scan and review the acquired images, making adjustments to ensure the images are of high quality and are capturing the areas and features of interest.

Once the scan was completed, the radiologist would interpret the images for clinical features, drawing on the patient's clinical information, including their symptoms, age, sex, and results of blood tests or other investigations. If necessary, the radiologist would make decisions about the need for further investigation before a diagnosis and treatment plan was prepared. They would then prepare a report including a provisional diagnosis and be available to discuss the report with the clinician. Based on the radiologists report and recommendations, the responsible clinician would decide whether a diagnosis could be made, or whether other investigations were necessary.

Use of the new AI-based system

Use of the new system is initiated by a clinician inputting their reasoning for requesting a scan directly into the system, along with supporting clinical data and an indication of what they are concerned about. Based on that information, the system recommends the most appropriate type of scan and modality to achieve the clinician's objective. Once the proposal has been approved by a radiologist, a radiographer physically looks after getting the patient into and out of the scanner and confirms that the system is set up to conduct the scan as planned. The system has the capability to complete everything else, to the point of preparing a diagnosis and a recommended treatment plan, autonomously.

The system has been designed to be given the authority;

a. to review the clinician's request and, drawing on its' extensive knowledge base and clinical reasoning capability, advise on the optimal scanning routine;
b. to set up and control the imaging sources to conduct the scan(s);
c. to interpret and evaluate the quality of the image produced;
d. to adjust the ongoing scan based on the system's interpretation of the images being acquired. That can include changing the modality or type of scan until images have been acquired that answer the clinician's question;
e. to terminate the scanning process, (either on completion or if it is stopped by the radiographer), and pass the patient back to the care of the radiographer;

f. to interpret the acquired scans and make recommendations about the presence and nature or injury or disease processes identified;
g. to prepare a recommended treatment plan or make suggestions for further investigation.

The system has several different modes of operation. As well as modes intended to be used in conducting scans, these include modes used only for maintenance and for calibrating the equipment. It is also possible for scans to be performed under manual control, rather than being controlled by the system.

With the new system, real-time images are capable of being produced and displayed much more quickly than when they are prepared manually. The rate of automatic image production is beyond the ability of the radiologist to monitor in real-time without interfering with the scanning process. Though the radiographer can initiate a pause in the process to examine images individually if they feel they need to.

Staff have been led to understand, both during training and throughout the implementation process, that the system's reliability and performance will be significantly better than the previous manual system. They have also been advised that, should they have doubts or concerns about anything the system is doing, they must intervene and either stop the process or take over control of the scan. They will retain legal responsibility for any adverse consequences to the patient or hospital should an incident occur during the automated scanning process.

The system has been in clinical use around the world for some time. Although the sample of patients and the range of conditions it has been used for are still limited, it has been found to be highly effective and reliable. Experience has shown that steps (a) to (f) listed above are all typically performed with little need for human involvement. Data to-date suggest the rate of missed indicators of disease, or misdiagnoses made by the system are substantially reduced compared with previous clinical experience.

The systems manufacturers are however aware that there are still conditions and some types of imaging situations where either the systems reliability has not been proven, or that are known to be outside the scope of the systems capability.

Impact on the Radiographer, Radiologist, and Clinician

With the new system, once the proposed scan has been approved and the system is engaged to begin, the radiographer is expected to remain "hands-off". They are though expected to monitor both the state of the patient and what the system is doing in real-time using a combination of what they can see of the patient, the imaging plan, the images acquired as well as the system settings displayed as the scan proceeds. They can intervene to adjust the parameters being used by the system, to take over manual control of the scanning process, or to stop the scan at any time.

The radiographer is expected, for example, to ensure that the correct area of the brain is being scanned, that significant features are being adequately explored, and that the quality of the final images is sufficiently high. They also need to ensure the system is not intending to do anything that may cause harm to the patients, such as planning to deliver a radiation dose that exceeds safety limits. And they remain responsible for ensuring the patient is safe, comfortable, and not showing signs of distress.

As with the radiographers, the roles of the radiologist and clinician have both changed with the new system: their task is now largely to review the interpretation of clinical issues produced by the system, as well as the provisional diagnoses reached and the system's recommended management plan. They must decide whether to confirm and adopt these, to initiate other investigations, or make changes. The lead clinician retains legal responsibility for the decisions made.

Of the three roles, the radiographer, radiologist and clinician, it is the impact on the radiographer that is the principal focus of this book. It is the radiographer who will be expected to adopt the role of a human supervisory controller of the real-time scanning process. They are the ones expected to pay attention and monitor the system in real-time and, if necessary, intervene to take control or adjust the system if they consider it necessary.

The psychological challenge: whether to intervene?

Unlike the radiographer, neither the radiologist nor the clinician's role in this hypothetical system is performed in real-time: they both begin after the system has completed its work, interpreted the images acquired and produced and recommended a diagnosis and treatment plan. As with the radiographer, however, and despite also having been told about the systems reliability and expected superior performance, both roles are aware of their ultimate responsibility for the patients care and to intervene if they have reason to doubt what the system is advising.

Even though the radiologist and clinician are not performing in real-time, introduction of the system has created a new type of challenge, putting them in a position which is psychologically challenging, and which needs to be understood and managed. That position is one in which they need to make judgements about whether to trust the system and to accept its recommendations or override them: while the radiographers do this in near real-time, the radiologists and clinicians work in (relatively) slow time. But all three roles are faced with the challenge of deciding whether they have sufficient concerns about the performance of the system to justify intervening. And all three are under professional and legal, not to mention economic, time and organizational pressures if they get it wrong: if they decide to over-ride the system when they did not need to, or if they decide not to intervene when it turns out they should have.

Clinicians are used to making diagnostic decisions in the presence of uncertainty. Sometimes that uncertainty comes from the information available to them about the patient's condition, or how they may respond to different treatments. Sometimes it is from the limits of their own knowledge and experience and the information they can easily access to support their decision making. With the introduction of the automated neuroscientist, there is an expectation that, because of the AI algorithms and the quantity of data used in its machine learning algorithms, the system will reduce the uncertainty about the diagnosis and the suitability of the treatment plan.

It is also however known that the system has limitations: both in terms of the experience it can draw on and on the technical limitations of the technology used to create and interpret some types of images. The psychological challenge, to the radiologists and clinicians as much as to the radiographers, is whether they are prepared to back their own reasoning and professional judgement, knowing the consequences if they were to make a mistake, against a system which comes with an impressive track record, credentials, and high expectations.

Note that, once the images are acquired and the treatment plan is prepared, the system is effectively open loop: although the system is in use around the world, there is as yet no process for feedbacking back information about the success or otherwise of the diagnosis and proposed treatment plan in managing individual patients. Although the system learns from published research, it is unable to learn directly from its own performance.

Although the roles of the radiologist and clinician in this hypothetical system are largely outside the scope of this book – at least they are not the central theme – the decision they face about whether to intervene to override the automation is very much at the heart of it. Psychologically, it is essentially the same decision as that faced by the radiographers. And it is the same judgement the driver in control of an automated vehicle faces when deciding whether to take over control of the car; or a driller on an oil rig faces when deciding whether to allow the automated drilling system to carry on drilling in the face of uncertainties about the nature of the formation they are drilling into.

By contrast with the radiographer in the role of human supervisory controller, the radiologist and clinician do not have to make decisions in real-time. But the context of the decision they are faced with is similar: in the presence of uncertainty, do the benefits of intervening and overriding the technology outway the costs of intervening and either being wrong or making things worse? Uncertainty might arise through a lack of awareness of the full picture the system was considering in its decision making. Equally, they may be disinclined to question the system's recommendations if they come to develop an overly high level of trust in the system over time, believing, for example, that it is "all-knowing".

To summarise the key aspects of this hypothetical system in terms of its relevance to the scope of this book;

1. The introduction of the automated neuro-scientist system has introduced change to the role of all three roles involved in conducting brain scans: radiographers, radiologists and clinicians.
2. Radiographers have become real-time human supervisory controllers of the system, while the radiologists and clinicians work in non-real time;
3. All three professions are faced with making the decision whether to intervene to assist or take over from the automation in the presence of uncertainty, and with significant implications if they either fail to override the automation when they should have or decide to override the automation when they did not need to;
4. While all three understand that responsibility lies with them, they are also aware that they need to have a lot of confidence in their judgements to justify over-riding the system and interfering with the anticipated benefits of the automation.

STRUCTURE OF THE BOOK

The remainder of the book is structured into three Parts:

- The three chapters in Part 1 recount the story of my experience taking ownership of a new car which, through it's Automated Driving System, is capable of driving autonomously. The chapters describe many examples of my experience as I explored and learned about the car's features and ability to drive itself, and my reflections on the demands it was placing on me as well as my new role as supervisor of the automation. The chapters recount the process I went through trying to understand how the automation works, what it was capable of and how and when I could trust it. Based on that experience and reflections, the Part extracts some general lessons likely to apply to most situations where the role of a human supervisory controller is being introduced.
- Part 2, comprising Chapters 6–9, delves into some of the psychology behind the lessons extracted from the experience set out in Part 1. The Part starts by reviewing what can be learned about human supervisory control from information that is in the public domain from eight incidents involving automated systems – some, though not all, disasters. Chapter 7 presents some basic concepts in human supervisory control and reviews models published over the past 50 years by three widely respected thinkers and researchers active in the field. Chapter 8 explains why, drawing on my own lifetime's learning and experience as a Human Factors professional, as well as additional knowledge from the science base, I concluded that even the most recent of the models discussed in Chapter 7 is some way short of providing a satisfactory account of the psychology involved when a human is expected to fill the role of a supervisory controller. Chapter 9 suggests a more comprehensive model of human supervisory control behaviour that builds on the models discussed in Chapter 7, adding features that Chapter 8 argued were missing.
- The third and final part of the book focuses on organisations involved in developing automated real-time control systems who need to demonstrate that they have done everything they reasonably can to support people filling the role of a human supervisory controller of their new system. Using the comprehensive model developed in Chapter 9 as a framework, Chapter 10 reviews six principal risks associated with human supervisory control. The chapter explains the nature of each of the risks and makes suggestions about the kind of questions that should be addressed to be confident the people involved will be able to fill the supervisory role effectively and reliably. Chapter 11 considers the role that the task of being proactive in monitoring for threats to the automation's performance – a task, referred to as "POM", that is at the heart of human supervisory control – can play as a defence, or "Barrier", against serious adverse events. Chapter 12 presents a "simple little tool", referred to as the Proactive Operator Monitoring Assessment Tool for supervisory control (POMATsc), that can be useful in evaluating

how well the combination of the design of an automated systems and the context and situation it is expected to be used in are likely to support reliable proactive operator monitoring. The method includes a means of calculating a numerical score (a "PMQsc score") that can give an indication of the quality of proactive operator monitoring likely to be associated with a highly automated or autonomous system.

NOTES

1 With the exception of a few omissions, the list of participants could hardly be more impressive, then or now: Tom Sheridan, Renwick Curry, Baruch Fischoff, Neville Moray, Richard Pew, Jens Rassmussen, Willaim Rouse, Henk Stassen, David Woods.
2 Chapters 2, 7 and 9 go into much more detail on the nature of human supervisory control.
3 Which may be a human or a "machine".
4 Figures 2.1 and 2.2 in Chapter 2 illustrate the use of technology to perform the supervisory control function in industrial process control systems.
5 The comprehensive model of human supervisory control set out in Chapter 9 explains what those core functions are.
6 A definition of what I mean by a "machine" as opposed to "automation" is included in Chapter 2.
7 The potential use of the Theory of Signal Detection (TSD) and the important role it can play in thinking about and assessing the risk associated with Human Factors in highly automated systems is explored in Chapter 7.
8 Though, as Chapter 5 recounts, I discovered a new feature only this week.
9 Note that this feeling of confidence in the reliability of the vehicle is entirely due to self-exploration. None of it is due to the car's User Guide or in-car help system, which were of very little help. See Chapter 4.
10 The hypothetical system to be described is entirely a fiction of my imagination. I am grateful to Dr Calum Nicholson and Dr Wendy Russell, from University Hospital Crosshouse for helping me understand the roles and processes involved in brain scanning in Scotland.

REFERENCES

1. National Academy Press. Research and Modelling of Supervisory Control Behaviour: Report of a Workshop. Committee on Human Factors. Commission on Behavioural and Social Sciences and Education. National Research Council. 1984.
2. Sheridan, T. Adaptive Automation, Level of Automation, Allocation Authority, Supervisory Control, and Adaptive Control: Distinctions and Modes of Adaptation. *IEEE Trans. Systems, Man and Cyber. Part A: Systems and Humans*. 2011 July;41(4).
3. CIEHF. Human Factors in Highly Automated Systems. Chartered Institute of Ergonomics & Human Factors. 2012. Available from: https://ergonomics.org.uk/resource/human-factors-in-highly-automated-systems-white-paper.html

2 Key concepts and terminology

SUPERVISORY CONTROL – THE BASICS

> ...the human activity involved in initiating, monitoring and adjusting processes in systems that are otherwise automatically controlled. [1]

Generically, the term "supervisory control" refers to any higher-level control system that oversees and manages the operations of lower-level systems or processes. Supervisory control involves monitoring, guiding, and adjusting the activities of those lower-level systems to ensure they operate correctly and efficiently. The concept has been in use in industrial control systems for many decades to describe the role of fully automated elements that monitor and automatically control lower-level controllers, especially where multiple processes or machines need to be coordinated.

The original concept was derived by analogy with supervisors in a work setting interacting with staff members. A supervisor is someone who sets goals and gives instructions within the scope of responsibility given to them by a higher authority. Those goals and instructions are translated into action by sub-ordinate staff members. Staff members carry out their tasks and aggregate and transform information about the state of the organisation or business into summary form for the supervisor to review and, if necessary, act on. Among many other considerations, the supervisor's willingness to delegate tasks, as well as how they act to monitor and check each staff members work is influenced by their judgements about the degree of competence, ability and reliability of each of the staff members.

From at least the mid 1970s it was recognised that the role of the human operator in real-time industrial process control systems was changing. It was moving away from the human operator being someone who actively monitored the process, compared what was happening with what was desired, and if necessary, acted. And it was moving towards a role whose main function was to monitor the automation, make minor adjustments to limits or process parameters though without taking over control, but to be ready and able to intervene and re-take manual control should problems arise with the automated control system.

Manual control

Figure 2.1a illustrates a simplified generic model of the human operator as a manual controller of a real-time process. This simple model can be applied to many situations where the human actively controls some operation in real-time: whether driving a car or aeroplane, controlling an industrial or agricultural process, remotely controlling robots or drones, or manually controlling the anaesthetics given to patients during surgery.

DOI: 10.1201/9781003679479-2

Sensors detect the state of key process parameters and provide information to the operator via some form of display devices. Sensor data might also generate alarms if variables exceed pre-set limits. The operator may also have direct perception – via sight, sound or even smell – of information indicating the state of the controlled process. The figure allows for the possibility that other people can also provide information to the human controller; a passenger in a car or another crew member in an aircraft drawing the driver or pilots attention to something they may not have been aware of.

In this simple model, the human monitors the information available to them and influences the process by acting on controls either through a computer interface, or by applying force directly to a control device; moving levers, opening or closing valves, changing pumping rates, applying brakes or turning a steering wheel.

In the case of industrial process control systems, such as refineries and chemical plants, the generic manual control model shown on Figure 2.1a needs to be slightly modified to reflect the reality of experience in these industries. Figure 2.1b shows a generic manual control model for industrial process control systems.

Industrial processes such as refineries and chemical plants have been using PID (Proportional, Integral, and Derivative) and other types of automatic controllers since at least the 1930s. Sometimes referred to as "regulatory" controls, these were typically pneumatic until the 1950s, when they were replaced by electronic controllers. In the mid 1970s, digital controls started to replace remaining pneumatic and electronic ones.

Each PID controller automatically controls a process variable which has a real-time sensor on the plant: temperature, pressure, flow, composition, etc. The human operator pre-defines the required set-point for each PID control and can adjust them according to changing priorities and targets. The PID controller continuously compares the value of the process variable detected by sensors against the operator entered setpoint. When a difference is detected, the PID controller automatically

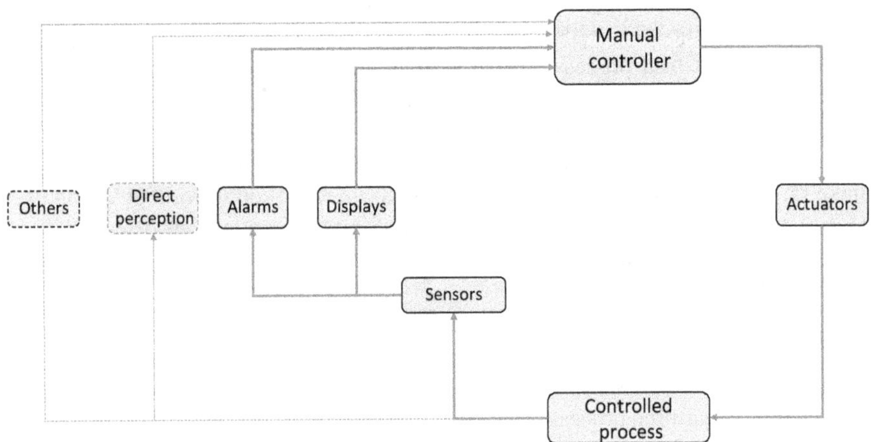

FIGURE 2.1(a) Simplified generic model of the human operator as a manual controller of a real-time process

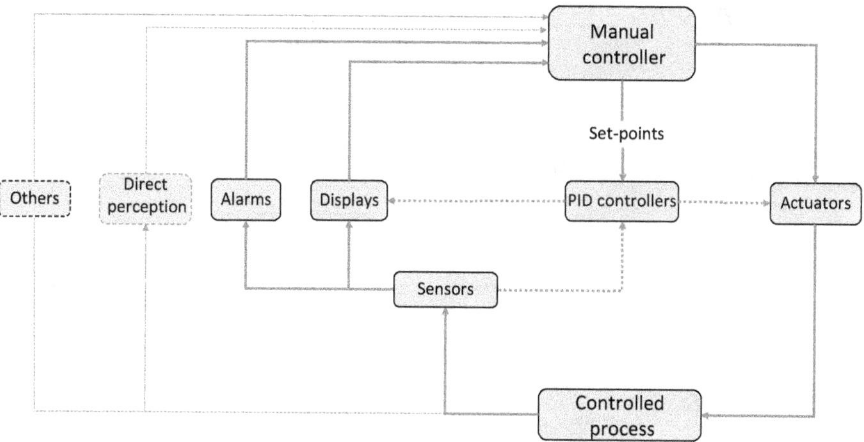

FIGURE 2.1(b) Generic manual control model for industrial process control systems

adjusts its output to manipulate the associated actuator (such as a control valve) to return the parameter to within the required setpoint.

Regulatory controls relieve the operator of the need to continually adjust control elements by hand in real-time by requiring only intermittent changes to set points. As such, they include a degree of simple automation. Though as the controllers are doing no more than attempting to return process parameters to set points that are pre-defined and can be adjusted by the operator at any time, these are still best thought of as manual control systems.

Even simplified manual control system such as those shown on Figures 2.1a and b quickly become complex depending on the dynamics of the system, including factors such as system gains, time delays, and couplings. For our purpose, the key point about these simple models is that the control system is closed through the operator: the PID controllers are not capable of ensuring the safe and effective performance of the controlled process without the human defining and managing the set-points in real-time.

Supervisory control

By contrast to the generic manual control system shown on Figures 2.1a and b, the concept of a generic supervisory controller is illustrated on Figure 2.2a. A digital control system has now largely replaced the operator in directly controlling the process (at least as long as it remains within the designed operating range). Sensors supply electronic data directly to the digital control system, which uses algorithms to compare the state of process variables (and derivates of them such as rates of change) and make predictions of future states against target states. Based on detected or predicted differences, the system automatically makes decisions and takes action to keep the process within its target goal conditions. As long as the system remains within the system design limits, the process is controlled automatically, with no direct human intervention.

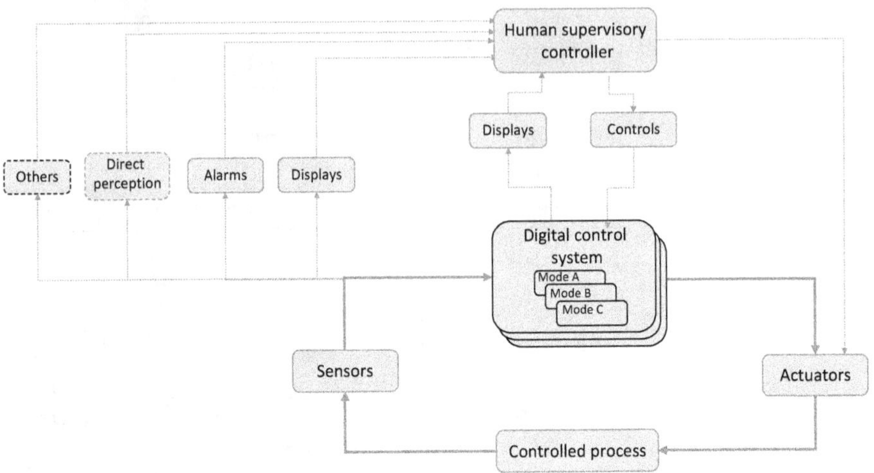

FIGURE 2.2(a) Generic model of the human operator as a supervisory controller

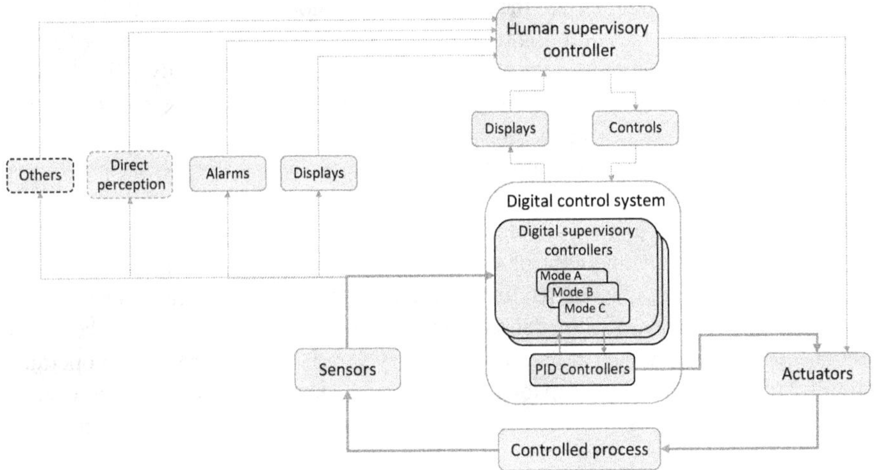

FIGURE 2.2(b) Generic human supervisory control model for industrial process control systems

As with manual control, in the case of industrial process control systems, such as refineries and chemical plants, the generic supervisory control model shown on Figure 2.2a needs to be slightly modified to reflect the reality of experience in these industries. Since the 1970s, digital supervisory controls have been introduced to many controlled processes. Figure 2.2b illustrates a generic human supervisory control model for industrial control systems.

Digital supervisory controllers use advanced and often highly sophisticated mathematical models of the controlled process and optimisation constraints to automatically define set-points independently of the human. The calculated set points are then used as inputs for the regulatory PID controllers. Digital supervisory controls can take many forms, including;

- "Best operator controls", that attempt to simulate how the best operator would adjust the set points;
- "Advanced regulatory controls", more sophisticated than standard regulatory controls, are typically used to control more complex variables such as reflux ratios, heats of combustion or reaction, temperature or pressure-based composition controls, etc;
- "Economics based controls" based on economics programming models that seek to adjust set-points in a way that optimises economic objectives within defined constraints.

Despite the sophistication and success of digital supervisory controllers, in most cases the human operator will still be provided with the option of overriding all or some of the digital control system and reverting to a manual control mode (though nowadays an operator would typically be required to justify to their supervisors why they had overridden the digital control system).

Note that the human in Figure 2.2a and b is still far from being redundant. Their role has changed to one where they are expected to monitor what is happening and to be ready and able to intervene either to adjust the parameters representing system goals and boundaries, or to intervene and take action to manually control the process if that was necessary. They also often perform a range of tactical and strategic tasks: adjusting goals, targets, and plans in response to operational and other influences; and adjusting the control system's objectives.

The human supervisory controller illustrated on Figure 2.2a and b now has - logically, though not necessarily physically - at least two sets of displays and controls to attend to: one showing the status of the process and allowing interaction with it; and another allowing interaction with the digital system The human supervisor not only needs to maintain awareness of the state of the process, they also need to be able to understand the state of the digital control system: what is it doing, why is it doing it, what is it likely to do next, etc. They need to be able to retain the knowledge, skills, and confidence to be able to interact with both systems if needed. And, perhaps most crucially, they need to have sufficient confidence in their own abilities to know that, if they do need to exert control, their engagements will be effective and not make matters worse. This is especially critical when the process moves into states that are abnormal, and when the abnormality is sudden and unexpected.

The supervisory controller models in Figure 2.2 also illustrates that most sophisticated control systems have several distinct different 'modes'. A 'mode' exists when the system behaves in different ways under different conditions, or when a control has a different effect depending on what state the system is in. Modes are common in modern computer systems. They have frequently been identified as a cause of users being confused about the behaviour of a system or making "errors" interacting with it. Not only does the human supervisor need to know what state the system is in at any time, but, if the system has different modes, they need to know what mode it is in, and what that means in terms of how the system will behave and how to interact with it. And all of this when the human is likely to be 'hands-off:' observing and monitoring what the system is doing, and while engaged in tactical and strategic activities. The demands on the human operator can be extremely challenging.

TERMINOLOGY

It is necessary to try to be clear about how I am using a few terms. This section explains how the following terms are used throughout the book;

- the difference between "machine", "automation" and "autonomy";
- the meaning of the terms "stages" and "levels" of automation;
- the meaning of the terms "system", "system boundary", "operational design domain", "task", "sub-task" and "function";
- The concept of "threats" and the difference between "internal" and "external" threats;
- The importance of cornerstone concepts of "ability", "authority", "control" and "responsibility".

MACHINE OR AUTOMATION?

Perhaps the most fundamental distinction this book needs to try to be clear about is the difference between what is a "machine" and what is "automation". The term "automation" lacks precision and can mean many different things depending on the context and the user's perspective and objectives. At one extreme, "automation" can be relatively trivial: using technology to perform essentially straightforward, limited tasks that can be fully specified in advance, have no uncertainty about them, and have clearly defined outputs or end points. At the other extreme, "automation" can refer to systems integrating the most state-of-the-art techniques in Image Processing, Machine Learning, Artificial Intelligence and Robotics to perform tasks that can be hard to specify in detail, are associated with a lot of uncertainty and complexity, and may need to be performed under a wide range of conditions that cannot be fully specified in advance, and that can be well beyond the abilities of human perception and cognition.

Automated Teller Machines (ATMs) have been ubiquitous around the world for at least four[1] decades as a means of automating transactions previously performed by human bank tellers.[2] While they can perform a variety of functions other than simply dispensing cash, the tasks involved, the processes to be followed, and the end states to be achieved can all be specified precisely with little or no uncertainty. The same is true of things like automatic washing machines, however sophisticated their programmes and settings. Although their abilities are more limited and dependent on the state of the external environment, things like cruise control in vehicles and auto-pilots in aircraft are similar: provided the world stays within acceptable pre-defined limits, the tasks involved can be defined precisely, with clearly defined goals and little or no uncertainty. The degree of uncertainty these systems can cope with, and the ability of the system to deal with variation is well defined by the design envelope they are intended to perform in.

Although ATMs perform tasks, physically manipulate cash and bank cards, and electronically make changes to computer systems and records, they do not actually control the real-time performance of anything other than themselves. ATMs are also open loop: they don't themselves look for or try to respond to feedback about whether the customer has achieved their goals.

Other people have tried to pin down this difference between "machine" and "automation". For example, in the context of automating drilling operations in the oil and gas industry, Federico Amezaga and his colleagues offered the following distinction:

> Mechanization is the process of introducing a machine to perform a task that was previously done by a human. Automation is the process of making the machines perform and complete a set of tasks automatically, with minimal or no human intervention in making real-time operational decisions based on operational data. [2]

An ATM is clearly a machine under this definition. A washing machine (which uses feedback about water temperature for example to make "decisions"), or cruise control in a car (to the extent that the car varies the engines output based on the load imposed by the driving conditions) could both be considered as "automation".

But we are interested in much more than that. The issue of whether technology is not only capable but *allowed* to make important decisions without human input – i.e. the extent to which it can genuinely act "autonomously" - is clearly central. Both the washing machine and cruise control act autonomously within their extremely constrained domains: but, in the scale of the kind of systems we are interested in, those domains are essentially trivial. So how to pin down this distinction between what is a "machine" and what is genuinely "automated" – not to mention "highly automated" or "autonomous"?

What Amezaga et al's definition misses about the kind of highly automated systems of interest in this book are three things:

i. the extent to which the operational tasks being automated need to be capable of being fully specified in advance (i.e. during design);
ii. a recognition of the limits or boundaries within which the automation is designed to perform;
iii. the ability to deal with uncertainty or unplanned variability in the world in which the automation operates.

When we wrote the CIEHF White Paper on HF in highly automated systems [3], I produced a description of what I meant by the difference between machines and automation. While it is perhaps a little complex, it seems to have found acceptance among my peers, so I will adopt the same distinction here. To quote from the CIEHF White Paper;

> Machine: If the designers have very little uncertainty about the details and variability of either the domain the product or system is expected to operate in, or exactly what the automation needs to do, or if the system does not have the capability to deal with unplanned variability in the domain, then the system is better thought of as a 'machine', rather than 'automation'.
>
> Automation: By contrast, if there is significant uncertainty or unplanned variability about either the domain the product or system is expected to perform in, or the way functions are to be performed, but the system is capable of dealing with those uncertainties with little or no reliance on a human, then the system is considered as having

> 'automated' those functions. It not only has the ability, but it is given the authority to behave autonomously in performing one or more of the core functions without relying on human input.
>
> The essence of this distinction is that to be considered as automation rather than a machine, the product or system must have the ability to detect and understand changes in the environment or circumstances it is operating in, and to adapt its behaviour accordingly. That is, it must have some degree of autonomy. Automation does not necessarily operate under a wider range of conditions than machines; but it has a sophisticated ability to detect changes in its environment, and to vary its actions in response to those conditions, that machines do not possess [3]

The term "automation", as I will use it, therefore means that the system has some ability to deal with uncertainty in achieving its goals that is independent from the individuals immediately tasked with supervising and monitoring it.

Automated and autonomous vehicles

In 2022, the Law Commissions of England and Wales and the Scottish Law Commission issued a joint report reviewing the status of UK law as it relates to automated vehicles [4]. The report adopted the terminology defined by SAE/ISO in its guidance on automated driving technology, J3016 [5], reserving in UK law the term "Automated vehicles" to be used as a general term "…to describe vehicles which can drive themselves without being controlled or monitored by an individual for at least part of a journey". Referring to the performance of the Dynamic Driving Task (DDT) as being "*…the real-time operational and tactical functions required to operate a vehicle in on-road traffic*", the report distinguishes between "Automated Driving System" (ADS), "Automated Vehicles" (AV) and "Driver Support Systems" (DSS). An "Automated Driving System" is defined as a system that performs the entire dynamic driving task on a sustained basis. "Automated Vehicles" are vehicles equipped with an ADS. Driver Support refers to features such as adaptive cruise control and lane keeping, that support the driver but are not capable of performing the entire driving task.

Crucially, the Law Commissions propose that; "While the ADS feature is engaged the user-in-charge is not responsible for dynamic driving. They cannot be held liable for criminal offences which arise from these activities". Automated Vehicles, where there is no legal requirement for the "driver" to monitor or pay attention to the movement of the vehicle, are therefore distinguished from vehicles that incorporate "Driver Support Technologies" where the user retains legal responsibility for the vehicle.

As this book is intended to apply to automated and autonomous systems across a range of disciplines, I have throughout used the term "autonomous vehicle" and related terms such as "autonomous capability" in a more general sense, without attempting to refer to legal responsibilities. In my usage, these terms are intended to refer to systems that have been designed with the capability to perform all or part of a continuous real-time activity on a sustained basis, at least under well-defined conditions, with no human input: they have been designed with the capability of performing control functions autonomously from human input, whatever the legal or regulatory context.

STAGES AND LEVELS OF AUTOMATION

Anyone interested in looking at the history of the published research into Human Factors aspects of highly automated systems will quickly find that a significant amount of effort has addressed itself to two questions;

i. *what* to automate. and;
ii. *how much* to automate?

That is, which type of functions best lend themselves to being automated without leading to adverse consequences on the ability of humans to support the system; and, to what extent should those functions be automated: from no automation, where the best option is to allow the human to do everything. to complete automation, where the automation decides and does everything without recourse to human intervention. Over recent decades, researchers and those who fund their work, have shown considerable dedication to answering these questions.

Within this general interest, one line of research has sought to determine whether it is possible to define an optimum balance of stages and levels of automation: i.e. a single combination that is generally universally the best. How much automation is too much? The answer, perhaps inevitably, is that it depends.

There is no need here to repeat the arguments and research findings on these topics. A brief mention of two published papers will serve both as an introduction to the literature for anyone interested in pursuing it, and will set up what is needed to support the remained of the book.[3]

PARASURAMAN, SHERIDAN AND WICKENS (2000)

In 2000, Raja Parasuraman, Thomas Sheridan and Christopher Wickens, all internationally respected leaders in the field, published an overview of the Human Factors issues associated with highly automated systems [6]. They provided a framework for investigating what types and what levels of automation to seek to implement for any system. Building on previous work by themselves and others, they recommended thinking in terms of four stages of general system functions that are amenable to automation:

- Information acquisition;
- Information analysis;
- Decision and action selection;
- Action Implementation.

Parasuraman et al pointed out, as had several previous authors, that these four stages map directly onto the stages used in simplified models of how the human brain processes and works with information that have been in wide circulation for many years. Figure 2.3 shows the traditional simple model of human Information processing, together with a slightly modified version[4] of the four generic stages amenable to automation. As Parasuraman et al point out, the simple four-stage information

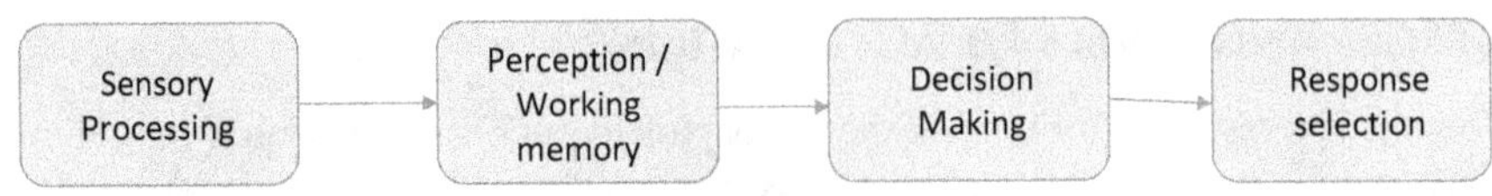

a) Simple 4-stage human information-processing model

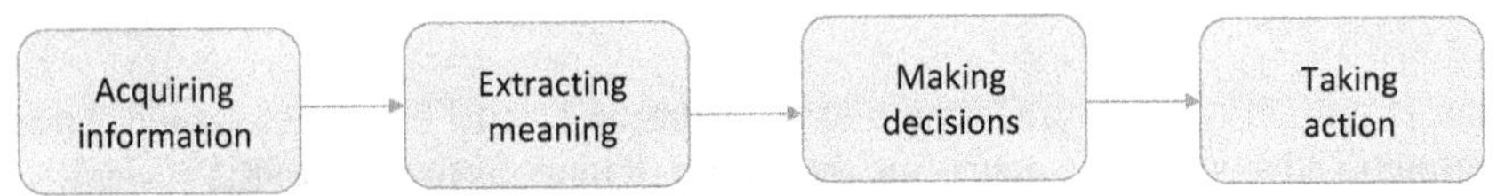

b) Four classes of function potentially amenable to automation

FIGURE 2.3 (a) Simple four-stage model of human information processing. (b) Four classes of function potentially amenable to automation

processing model shown on Figure 2.3a is certainly a gross simplification: how the brain goes about processing information is a lot more complex.[5] The simple model however serves as a useful point of reference.

As well as recommending the use of the four stages, Parasuraman et al also recommended four "primary evaluative criteria" based on human performance capabilities that could be used to assess the relative merits of seeking to introduce automation for a given application; (i) mental workload, (ii) situation awareness, (iii) complacency, and (iv) skill degradation. They also suggested two secondary, criteria that could be used to further screen potential automation candidates that met the primary criteria; (v) automation reliability; and, (vi) the cost of decision actions or outcomes.

There is little more that needs to be said about *stages* of automation for the purpose of this book. The essential point is that, when we talk about "automation", in any detail, it is important to be clear about which combination of one or more of the four generic stages shown on Figure 2.3b are being automated.

As for seeking to identify and describe *levels* of automation, there is no need in this book to go further than to recognise that functions can be automated to varying extents. For the purpose of the book, where it is needed to suggest different levels of automation, I will adopt a minor variation of the five levels used in [3];

1. None: entirely human, no automated support.
2. Low level automation.
3. Medium level automation.
4. High level automation.
5. Fully autonomous, performed with no human support.[6]

Complex systems generally automate more than one function. And each function can involve one or more of the four generic stages amenable to automation shown of Figure 2.3b. Which means that a single automated "system", will often comprise multiple functions, each automating one or more of the four generic stages to different levels.

Onnasch, Wickens, Huiyang and Manxey (2014)

The second paper worth drawing attention to, both as an introduction to the literature and for setting up the discussion later in the book, was published by Linda Onnasch and her colleagues in 2014 [7]. They were interested in determining whether it was possible to combine the dimensions of stages and levels of automation into a single metric: something they refer to as the "Degree of Automation" (DOA). They were interested in making comparisons across different systems that had been the subject of research studies, but that differed in the stages and levels of automation supported. They suggested that a DOA metric could provide a basis for making such comparisons across those different systems. Onnasch et al were particularly interested in testing the idea that the "more" a system is automated - the higher its DOA score – the better overall system performance would be. But if there were problems that relied on human intervention, they anticipated that systems with large DOA scores would prove more problematic than those with lower scores. Something they refer to as "the lumber jack effect" - the higher they are, the farther they fall.

Onnasch et al performed a statistical meta-analysis of 18 research studies that had looked at various aspects of human performance with a variety of systems automated to different extents. Their conclusion supported the expectation of the lumber jack effect: higher levels of automation helps when all goes well but can lead to considerable performance impairment if it fails. A risk that was seen to increase as the combination of stages and levels of automation increased. Specifically, and importantly, they concluded that:

> When DOA moves across the critical boundary from information acquisition and information analysis to action selection, the latter alleviating the human from some or all aspects of choosing an action, then the human is much more vulnerable to automation "failures." [7]

SYSTEM, OPERATIONAL DESIGN DOMAIN, TASK, SUB-TASK AND FUNCTION

Figure 2.4 illustrates the relationship between the terms *system*, *operational design domain*, *task*, *sub-task* and *function* as they are used this book.

System

Put simply, the term "system" refers to the entirety of things that perform tasks, interact, exchange information or data or otherwise collectively contribute to achieving some human purpose or goal. In most cases, certainly in the case of complex systems, that is going to mean a *socio-technical* system: achieving most complex human goals requires co-ordinating the activities, values and motivations of a combination of human and technological elements, often distributed in time and space.

Defining the limits of any even moderately complex system can sometimes seem an impossible task: wherever a boundary is drawn, it is not usually difficult to expand it to include additional, typically higher level, objectives. So for the purpose of doing any useful systems analysis of the Human Factors requirements

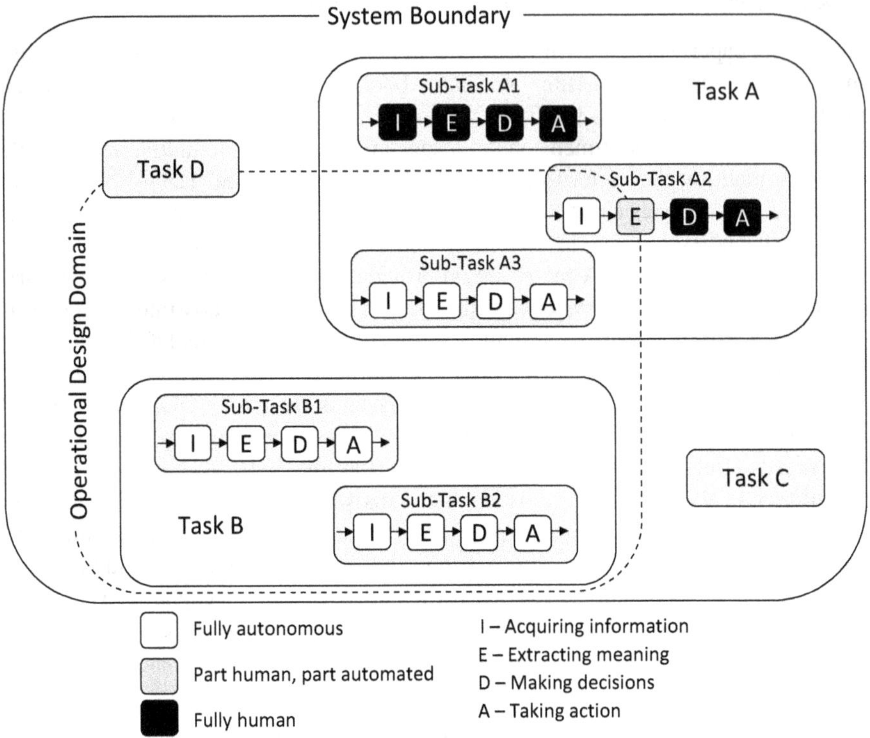

FIGURE 2.4 Illustration of system, tasks, sub-tasks and functions

and risks associated with the development of an autonomous capability, we need to start by being clear where the overall boundary of the socio-technical system we are interested in lies. That is what is referred to as the "System Boundary" as shown on Figure 2.4. However, again, it can be frustratingly difficult to clearly and unambiguously define a system's boundary: it can depend on the goals and objectives to be achieved, the nature of the activities and processes involved, as well as the environmental and other conditions associated with the system's performance. Sometimes, the best we can do is to insist that the description should be explicit, and as clear as possible.

Operational design domain

We need to be equally clear about the limits of conditions the automation is intended to be able to operate in and the tasks it is intended to be able to perform, whether with or without human support. For the purpose of the book, I'm going to adopt the term the *Operational Design Domain* (ODD) to describe the limits of the automation's intended capabilities. This is the term used by SAE recommended practice J3016 [5] to describe the conditions and limits automated vehicles are designed and expected to operate within. Generalising from the SAE definition to make it relevant to other types of systems, the ODD could be defined as;

> Operating conditions under which a given automation system or feature thereof is specifically designed to function, including, but not limited to, environmental, geographical, and time-of-day restrictions, and/or the presence or absence of certain characteristics of the environment. (amended from [5])

There is a neat illustration of the limits of a highly automated system's capabilities in the film "Good Night Oppy". The robotic rover Opportunity ("Oppy", or MER-B to give it its' Sunday name), was one of two rovers sent by NASA to explore the surface of the planet Mars in January 2004. Although its mission was planned for 90 days, in the end, with careful and intelligent management of its' power supplies (by its' human controllers 150 million miles away back on earth) Oppy continued actively exploring the planet until June 2018. As with the other five robots that have been sent to the planet under NASA's Mars Rover missions, Oppy used a variety of intelligent technologies that allowed it to navigate and explore the Martian surface autonomously, even making important decisions independently from its human controllers. Once the humans on earth gave it an objective for each day's exploration, the robot could plan its own route using machine vision and other technologies, to navigate its way to the day's objective.

Towards the start of the film, there is a sequence where Oppy comes to a stop in its journey across the Martian surface. Its machine vision capability has not been able to interpret images in the path ahead captured by its on-board cameras. The sequence is revealing both in terms of the limits of Oppy's autonomous capabilities, and in the nature of the cooperative relationship between the AI-enabled robot and its human controllers. Rather than risk a potentially disastrous collision (the robot and mission represented a US national investment running into billions of dollars) the robot stopped and waited for instruction from earth. After a few moments delay (in the film – the actual time delay for signals to travel to earth and back was around 26 minutes, plus time for the ground controllers to bring human intelligence to make sense of the imagery), a message appears on-screen saying "it's only your shadow. It's safe to continue". As impressive as the machine vision and other AI capabilities of the robot were, it had reached the limit of its ODD: it had encountered an image it was not capable of interpreting without human help. Of course, Oppy's designers knew that machine vision would not be able to recognise every image the robot's cameras would encounter on the surface of Mars without help and additional training. Though among the enormous number of possible images that could be encountered, it's unlikely they would have known that the robot's own shadow, in the specific orientation, would have been beyond its abilities at that moment.

Tasks and sub-tasks

Figure 2.4 shows details for two of the four tasks needed for this system to achieve its goals. Task A comprises three individual sub-tasks, while task B comprises two sub-tasks. A sub-task is defined as the lowest level of activity that requires each of the four basic information-processing functions[7] to be performed[8];

- Acquiring Information,
- Extracting Meaning;

- Making Decisions;
- Taking Action.

The figure shows that while Task B lies entirely within the ODD, task A lies partly inside and partly outside it.

Figure 2.4 illustrates that, within task A, sub-task A1 lies outside the ODD of the automation. The automation has no capability to support the sub-task, so it is performed entirely manually. Everything associated with sub-task A3 on the other hand lies entirely inside the ODD, and this sub-task can be performed fully autonomously – all the constituent functions can be completed entirely without human intervention. Sub-task A2 however is more complex as it lies across the boundary of the ODD: some of it can be automated, but some needs to be performed by people. Of the four functions in sub-task A2, acquiring information is fully automated, and the automation collaborates with the human in extracting meaning from the information (as in the Oppy example above). Making decisions and acting however are outside the ODD boundary and are performed entirely by the human without support from the automation.

By contrast to task A, task B can be performed fully autonomously. All aspects of the two sub-tasks involved lie within the ODD: the automation has the capability of performing both sub-tasks without human intervention.

The system illustrated on Figure 2.4 is of course highly idealised and simplified for a system even having the automated capability described. It does not, for example, attempt to capture any of the tasks or complexity associated with the human engaging, adjusting, monitoring or intervening to take over control from the automation should that be necessary. But as the purpose here is purely to illustrate the relative meaning of the terms, it is sufficient for the purpose of the book.

THREATS

The concept of "threats" is well established in the world of safety management. Conceptually, it derives from the aviation sector where it was first developed in the mid-1990s as a way of thinking about the performance of pilots. The concept of Threat and Error Management (TEM) was developed by Robert Helmreich and his colleagues from the University of Texas [8]. Under the TEM concept, and put simply, a threat was considered as anything that needs to be managed by the pilots to maintain flight safety. The concept is used here in essentially the same way. It means anything that could lead to automation being unable to provide effective control over the activity or process it is designed to control without the intervention of a human. Threats can be either specific events (such as the failure of a sensor or other technical component), or situations (such as several things happening in the world around the same time).

Note that I am using the term "threat", rather than "hazard". There can often be confusion about terms such as "risk", "hazard", "threat" and others, particularly among those involved in conducting risk assessments who are not risk management professionals. A Hazard is best thought of as something with the potential to cause harm. Hazards are often a source of energy - pressure, temperature, electricity, chemicals, etc – though activities can also be hazardous - such as performing open-heart surgery, or landing an aircraft in strong, blustering cross-winds. Threats here are considered in the sense of "a threat to the successful completion of the activity". Threats will often involve hazards, though not necessarily. Threats can involve things like not meeting targets such as completing operations on time, organs that might obscure a cancer site, inefficiency, not meeting financial goals or consuming excess resources.

To give an example, I recently attended a training session for mental health professionals that covered how to conduct risk assessments for patients convicted of a criminal offence who have mental health problems. The instructor asked us to consider a hazard where a disturbed patient had a knife. As politely as I could, I suggested that she was really referring to a threat, rather than a hazard: the knife (or at least the ability of its sharp edge to damage human flesh) is undoubtedly a hazard, though the combination of a disturbed patient wielding it is a threat, not a hazard.

Threats can be identified by detecting a specific piece, or combination of pieces, of information; an unexpected increase in pressure, or unexpected change in some other data; the unexpected rapid slowing of a vehicle ahead; or, indeed, a disturbed patient known to have mental health issues being in possession of a knife. They can also be identified by detecting a change in something over time, such as temperature or pressure moving gradually towards a limit, or a movement in an unexpected direction. The concept of threat implies the potential for risk (i.e. the potential to cause loss or harm). Identifying a threat implies some assessment of the likelihood that it may lead to a problem; i.e. of the associated risk.

To understand the role of a human supervisory controller, we need to distinguish between two distinct types of threat;

- ***Internal threats*** are things that form part of the design of the system that have the potential to prevent the automation from doing what it is designed to do should they not work correctly. Examples include technical components such as sensors, or activators going out of calibration, errors in software algorithms, or data becoming corrupted.
- ***External threats*** are events, conditions or objects that exist in the external world that do not form part of the design of the automation, but that represent risk to the ability of the automation to do what it is designed to do: conditions outside the automation's design envelope; a loss of information or data the automation relies on from external sources; or constraints on acceptable performance that the human is aware of, but that are not known to the automation.

Internal threats

Internal threats represent risk that the technology is not performing in the way it is expected to. In his book 'Taming Hal', Asaf Degani [9] describes a scenario where a commercial airliner exceeded it's permitted airspeed due to an issue associated with the design of the aircraft's automated vertical navigation system. The problem was only corrected when the pilots, in the role of human supervisory controllers of the automation, detected the threat and intervened. Here's a summary of Degani's description of what happened.

Shortly after take-off, the pilots engaged the vertical navigation mode of the aircraft's autopilot to achieve a flight level of 11,000 ft. The system automatically chose the most economical airspeed of 300 knots. Before reaching 11,000 ft, air traffic control instructed the pilots to reduce speed to 240 knots to maintain horizontal separation from aircraft ahead. This was achieved by manually adjusting the required speed, overriding the automation (but without taking over manual control of the aircraft). Shortly after reaching 11,000 ft, the air traffic controller instructed them to climb to 14,000 ft. The co-pilot again engaged the vertical navigation mode, setting the required altitude accordingly. Unknown to the pilots, re-engaging the vertical navigation mode also allowed the automation to again select the most economical speed – which was still 300 knots. It was only when the aircraft reached 290 knots – 50 knots faster than the speed restriction given by air traffic control – that the pilot noticed the divergence in speed and intervened, taking over manual control and reducing speed to the required 240 knots. The whole incident lasted around 15 seconds. The internal threat built into the design of the system allowed the automation to exceed the restriction on the aircraft's speed imposed by verbal communication between the air traffic controller and the pilots that was unknown to the automation.[9]

In 2024, Sydney Dekker and David Woods published a paper [10] discussing problems with what they referred to as "wrong, strong and silent" systems. The paper was concerned with systems where automation is given high degrees of both autonomy – being able to perform functions with little or no human involvement – and authority – being able to override the actions of human operators if the automation decides it is the right thing to do. Their interest was in how such systems can actively create problems when there is a mismatch between the world the automation believes it is in, and the world it is actually in: they describe this as systems "misbehaving";

> When assumed and actual worlds do not match, automated systems will misbehave, taking actions that are inappropriate and possibly dangerous. Some supervisory role has to recognize the behavior as inappropriate or dangerous and intervene by redirecting the automation.
>
> …the system will do the right thing—in the sense that the actions are appropriate given its model of the world, when it is in a different world—producing unexpected/unintended behavior and potentially harmful effects [10]

Asaf Degani uses the phrase "dumb but dutiful" to describe something similar in the behaviour of an autopilot [9]. Although the issues are relevant to any form of complex systems where automation is given high autonomy and high authority, Dekker and Woods were writing largely from the perspective of what has been learned from a series of aviation accidents. They drew on the two crashes of the new Boeing 737 MAX aircraft, in Indonesia in October 2018 and in Ethiopia in March 2019 as a case study to illustrate their argument.

In automated systems designed to exert control over objects or activities in the real-world, mismatches between the world the system "thinks" it is in and the real world can arise for many reasons. To give just two examples;

1. When automation relies on data from sensors or other external sources to compute solutions or perform actions, and that source data is wrong. This can occur, for example, if sensors are mis-calibrated, or otherwise fail to perform correctly. The "misbehaviour" occurs when the system continues to function, believing its model of the state of the world derived from the sensor data provided to it is correct, when it is actually wrong. This was the situation in the crashes of the two Boeing 737 MAX aircraft in 2018/2019. It is different from the events that caused the loss of the Air France flight over the Atlantic in 2009. In that case, when the aircraft's pitot tubes froze, and the flight computers had no input of airspeed data, the autopilot system simply stopped working and relied on the human pilots to re-take control. Though, tragically, the pilots did not realise what had happened and were not able to take over manual control.[10]
2. If the model the automation holds about the world it is designed to operate in is incorrect. When a car in which automation has been given the authority to control the vehicle autonomously encounters a sharp bend in the road, the automation will automatically slow the vehicle down to a speed where the bend can be navigated safely. The decision-making algorithms that decide how much to slow the car down by to manoeuvre the corner safely are based on a mathematical model built-in to the automation of the dynamics of the vehicle: the safe speed will vary between different models depending factors such as the car's weight, it's mass and the design of its chassis, its centre of gravity, and so on.[11] Models of the dynamic characteristics and behaviour of the object or process being controlled are held by every form of automation that performs real-time control. Models can be expressed in mathematical or other forms depending on the nature of the world the world the algorithms need to interact with.

   These types of models can, however, be wrong for a variety of reasons. In the case of the car taking the corner, the mathematical model of the car's cornering abilities must include assumptions about things like the quality of the road surface, as well as the state of the vehicle's tyres. If the road surface happens to have a covering of black ice, or a layer of oil following a spill, or if the depth of tread on the vehicle's tyres are well below the legal limits reducing their ability to grip the road surface, the model held by the automation of the car's cornering capabilities, and therefore the speed to take the bend may turn out to be significantly too high. A human manually driving the car might have a chance of detecting the problem and reacting to these situations. The autonomous vehicle would need to rely on the driver-as-supervisory-controller detecting the threat in time to act.

The potential for systems with high autonomy and high authority to carry on functioning, believing it is doing the right thing when, in reality, it is creating a problem is especially challenging for the human supervisors we are focused on in this book. Detecting and reacting to internal threats leading to systems misbehaving can be extremely difficult in complex systems that have high levels of autonomy. As Dekker and Woods put it;

> Some supervisory role has to recognize the behavior as inappropriate or dangerous and intervene by redirecting the automation. Stepping from a monitoring role into an active role in a developing non-normal or abnormal situation is a difficult shift. If the behavior and configuration of automated systems is opaque (What is it doing? What will it do next? What is the configuration currently controlling key parameters or processes?), and if the mechanisms for re-directing parts of the suite of automation are clumsy, then the integrated system has a built-in vulnerability to breakdowns where the automation is strong, silent and wrong. [10]

External threats

Here are some everyday examples to illustrate the concept of external threats;

- In recent years, I've noticed what seems to be a growing trend in the UK to do with drivers on slip roads seeking to enter the main carriageway of a dual carriageway or motorway. When I was learning to drive (c.1976), we were taught, in that situation, to slow down and wait for a clear gap in the traffic on the motorway. And, once a gap appears, to gently accelerate if needed to merge into place in the line of traffic on the main carriageway. Perhaps it is a sign of my ageing, but I've noticed in recent years that, being a driver on a motorway with a vehicle merging from the left (in the UK), drivers on the slip road frequently believe the correct approach is to accelerate to get in front of the traffic already on the main carriageway, forcing those drivers to brake to make space. Cars accelerating on an interception path from the left are a threat to the safety of my vehicle.[12]
- Another example happened to me as I was cycling to our bridge club recently. While there was no automation involved, the mental process I went through is essentially the same as that which frequently faces the human supervisor of an automated system. My route to the bridge club takes me through two suburban villages in the south of Glasgow. They are busy roads, lined with shops, restaurants, and commercial properties. As I approached the middle of the first village, I could see two pedestrians about 250 metres ahead of me step off the pavement and start to walk on the road in front of me. My first impression was that they had stepped off to pass some obstruction on the pavement. I quickly realised however that they were walking at a shallow angle to the pavement, gradually getting further into the road – and directly into my line of travel. So, I was faced with the decision, with rapidly reducing time available to make it, whether to slow down and stop (something that, as a 67-year-old cyclist having invested the physical effort in getting up to speed, you have a bit of a reluctance to do if you don't have to), or to run the risk of having to divert my route into the busy traffic. In the end, after frantically sounding my bike's bell to alert them to my presence, the pedestrians turned towards me and realised what they were doing. The behaviour of the pedestrians was a threat to the safety of my journey. A threat that I had to monitor to the point where I had to decide what action to take.
- As a final example, again to do with driving, I was recently driving at about 50mph on an 'A' road when I closed on a white van[13] ahead of me. I noticed

the van was extremely close to the car immediately in front of it – it looked to me no more than about 2 metres - though the road ahead was clear for some distance. The van's driver seemed to be impatiently looking for an opportunity to overtake but was being repeatedly frustrated by a combination of the curvature of the road and oncoming traffic. After watching for some time, I noticed yellow flashing lights moving slowly in our direction some distance ahead of the two vehicles. These turned out to be hazard warnings lights on a tractor moving at less than half our speed. Fearing the behaviour of the driver of the white van had the potential to lead to a need to take emergency action, if not causing a collision, I decided to slow my speed to keep well clear. The van eventually overtook, and the situation passed. What I saw as the reckless behaviour of the driver of the white van, combined with the fact I could see a slow-moving vehicle some distance ahead had led me to feel sufficiently at risk that I needed to take precautionary action: the combination was a threat to my safety, leading me to act. Reflecting on this event in the context of automated vehicles, I realised that, had my vehicle been in its' self-driving mode, this was an example of a threat that I, as the vehicle's supervisory controller could detect but that the automation could not. Enough of a threat to lead me to intervene to take control over the automation and move to a safe distance behind the potentially unfolding events ahead of me.

ABILITY, AUTHORITY, CONTROL AND RESPONSIBILITY

The academic Frank Flemisch has spent many years researching aspects of shared and cooperative working between people and automation. In 2012, he and his colleagues published a paper, building on earlier work, that provides a useful framework for thinking about and analysing the relationship between people and highly automated systems [11].

The framework is based around four "cornerstone concepts" that need to be understood and applied in the design of highly automated systems: ability, authority, control and responsibility. As they are central to thinking about and understanding the relationships between people and technology, we need to be clear about what they mean;

- **Ability** means possessing the means or skill to do something.[14] For our purposes here, it refers to whether the human or the technology possess the ability to perform some real-time activity;
- **Authority** is defined as the power or right to give orders, make decisions, and enforce obedience. It governs what the human and technology is, and is not, allowed to do. Authority needs to be given to the human or technology by some higher authority in advance of exerting control: neither should act without the authority. Flemisch et al distinguish between two types of authority: the authority to exert control over a function (control authority); and the authority to change where the control authority lies at any moment (which they refer to as (Control) Change authority). Note that, in principle,

both the human and the technology could have authority to exert control at the same time. This raises questions about where the governing authority lies;

- **Control** means having the power to influence or direct the course of events. For our purposes, we are interested in where control lies at any moment: whether it is the human or the technology that is in control of the activity at a specific moment;
- **Responsibility** means being accountable for something in the sense that you can be blamed if something untoward happens. Responsibility can be a difficult concept to pin down clearly. It may or may not imply legal liability: an individual's values, religion and self-identify, as well as social pressures, can lead people to feel a responsibility to do something or act in a certain way, even if they don't hold legal liability. There can also be differences between what an individual believes they hold responsibility for, and what their employers, peers or society are likely to hold them to account for. It is sufficient here, for the purpose of understanding the role and behaviour of a human supervisory controller, to view responsibility as being what the individual believes they will be held accountable for either by their employers, or others whose opinion matter to them.

There is nothing original in recognising the importance of these cornerstone concepts. They had been recognised in the research literature for a long time before the Flemisch et al paper was published. For example;

> Responsibility should always be accompanied with suitable authority. [12]
>
> …the analysis of problems with highly automated aircraft has shown that where responsibility is ambiguous or poorly indicated in the control station, several problems arise…. less than perfect understanding of each other's abilities and characteristics could lead to misunderstanding between operator and machine, and hence competition for control. [10]

A2CR CONSISTENCY

Figure 2.5 shows the expected relationships between the four cornerstone concepts of ability, authority, control and responsibility. For clarity, it is worth expressing these relationships (which Flemisch et al refer to as "double and triple binds") in language;

- the responsibility someone has should not be greater than both the authority assigned to them and the ability they have;
- the authority given to an individual should not exceed their ability;
- having the ability and the authority allows an individual to exert control over a function;
- having the authority to perform a function suggests a responsibility for the performance of that function (but does not necessarily impose it);
- having responsibility should provide assurance over how well control is exerted;
- exerting control over a function implies responsibility for the performance of that function.

The concept of A2CR consistency refers to the extent to which the relationships shown on Figure 2.5 have been maintained in a system once it is operational;

> A2CR consistency means that the double and triple binds between ability, authority, control and consistency are respected, e.g. that there is not more responsibility than would be feasible with the authority and ability, that there is enough ability for a given authority, that the control is done by the partner with enough ability and authority and that more responsibility is carried by the representative who has more control. [11]

Failure to maintain A2CR consistency can create issues in many ways;

- an agent (human or automation) that has authority to control a task, but that lacks ability is clearly in a risky position;
- an agent that holds legal responsibility for the performance of an activity but does not have authority or ability needs to have other means available to ensure it is able to exert control;
- an agent that has ability but does not have authority will need to be supervised.

A2CR consistency can be a useful concept in thinking about the design of systems that rely on human supervisory control. A lack of A2CR consistency in the design and deployment of an automated system does necessary imply that a solution is unacceptable. But it does require that other means or controls are put in place to ensure consistency is maintained in the relationships between the concepts.

Understanding what is needed to ensure consistency in the relationships between these cornerstone concepts is self-evidently important in the implementation of highly automated and autonomous systems. They have implications not only for employers, but for governments that legislate and provide permission for these systems to be sold and use in their jurisdictions. It is also critical that the concepts are consistent with the beliefs and expectations of the individuals filling the role of human supervisory controllers. Lack of confidence in their abilities, uncertainty about whether they have the authority to act to support or take over from automation, or concerns over their responsibilities and liabilities if they do - or do not - decide to intervene all have the

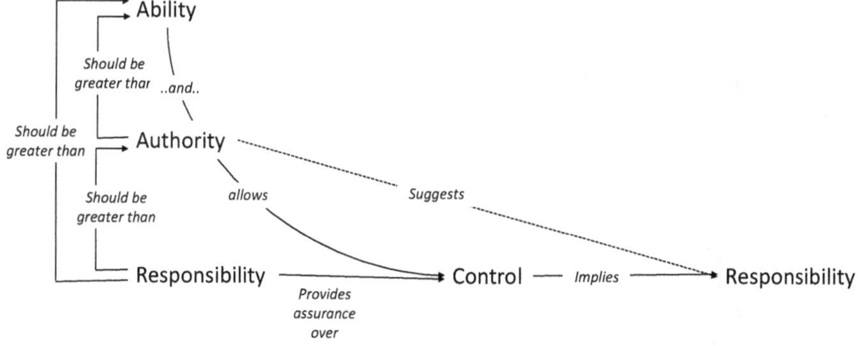

FIGURE 2.5 Relationships between ability, authority, control and responsibility (my variant after Flemisch et al. [Modified from 11])

potential to influence the effectiveness of someone filling the role of a supervisor. Which is why they are important to the comprehensive model of human supervisory control performance which is described in Chapter 9.

The A2CR framework is based on recognition that in other than the simplest systems, there can be a wide range of different relationships between what people do and what the automation does. Where the balance lies can change from moment-to-moment, depending on what tasks the overall system is engaged in, and what the relative strengths of the human and technological elements of the system are.

A2CR DIAGRAMS

Figure 2.6 illustrates diagrammatically the core idea of the relationships that might exist between people and technology in a shared or cooperative system [11]. The figure captures the entire theoretical range of ability available to control some function in a specific situation. The extreme left-hand side shows the situation where the human can control the entire function, while the technology has no ability. In this area, the function needs to be entirely manually controlled. The extreme right-hand side shows the opposite: the technology has complete ability to control the function in this situation while the human has no ability. This area represents autonomous control, where the technology controls the entire activity with no human support. There is of course nothing new in the basic concept shown on Figure 2.6a. It is a variant of the issue of levels of automation that researchers have been studying for decades.

Within the range of theoretically possible options for sharing control of a function between the human and the technology illustrated on Figure 2.6a only a small

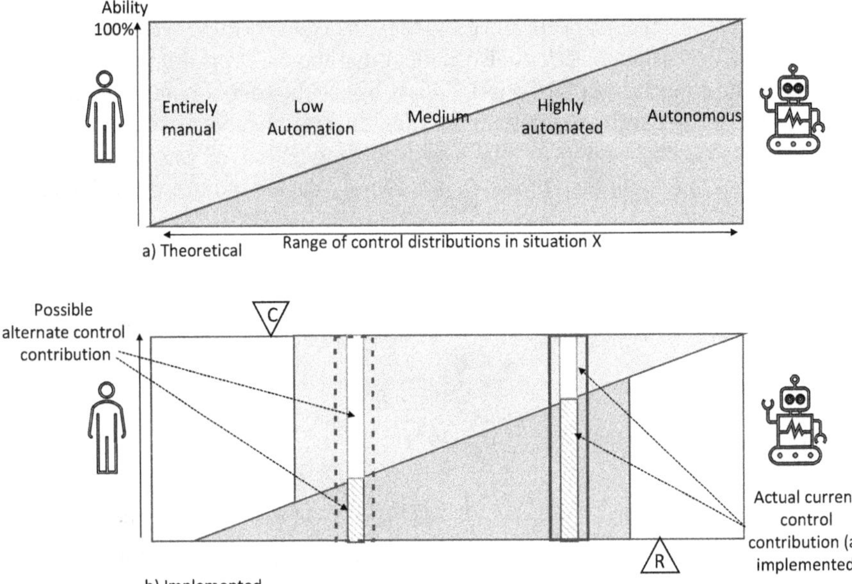

FIGURE 2.6 Illustration of theoretical range of options for shared cooperation between human and technology (Modified from [11])

number, perhaps no more than one or two, will be implemented and available for use at any time. An example of this is illustrated on Figure 2.6b which shows a system where, within the theoretically possible range, two actual possible control options have been implemented. In the first – towards the right – technology provides most of the control, with some human support. In the second option, towards the left, the human exerts most of the control, supported by the technology. Clearly, only one of these options can be in use at any one time. On the figure, the option to the right, which is drawn with a solid border is currently in use. The alternate option, shown with a dashed border, is available, but not currently in use.

Note that on Figure 2.6b neither the human or the technology has the ability to exert control across the full range of theoretical control options: in both cases, there are areas that either the human or the technology has no capability. Note also that in this case – fortuitously - between them, the combination of human and technology cover the full range of ability needed. In many real-world situations, that may not be the case.

An A2CR diagram such as that illustrated on Figure 2.6b can also convey information about where both *notional control* and responsibility for the performance of the activity lies. These are illustrated on Figure 2.6b using the "C" and "R" tokens. The location of the "C" token towards the human end of the figure indicates that, even when authority to control the activity has been given to the automation, notional control, exerted through the role of the human as supervisory controller, still lies predominantly with the human. On the other hand, the location of the "R" token towards the technology end indicates that responsibility for the performance of the system, in this case, lies predominantly with the people or organisation behind the use of the technology.

The concept behind A2CR diagrams doesn't stand overly rigorous inspection (for example, the suggestion of a continuous range of both ability and control implied by the horizontal and vertical axes quickly breaks down). Nevertheless, the diagrams can be a useful way of thinking about and communicating the potential relationships between ability, authority, control and responsibility, as well as the range of options that could be available to project teams involved in conceiving and developing highly automated systems.

SUMMARY

In today's world, the terms "Artificial Intelligence", and "AI", seem to have become ubiquitous while at the same time close to being meaningless. Their use in the general media, and by politicians and others often conveys not much more than a desire to suggest an association with the leading edge of extremely clever technologies. Outside of well-informed forums, there is rarely any attempt to be clear about precisely what the user means by these terms. The situation around automation and autonomous systems, while not being quite so opaque, has similarities. Terms are used in ways that, while they might be clear and meaningful to the authors, are somewhat less meaningful to the audience.

This chapter has sought to try to bring a degree of clarity about the way this book uses a number of terms and concepts that are central to talking about the role of people as human supervisory controllers of highly automated and autonomous systems.

The chapter has tried to be clear about the difference between "machines" and "automation" and when an automated system has "autonomy". It has explained how the concept of stages and levels of automation are used in the book, and what is meant by the terms system, system boundary, operational design domain, task, sub-task and function. The chapter has tried to be clear about where the boundary lies between supervisory and manual control, the different kinds of threats the human supervisor is expected to monitor for, and the importance of the cornerstone concepts of authority, ability, control and responsibility.

While others may take a different view on the meaning of some of these terms, the chapter has tried to be clear what is meant when they are used in the remainder of the book.

NOTES

1 According to Wikipedia, the world's first cash machine was put into use by Barclays Bank, Enfield, north London in the United Kingdom, on 27 June 1967.
2 Although their ubiquity seems to be declining as societies around the world move towards cash-less economies.
3 There are additional references to this literature in Part II.
4 The modifications in Figure 2.3b are concerned only with the labelling of the four stages. The labels used here are consistent with [3].
5 Chapters 8 and 9 discuss and emphasises the concept of "macrocognition" that builds on this point.
6 Ref [3] describes level 5 as being "Fully automated". I have changed this to "autonomous" to capture the distinction between automated systems that rely on some degree of human input to support their performance, and autonomous systems that are intended to achieve their entire purpose – assuming all necessary pre-conditions are met and the system remains within its design limits - without human support.
7 Note that when performing a Task Analysis of tasks assigned to humans, the lowest level of human behaviour of interest is sometimes referred to as an Activity. When they are referred to generically – without a specific analysis to hand - there can be a degree of lack of clarity about these terms. For the purpose of this book, the term "function" is reserved for this lowest level, on the grounds that it may be performed by combinations of hardware, software and human effort.
8 Note that complex systems may have many more layers of tasks between the top-level tasks and the lowest level requiring the four information processing functions.
9 Degani also provides an analysis of why the overspeed happened, though that is beyond our needs here. See Chapter 14 of [9].
10 Chapter 6 includes summaries and a discussion of what can be learned about human supervisory control from both of these incidents.
11 I used examples of this situation a few times in the narrative of my experience transitioning to owning a car with autonomous capability in Part 1 of the book. I mentioned my experience of times where the speed the car thought was safe to take a bend in the road was significantly above the speed I felt comfortable at. While I don't doubt the manufacturer's understanding of the dynamics of their vehicles, my level of unease was such that I would frequently take over control and reduce the speed to a level within my risk tolerance level.
12 I have not yet developed the confidence to find out at what point my car will detect the accelerating vehicle approaching from the side, and at what point after detecting the approaching car, it would take avoiding action.

13 In the UK at least, the "white-van man" has, in recent years, become something of an item of cultural folklore synonymous with thoughtless driving behaviour that causes frustration to other road users. I first came across the term when I worked for Shell International and attended a presentation given by one of our contractors. The presentation concerned a study of driver behaviour to support thinking about user's design needs in future Shell retail outlets. Reference to the "white-van man" was sufficient to allow the audience to recognise they were referring to drivers who would park or behave in ways not intended in the design, but that could interfere with or inconvenience other users of the retail station going about their business. Drivers of other colours of vans may behave in similar ways. But in the UK at least, there seem to be many more white vans than other colours.

14 Definitions are from Oxford Languages web pages in February 2025. They are identical to those include in the Flemisch et al paper.

REFERENCES

1. National Academy Press. Research and Modelling of Supervisory Control Behaviour: Report of a Workshop. Committee on Human Factors. Commission on Behavioural and Social Sciences and Education. National Research Council. 1984.
2. Amezaga F., Cummins T. & Hooker J. Automating Tubular-Running Operations to Increase Rig Efficiency. IADC/SPE International Drilling Conference and Exhibition; OnePetro. 2020.
3. CIEHF. Human Factors in Highly Automated Systems. Chartered Institute of Ergonomics & Human Factors. 2012. Available from: https://ergonomics.org.uk/resource/human-factors-in-highly-automated-systems-white-paper.html
4. Law Commission. Automated Vehicles: Joint Report. Law Commission No. 404, Scottish Law Commission No. 258, HC 1068, SG/2022/15. 2022.
5. SAE. Surface Vehicle Recommended Practice: Taxonomy and Definitions for Terms Related to Driving Automation Systems for On-Road Motor Vehicles. J3016. SAE International. 2021.
6. Parasuraman, R., Sheridan, T.B. & Wickens, C. A Model for Types and Levels of Human Interaction with Automation. *IEEE Transactions on Systems, Man and Cybernetics: Part A: Systems and Humans.* 2000 May;30(3):286–297.
7. Onnasch, L. Wickens, C.D., Huiyang, K. & Manzey, D. Human Performance Consequences of Stages and Levels of Automation: An Integrated Meta-Analysis. *Human Factors.* 2014 May;56(3):476–488.
8. Helmreich, R.L., Klinect, J.R. & Wilhelm, J.A. Models of Threat, Error, and CRM in Flight Operations. In Proceedings of the Tenth International Symposium on Aviation Psychology (pp. 677–682). Columbus, OH. 1999.
9. Degani, A. Taming HAL: Designing Interfaces beyond 2001. Palgrave Macmillan. 2001.
10. Dekker, S.A. & Woods, D.D. Wrong, Strong and Silent: What Happens When Automated Systems with High Autonomy and High Authority Misbehave? *Journal of Cognitive Engineering and Decision Making.* 2024 April. DOI:10.1177/15553434241240849
11. Flemisch, F., Heesen, M., Hesse, T., Kelsch, J., Schieben, A. & Beller, J. Towards a Dynamic Balance between Humans and Automation: Authority, Ability, Responsibility and Control in Shared and Cooperative Situations. *Cognition, Technology & Work.* 2012;14:3–18.
12. Moray, N., Inagaki, T. & Itoh, M. Automation, Trust and Self-Confidence in Fault Management of Time-Critical Tasks. *Journal of Experimental Psychology: Applied.* 2000;6(1):44–58.

Part 1

Transitioning to a self-driving car

A learning experience

INTRODUCTION TO PART 1

The three chapters in this part of the book set the scene. I'm going to reflect on my experience, as someone who has spent a career of more than 40 years specializing in the Human Factors associated with complex systems, becoming the owner of a new car, a Lexus NX450h+. A car that, to my surprise when I took ownership of it, offers a degree of autonomous capability. And I'm going to use that experience as the basis for identifying, reflecting on and explaining some lessons that those responsible for the development and use of these highly automated, and especially autonomous, systems need to take much more seriously: governments, regulators and insurers, as well as manufacturers and vendors.

The content of these chapters is not intended as a specific critique of Lexus or their products. There is no reason to believe that the new product development processes followed by Lexus are any better or worse than other global manufacturers: indeed, they might be expected to be among the leaders in these areas. Rather, the chapter is intended as a critique of process: both the process by which a global product having autonomous capability is developed and introduced, as well as the way society allows that kind of product to be sold and used.

The key lesson is one which has been recognised in the scientific and research communities for some decades: introducing autonomous capability changes the role of the user (in this case, the driver), from a manual controller to a supervisor.[1] Supervisory control relies on the driver being effective as a *proactive* monitor of the system: someone who actively seeks information about the status of both the activity

DOI: 10.1201/9781003679479-3

(the movement of the vehicle in the world) and the performance of the automation in advance of problems arising or the driver needing to intervene. *Proactive* operator monitoring is contrasted with monitoring *reactively*: simply responding to alarms or alerts from the system intended to attract the driver's attention.

To set out the argument as clearly as I can;

I. There is good reason to be confident that self-driving vehicles will become widespread, possibly even the norm, at some point in the future, at least in some countries, and in well-defined areas of the road network.

II. That time is, however, still some time in the future and relies on advances in technology, together with engineering and regulatory developments in the design, implementation and maintenance of the road infrastructure. There are going to be generations of vehicles designed, developed, used, and that become obsolete, before self-driving vehicles are the norm in any meaningful sense. Over that transition period, the human driver is going to continue to be required, both by vehicle manufacturers and the law, to be responsible for the safe passage of their vehicle.

III. Advances in technology or engineering are never going to change the fundamental abilities of the human brain to maintain sustained concentration over extended periods of time when the driver has little or no active engagement in a task. That is a result of evolution and would only change – assuming there was some evolutionary advantage (it is not easy to see what that would be) - over timescales that are orders of magnitude longer than the development of self-driving vehicles.

IV. The solution over the transition period must rely on the manufacturers of these vehicles taking every opportunity to use design and engineering to optimise the way the vehicle actively supports the driver in their new role as a supervisory controller. That means re-thinking the design of the driver user interface to autonomous capabilities and developing solutions that help the driver develop and maintain awareness of what the system is doing, what it thinks the world around it is like, and what is expected to happen soon. As well as helping them sustain attention and retain driving skills when they have little to do over sustained periods.

The path towards resolving this challenge will only really begin when the designers of systems having autonomous capabilities properly analyse the information needs and psychological demands on the users of the systems in their new role as a human supervisory controller reliant on proactive monitoring. What is clear, is that a mindset that believes that simply adding controls and information needed for autonomous functions to an interface designed to support manual control is not the route to effective long-term solutions. What is needed is a change of mindset, and a recognition of the changes that introducing autonomous capability brings to a system's users.

Putting Part 1 in context

This Part of the book is a narrative, based on my personal experiences over a period of 3–4 months.

It is, of course, bad practice, especially in the psychological sciences, to try to extract general lessons based on a sample of one user and one product. Anyone that wanted to defend the industry against the arguments I make in this Part of the book could readily argue that it is not based on "research": it is not the result of a structured investigation, carried out under controlled conditions, or been subject to peer review. It is no more than one person's opinion.

The interested reader who is new to the topic of Human Factors in Autonomous Vehicles (AVs) should be aware that there exists a very substantial, generally high quality and rapidly growing body of research evidence, much of which is easily accessible via internet searches. That evidence has been generated both by academic as well as industry-based research laboratories around the world. Much of it aims to support, guide and help validate the development of future generations of AVs.[2] Frequently it supports the development of national and international regulations around licencing emerging AV capabilities.

The published research literature into Human Factors and automated vehicles includes large and growing bodies of both objective and subjective data gathered under more-or-less controlled conditions. Much of this data is gathered using high-fidelity vehicle simulators. Where possible, studies use large enough samples of human subjects to able to carry out statistical analyses with the aim of being able to identify research findings unlikely to have arisen through chance. Objectively measured data include things such as where people were looking at different times in experimental scenarios, how, when and how frequently they interact with the automation, and how much time they spend engaged with non-driving activities. Researchers often use EEG measurements of brain activity as a way of estimating the cognitive demands involved in monitoring an automated vehicle. Subjective measurements often include measures of the levels of mental effort people felt they experienced monitoring and interacting with automated systems, levels of situation awareness, as well as various subjective reactions to the experience of "driving" automated vehicles.[3]

Perhaps the most important and influential body representing Human Factors as a professional discipline active in influencing developing regulations around autonomous vehicles is the HF-IRADS (Human Factors in International Regulations for Automated Driving Systems) committee. HF-IRADS operates under the auspices of the International Ergonomics Association (IEA), the principal body representing the professional interests of professional Ergonomists around the world.[4] The committee brings together Psychologists and other Human Factors specialists from around the world, including representatives from government bodies, academia, and industry. The committee supports the work of WP.29[5] of the United Nations Economic Commission for Europe (UNECE) on the safety of automated driving systems. Among other important contributions, in 2022 the HF-IRADS committee prepared a set of Human Factors principles to guide the development of future AVs [1]. While there is little new or original in the principles themselves in terms of Human Factors content, the fact they were prepared and submitted to such an influential international body under the auspices of the IEA is important.

Against that background of a substantial, high quality and growing body of research evidence, as well as internationally co-ordinated activity at the highest levels that draws on both research as well as the views of those actively engaged in it to advise future regulations and policy, what is the value of three chapters providing a narrative of one person's experiences? The answer comprises two parts.

The first is that you will not find the kind of experiences I describe in these chapters in any published research. That is hardly surprising: it would be inconceivable for any research organisation to set up experimental conditions and conduct and publish research that was equivalent to my experience as a buyer in the real world. No safety and ethics committee in any research organisation anywhere in the world would allow a research team to hand a driver the keys to a vehicle and allow them, unaccompanied and unobserved, to engage an autonomous driving system on a busy motorway with no training or supervision. Yet that is exactly what happened to me. And it is what must be happening every day all around the world as vehicle manufacturers transition to selling, and governments allow the use of, vehicles with ever increasing autonomous capabilities. All with little or no recognition of the need to properly design for and support drivers as they transition from entirely manual to increasingly autonomous vehicles. So the first part of the answer is that it is highly unlikely that any institute, academic, industrial or government-based, could conduct and publish work describing the experiences I, simply as a member of the car-buying public, went through.

The second part of the answer is to do with the concept of what Psychologists and Human Factors professionals refer to as "situated rationality". The concept of situated rationality captures the idea that, when we seek to learn about how or why people made the decisions or behaved and acted in the ways they did in the events leading up to any sort of adverse event, we should not work backwards from a knowledge of what they actually did (or did not do) to try to understand and learn from why they did it. Rather, we should try to understand the world as it appeared to the people involved, ***in the situation and the context they were in*** at the time they made decisions and acted[6]: we should try to put ourselves in their shoes and understand the world as they understood it at the time.[7]

The narrative in these three chapters represents something close to the situated reality I experienced in near real-time over the period of around 3–4 months while I was learning about the autonomous capabilities of my new car. A period when I was transitioning from someone who, while I had years of experience using cruise control, would never have conceived of taking my hands off the wheel and allowing my car to drive me at 70 mph on the outside lane of a motorway. The content of the narrative was captured contemporaneously with the experiences described, either by voice recordings made at the time, or through written notes made usually on the same day as the experiences occurring.

With hindsight, had I realised before I first sat in my new car that I was going to spend hours documenting my experiences over the coming months, and that I would eventually see those notes as capturing something of the reality of the situation I was in at the time, I would have prepared better. I would, at least, have kept a detailed diary. And I would have put effort into being ready to verbally (perhaps even visually) record what I was experiencing, what I thought about it at the time and why I made the decisions and took the actions I did. I might even have been systematic about making sure I properly captured the situation (weather, time of day, speed, volume of traffic, route, etc) and the context (was I in a hurry, what had I just been doing, what was on my mind beforehand, etc) I was in each time I engaged or had issues with the self-driving capability. Unfortunately, that opportunity was missed. Though if there are any other experienced Human Factors or Applied Psychology professionals about to embark on a similar journey of transition to an autonomous vehicle, and interested in capturing their experience, hopefully they will be better prepared than I was.

As an example of how this can be done, in 2017, Mica Endsley published a paper documenting the results of a "naturalistic" driving study in which she documented her own experiences driving a Tesla Model S over a six-month period [2]. Unlike my experience however, where I found myself in control of a car with self-driving capability unintentionally, with no attempt to prepare to record my experiences and with no training at all on the car's automated features, Endsley not only received some (though not very good) initial training from the service representatives, but was prepared for her study. At the end of each trip, she collected data on her experiences (such as the distance driven, weather conditions, the automation used, and any issues encountered) and completed a 5-point self-rating covering satisfaction, trust in the automation, and perceived workload. She also made regular assessments of her situation awareness and workload at the moment the assessments were made. Being a good scientist, Endsley also recorded the same data in a control condition over a one-month period in a different car prior to driving the Tesla.[8] The purpose here is not to compare Endsley's findings with my own experiences. Though it is worth noting the extent of similarity between the problems and issues we both faced.

A word on "autonomy"

Throughout this Part of the book, I am going to refer to my vehicle as having, under some circumstances, self-driving or "autonomous" capability. The use of that term may seem contentious in some quarters, most especially among those who manufacture and sell vehicles having the capabilities I refer to. It is clear from the literature surrounding the ongoing development of national legal frameworks and attempts to agree international regulations surrounding the development and introduction of automated vehicles that, while vehicle manufacturers are keen to develop and offer autonomous capabilities to their customers, they are significantly less keen to accept that their vehicles have autonomous abilities. The reason seems to be the desire to avoid their vehicles being classified as Level 3, rather than Level 2 in terms of the SAE/ISO J3016 framework for levels of vehicle automation [7]. The significant difference being that if systems are classed as Level 2, the driver maintains legal responsibility for the safe performance to the vehicle. At Level 3, manufacturers are likely to need to accept legal liability for accidents.

Probably the most prominent and high-profile example of this is the way car manufacturer Tesla's Chief Executive Officer Elon Musk refers to the "Autopilot" functionality available in Tesla vehicles. Tesla's website repeatedly makes clear that "Autopilot" does not make the car "autonomous".

> Autopilot, Enhanced Autopilot and Full Self-Driving capability are intended for use with a fully attentive driver, who has their hands on the wheel and is prepared to take over at any moment. While these features are designed to become more capable over time, the currently enabled features do not make the vehicle autonomous…. The currently enabled Autopilot, Enhanced Autopilot and Full Self-Driving features require active driver supervision and do not make the vehicle autonomous.[9]

On the other hand, it is clear from many pronouncements made publicly by Elon Musk himself that he, and therefore presumably the company, consider the vehicles as having autonomous capability. In 2016, he told a RECODE conference that "The model X and Model S at this point can drive autonomously with greater safety than a person". And in a TED talk in 2022, he stated that "The car currently drives me around Austin most of the time with no interventions".[10] It seems clear that a vehicle that usually drives Mr Musk around Austin with almost no human interventions is performing in a way that would be considered "autonomous" under any reasonable definition.

NOTES

1. See Chapters 2, 7, 8, 9 and 12 for a detailed review of the characteristics of supervisory control and proactive operator monitoring.
2. In [2], Mica Endsley includes a summary of key research findings into the Human Factors of autonomous vehicles, as well as summarising the three key issues: Attention and trust, Engagement and Workload, and Mental Models.
3. There are a number of recent books describing bodies of research into Human Factors of AVs. See for example [3].
4. Full disclosure: in 2022 I was elected a Fellow of the IEA.
5. Working Party on Automated/ Autonomous and Connected Vehicles (GRVA).
6. See [4] for definitions and an introduction to the importance of distinguishing between Situation and Context when trying to understand human performance. Chapter 10 also explains the concepts.
7. For an extended discussion of the concept of situated rationality, see [5]. In [6], Sydney Dekker also provides a clear explanation of the need to understand human behaviour in terms of the context in which it occurs.
8. Coincidentally, that car was a Lexus 350ES.
9. https://www.tesla.com/en_gb/support/autopilot. Accessed 20 February 2025.
10. Elon Musk: Elon Musk talks Twitter, Tesla and how his brain works — live at TED2022 | TED Talk.

REFERENCES

1. UNECE HF-IRADS. Human Factors Principles to Guide the Design, Standards and Policies for Automated Driving Systems. Informal Document No. 2. Economic Commission for Europe. Inland Transport Committee. Global Forum for Road Traffic Safety. 2022.
2. Endsley, M.R. Autonomous Driving Systems: A Preliminary Naturalistic Study of the Tesla Model S. *Journal of Cognitive Engineering and Decision Making*. 2017;11(3):225–238.
3. Young, M. & Stanton, N. Driving Automation: A Human Factors Perspective. CRC Press. 2023.

4. CIEHF. Human Factors in Highly Automated Systems. Chartered Institute of Ergonomics & Human Factors. 2012. Available from: https://ergonomics.org.uk/resource/human-factors-in-highly-automated-systems-white-paper.html
5. Townley, B. Reason's Neglect: Rationality and Organizing. Oxford University Press. 2008.
6. Dekker, S. The Field Guide to Understanding Human Error. Ashgate. 2006.
7. SAE. Surface Vehicle Recommended Practice: Taxonomy and Definitions for Terms Related to Driving Automation Systems for On-Road Motor Vehicles. J3016. SAE International. 2021.

3 Taking ownership of my new Lexus

I recounted in the Prologue the feelings of surprise and confusion I experienced after being handed the keys to my new Lexus NH450h+, with no training or briefing of any sort on the automated features of the car, and let loose into the rush-hour traffic in central Glasgow. That surprise and confusion started shortly after I left the showroom, joined the M8 motorway, and tried to turn on the car's cruise control - a function I had grown to rely on in all the cars I had owned over at least the previous 15 or so years. The layout and appearance of controls on the steering wheel looked - superficially it turns out – nearly identical to other models of car I was used to. However, the cruise control would not turn on. And I had no idea why: I kept getting a text message on the drivers display saying cruise control was not available.

I clearly recall the confusion and mental effort involved as I tried to divide my visual attention between the world outside the vehicle as I drove down the motorway, and the information cluster and steering wheel controls in front of me inside the car. I didn't have time to take my eyes off the road for long enough to properly locate and study the numerous icons and controls. I didn't understand what the text message displayed on the user interface meant (I found the text difficult to read, and the abbreviations used were meaningless to me as a new owner). I had no idea why cruise control would not turn on, or what I needed to do to make it work. Fortunately, I did remain in control of the car.

Over the next few days, I spent time studying the car's Information System and User Guide to try to understand how to use the cars many features. That itself turned into a learning experience. Chapters 3, 4, and 5 explore in some detail the kinds of issues I faced transitioning to owning my new car over the coming weeks and months. They describe my reflections on them and lessons that can be learned, not only for autonomous vehicles, but wherever autonomous systems are being introduced to control previously manual activities.

DRIVER-ASSIST FEATURES ON THE LEXUS NH450H+

To provide context for this and the next two chapters, a little background on the assisted driving features available in my version of the Lexus NX450h+ is necessary. My model comes with three driver assist features[1];

- Cruise Control: the vehicle maintains the speed set by the driver, but the driver retains control of direction and lane keeping as well as the distance from vehicles in front;

DOI: 10.1201/9781003679479-4

- Speed Limiter: the driver controls the vehicles direction, lane position, as well as speed and distance from vehicles in front, but the vehicle limits the maximum speed;
- Adaptive Cruise Control (ACC): the vehicle controls speed, as well as distance from vehicles in front.

In addition, the vehicle is provided with Lane Tracing Assist (LTA): the vehicle's radar detects lane edge markers and automatically acts on the steering wheel to maintain the vehicle's position in the lane. The driver can temporarily override the LTA – for example to change lane or turn onto a different road – but, unless the system is disengaged, the vehicle takes over again once the driver stops applying force to the wheel.

Only one of Cruise Control, Speed Limiter or ACC can be engaged at any time. Though LTA can be engaged in combination with any one of the other three features. None of the assisted features, nor the LTA, could on their own be considered autonomous: the driver retains real-time manual control over either the vehicle's speed (including obeying speed restrictions and maintaining a safe distance from the vehicle in front) or lane keeping.

However, if the combination of LTA and ACC are both engaged simultaneously (I will refer to this state as "LTA/ACC"), the vehicle can be considered to be acting autonomously of the driver: in this situation it is genuinely "self-driving". This is a fundamentally different state from any previous form of driving. With both features engaged, the driver can[2] take their hands, feet, and even attention away from controlling the vehicle. The car will drive itself (at least within constrained conditions of maintained "highways or expressways:"[3]), staying in lane and maintaining a minimum distance from the vehicle in front with no driver input at all: if the vehicle in front slows down or even stops, so will the Lexus. And if the road and driving conditions remain within the design limits of the vehicle, it will continue to do so safely until the driver takes over control. Though if the road conditions go outside the design limits of the vehicle (for example, if lane markings disappear) and the driver does not take over control, disaster beckons.

Figures 3.1 and 3.2 show the information displays and controls available to the driver to support these driver-assist modes on my Lexus. To enter one of the assisted modes, the driver must first press the 'Driving assist mode select switch'[4]: each press toggles the current selected mode between Cruise Control, Speed Limiter and Adaptive Cruise Control. The associated information icons appear in the Multi-Information Display on the information cluster (Figure 3.2) using off-white characters. It is only when the driver presses the "Driving assist switch" that the currently displayed mode is activated. At this point, the symbology in the Multi-Information Display becomes colour-coded (predominantly green lines) and of higher luminance, increasing their visual prominence. If no action is taken within around 10–15 seconds of selecting the mode, a text message appears in the information area saying "Push [driving assist switch icon[5]] to Activate Driving Support System".

Figure 3.3 illustrates the icons associated with these features. These icons appear both on the right-hand side of the steering column and on the multi-instrument display area of the information cluster.

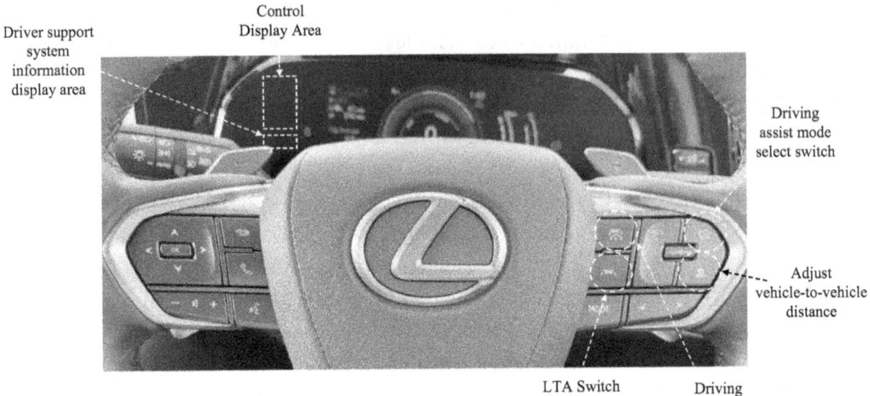

FIGURE 3.1 Close-up view of driver assist controls on the steering column of my Lexus NX450h+

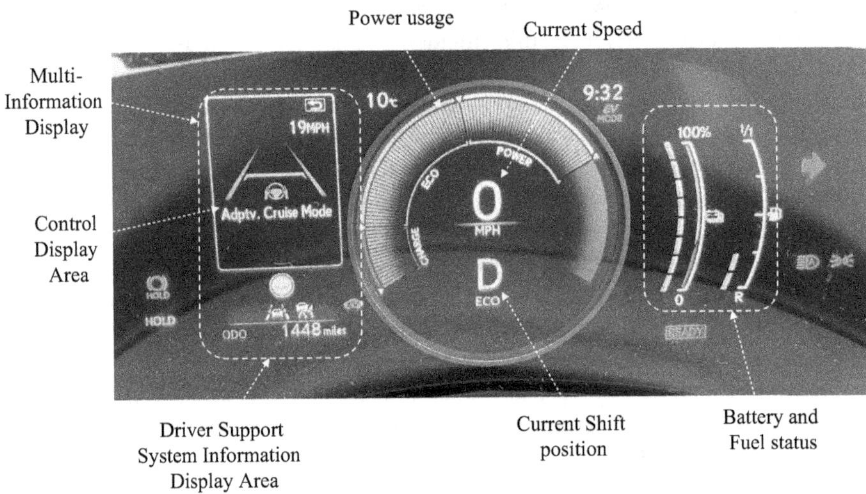

FIGURE 3.2 Close-up view of the instrument cluster

It had been more than six months since I had researched and placed my order for the car. And I had been given no briefing at all on the driver assist system before being handed the keys to my new car. In that context, the confusion I experienced during my first drive can be easily understood;

- the driver assist mode select switch is in the same location as the cruise control switch was in my previous car. The graphical icon associated with it also looked similar, at least at a glance. I assumed that pressing the switch would initiate cruise control, as I had long been accustomed to;

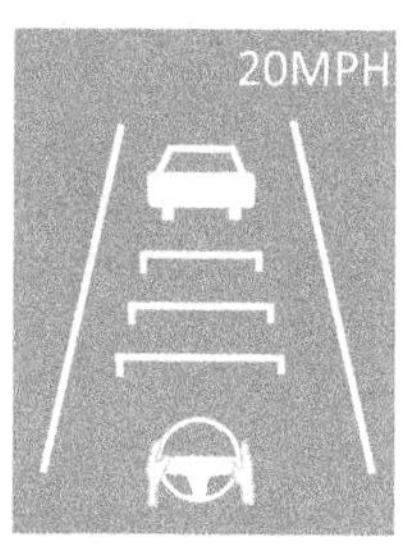

Combined ACC<A
(Information cluster only)

Driver Assist Switch

Driver Assist Mode Select

Lane Tracing Assist (LTA)

Vehicle-to-Vehicle Distance

FIGURE 3.3 Illustrations of driver assist icons on the Lexus NX450h+

- in fact, on the Lexus, pressing the switch toggles through the three possible driver assist modes but does not actually activate any of them. The mode only becomes activate when the driving assist switch is pressed. I didn't know this;
- at this point, at least four factors were combining to create my feelings of surprise and confusion;
  i. having pressed what I assumed was the cruise control switch, the system did not take control of the vehicle's speed. I did not understand why or what I needed to do;
  ii. I was not aware that the vehicle had two other driver assist modes (other than manual and cruise control);
  iii. although I could read the text message, I could not make out what the icon embedded in the message was supposed to be: it is too complex a graphic to be uniquely identified and understood at a glance – certainly by someone having no knowledge of how the system works, or prior exposure to the icon. Not to mention someone driving in motorway traffic, in low light conditions, having to maintain visual awareness of the state of the busy road outside while trying to read and understand the message and scan the vehicles controls.
  iv. as well as maintaining attention to the vehicle's movement on the road, and in low light conditions making it difficult to make out details on the steering wheel, I was trying to visually scan the icons on the steering wheel and compare them with the icon that was barely legible on the visual display. To a first-time user, at a quick glance the driving assist switch icon looks like the driving assist mode select icon embedded on the steering column.

I quickly gave up trying to use cruise control and reverted to manual control for the journey home. But for the remainder of the trip I wondered why it was that I had not been able to figure out how to turn on the car's cruise control: a function I had been familiar with across a range of vehicles for many years and had never previously experienced a problem with. Indeed, a function I had usually found straightforward and easy to use, and which I had come to take for granted. My confusion was really down to my unfamiliarity with the driver user interface on the Lexus. It was not because the vehicle had autonomous capability. Though with the many advanced features, the driver interface is considerably more complex than I have been used to.

Once I safely arrived home, I opened the car's on-screen Information System and manual User Guide [1] and set about finding out where I had gone wrong. This, however, led to another unsettling discovery: if the dealership I bought the car from had not ensured I was adequately equipped with the knowledge I needed to operate the cars sophisticated features safely, and the user interface had not been sufficiently intuitive to allow me to work out what to do, the User Guide was not going to be of much help either.

I cannot pretend I have not been impressed with my car's abilities, because I have. In fact, despite the observations in these chapters, I'm delighted with it and find the driver assist features – at least now that I have learnt how to use them and understand some of their limitations - very useful.[6] Though a number of events occurred during this journey of discovery into the car's abilities and features that caused me to reflect on what seem to me to be significant Human Factors issues.

THE DRIVER USER INTERFACE

Before going into detail about my experiences learning about my new car, I need to say a few words about user interface styling: specifically, the use of screen space and symbols (icons), colours and fonts in the design.[7] Again, the purpose of this section is not to offer a critique of the design produced by Lexus for the NH450h+ model: my assumption is that other vehicle manufacturers will have approached the design of the user interface to their automated features in more or less the same way that Lexus had. Indeed, my argument here is not limited to vehicle manufacturers: from what I have seen and learned both through direct personal experience in my career, as well as researching for this book, the same approach is often true of highly automated systems being developed in many industrial sectors.

To provide a little context, user interface design has long been a core professional interest for me. The first major project I worked on, between about 1988 and the late 1990s, was as part of a large team developing the command system for the UK's new Type 23 Frigate, which was at that time under development. I had technical responsibility for a programme of Human Factors activities around the design of the operator consoles, as well as the human-computer interface between the system operators and its computers and other systems.

A major focus of this programme was to design and prove - to the customer's satisfaction - that all information presented on the consoles screens would be easy for the system operators to identify and understand (i.e. read text and understand the meaning of symbology). We put a lot of effort into developing and testing the legibility of different fonts, in different sizes, colours and luminance. We also developed around 80 graphical symbols needed to represent a whole variety of possible objects the operators needed to know about in the world around them (such as aircraft, ships, submarines or land). Operators needed to be able to uniquely identify every individual symbol, in any of four possible colours, at any of two possible luminance levels, and presented against any of four possible background colours. All of this work was done under the real-time pressures of a fixed price commercial system development project that had been won after a long period of ferocious competitive tendering. There were contractual acceptance tests, linked to major milestone payments, tied to our ability to prove the usability of the symbols, colours and fonts we used in the design. This was not a research programme. I'm pleased to say that, with the strong commitment of the Royal Navy to support the work, we comfortably passed all 110 Human Factors contractual acceptance tests.

With that experience as the first major project in my career, it is not surprising that screen layout and the usability and usefulness of text, symbols and colours used in user interface design has long been an interest of mine. Of course, when I first sat in a showroom model of the new Lexus NX450h+, I checked whether the information on the screen was clear. To my shame however, I did not do what any experienced Human Factors professional should always do: test the design under realistic task conditions – i.e. while driving on a busy road, and under the likely range of lighting conditions. During my test drive of the model, with everything else going on, including the salesman giving his pitch, I simply forgot.

The purpose here is to illustrate the key argument of this book: that from a Human Factors perspective the design of the user interface to highly automated systems cannot simply be treated as a variant of an existing manual system. Rather, it needs to be based on a proper understanding of the new role of the driver in supporting the automation and of what the user's requirements are to be able to fill that role. And a recognition of the demands on people as they transition from the traditional role of manually driving to the new role of being a supervisory controller of an autonomous vehicle. Fundamental to the driver being able to fill their new role effectively is the need to ensure they have the information they need, presented in a format that is easy to understand and use, when it is needed. The purpose of this discussion is to illustrate that argument.

The user's role and information needs

At the core of designing a good user interface is an effective mapping between, on the one hand, the information the user needs to support their various tasks and, on the other hand, the features available to the designer in laying out and presenting information to the user to support those tasks. Achieving an effective mapping requires those responsible for the design of the user interface to have a sound understanding of;

1. The role of the user in the overall system, the tasks they need to perform and the information they need from the user interface, as well as the relative importance of each of those items of information to be able to perform their role successfully.

2. The concept of "salience" in user interface design: that is, the extent to which the various features available to the designer (screen position, colour. size, shape, brightness, behaviour (e.g. flashing), etc) are effective in capturing the user's attention and conveying the relative importance of the information. Information that is highly salient, attracts the user's attention with little effort and is easy to understand and use of in performing tasks. Information that has low salience might require some time and effort to acquire and use. That time and effort can be acceptable provided the tasks the information supports are relatively low priority and risk.

The first event that really captured my professional interest occurred not long after I had the confidence to allow the car to slow itself as it approached traffic, eventually coming to a stop behind a stationary vehicle. Shortly after this – it may have been the same day – I found myself again approaching slow moving traffic: though this time I was surprised to suddenly realise that the car was no longer in ACC/LTA mode, but that I was in manual control of the car. And unless I took rapid action, there was going to be a collision. In her naturalistic report of her experience learning to use the automated features on a Tesla Model S [2], Mica Endsley reports having had the same experience of not being sure whether she or the automation was in control of the vehicle.

The psychology behind why I made this mistake is beyond the scope of this chapter, though fully understandable: highly skilled and practiced manual tasks (such as driving a car) can be carried out with little or no active conscious control – that is in the nature of what it means to be "skilled". In the situation I was in, although I was manually controlling the car as we approached the traffic ahead, my frame of mind was that the vehicle was still in ACC/LTA mode (as it had recently been), and I could therefore allow the car to slow itself.

Allow that my mistake could not simply be put down to carelessness in those moments, but that I was driving in the same pre- or sub-conscious state that any skilled driver would be in much of the time. I asked myself if there was anything in the design of the information cluster that was intended to help me maintain conscious awareness of what mode the vehicle was in, without simply relying on my awareness. Without relying on my knowledge and immediate memory, how would I know whether the car was in ACC/LTA mode, and acting autonomously? What was the car's information cluster telling me?

To begin to answer these questions, consider what information is available to the driver on the car's information cluster, and how it reflects their needs for information in the different vehicle states. Figure 3.4 shows the information cluster when the car is in manual mode, with ACC/LTA selected, but not enabled. From the driver's position inside the car, the most visually prominent features are;

- The power usage meter;
- Current speed indicator;
- Shift indicator (D for "Drive" in this instance).

These are the largest individual items in the cluster and are in the most visually prominent positions.[8] After these, the Control Display Area on the left-hand

FIGURE 3.4 Close-up view of the Instrument cluster. Vehicle is in manual control mode: Adaptive Cruise Control (ACC) and Lane Tracing Assist (LTA) are selected but not enabled

side (which is part of the multi-information display, that also includes the Driver Support Information Display Area) and the Battery and Fuel Status indicators on the right-hand side are probably of around equal visual prominence.

When in ACC/LTA mode, the Control Display Area shows the combined ACC/LTA icon including the current set cruising speed all high lit in a bright green colour. When in manual mode, when ACC/LTA is not engaged, this icon is drawn in a less bright, near white colour.

The Driver Support System Information Display area, at the bottom left of the multi-instrument display, is relatively less prominent. The icons in this area use the same colour coding policy as the larger ACC/LTA icon. There is a lot of detail in this small area: it is not easy to extract information from it when driving. This is, however, the only area that permanently indicates to the driver whether the vehicle is in autonomous mode.

Figure 3.5 shows how this area of the information cluster appears when ACC/LTA is enabled: the set cruise speed, steering wheel and lane edge markers are now all drawn in a bright green, increasing their visual prominence. There is also a symbol showing the vehicle has detected a vehicle in front. The set vehicle-to-vehicle distance is shown as a blue horizontal line.[9]

Note that the most visually prominent features of the information cluster remain the power usage meter, current speed and shift indicator - the same as in manual mode.

In the autonomous driving mode however, this is not the information that is most important to the driver at that time. The single most important thing the driver needs to know when the vehicle is in autonomous mode is that the vehicle is driving itself: everything else is secondary. In these moments, the speed the car is doing is not of primary importance to the driver. What is important – in fact critical - is that the driver is in absolutely no doubt what mode the system is in. Without simply taking over control, the only real way I, as the driver of my new Lexus, can tell that is by actively directing my visual attention to checking the large icon on the left of the

FIGURE 3.5 The car's information display with Adaptive Cruise Control and Lane Tracing Assist enabled: i.e. the vehicle is in ACC/LTA mode

information cluster. It is completely down to me as the driver to check and to realise the meaning of the *colour coding* on the icon (bright green meaning the system is active, off-white showing it is not). The design of the information cluster for this driving state does nothing to actively draw my attention to the different state the vehicle is now in. All it does is, passively, to show the system is in control by colour coding the icon, with a consequent increase in its luminance. It might be argued that if the driver is in any doubt, they should simply take over control. Which is probably correct. Although it raises the question of why provide the indicator at all if it is not needed? If it is needed, it should do the job required of it.

This situation gets worse: in exploring the drivers' facilities, I realised that the Control Display Area (Figure 3.2) is not dedicated to giving the driver a clear indication of the status of the driver assist system. The driver can choose to use this area to display three other types of information: driver information (such as range and fuel utilization), navigation data, and the settings of the audio-system (radio station or audio currently playing). An example of this is shown on Figure 3.6. Choosing to use this area to check my available fuel range was hiding the key piece of visual information that told me whether the vehicle was operating in its' autonomous mode.

So, even when in the ACC/LTA mode, with the vehicle driving itself, it is possible for the driver to choose not to display the large ACC/LTA icon. In this situation, the only way the user can confirm that the vehicle is in autonomous mode, is by looking at the much smaller icons towards the bottom left of the cluster: icons that, while it is easy to tell if they are engaged or not (because of the use of a bright green colour when engaged), are not easy to discriminate which driver assist state is active (LTA or ACC) without looking carefully at the icons. Even then, simply noticing that the systems are engaged does not, in itself, remind me what my current role is: manual driver or supervisory controller of the vehicle.

More could be said about the relationship between the content and design of the information cluster on the Lexus NX450h+ and the drivers information needs. But the above is sufficient. At this point in my experience with my new vehicle, and as

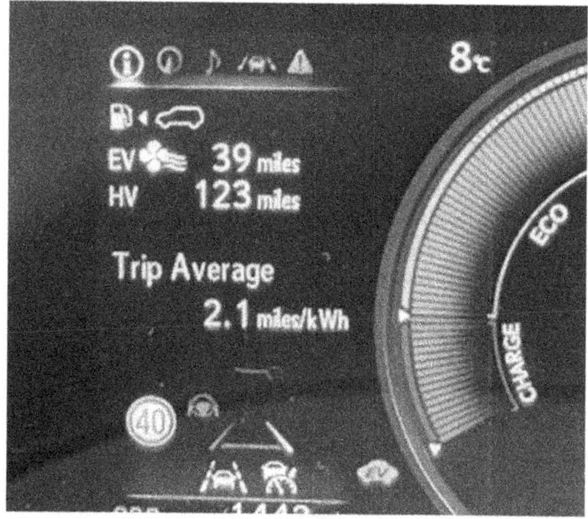

FIGURE 3.6 The Control Display Area with the driver information display selected. The large ACC icon is no longer displayed: the only indication of the vehicle mode is in the Driver Support System Information area at the bottom left. In this case, the driver assist systems are disengaged: the display would be the same, though colour coded, if ACC/LTA were enabled

someone with more than 40 years professional experience applying Human Factors in the design of complex systems, I had come to a clear conclusion: the design of the driver interface cluster to my new car had not been optimized to reflect the information I need when the vehicle is in autonomous mode. That is, the needs I now have as someone whose task has changed from manually driving the vehicle, to being a supervisor of the vehicle's automated systems.

What is the car thinking?

My third important learning as I explored the autonomous capabilities of my new car, was that I really needed to know what the car was "thinking". Consider the situation: provided I remain reasonably alert and attentive, I can see the road ahead and can anticipate the situation the vehicle will be in over the coming seconds, including any potential external threats that may need action: such as approaching slow moving traffic, unusually tight turns in the road ahead, lack of road markings, etc. Inside my Lexus however all I can see is information about the vehicle's status: current speed, shift position and whether the ACC and LTA are engaged. Although the ACC/LTA display shows vehicles directly in front or crossing onto or away from my vehicle's path, it tells me nothing about whether it is aware of the threats I can see that I know are going to need action in the coming seconds.

Here's an example of why this matters. On one of the motorways I use reasonably often, there is an especially tight bend. Roadside signs at the start of the bend advise drivers to reduce speed to 40mph. There are also crash barriers and black chevrons drawing attention to the hazard. The first time I approached this bend in my new car, the vehicle was in ACC/LTA mode, with speed set at 70mph. I could see the warning signs, and the approaching tight bend. Though there was nothing to tell me that the vehicle was aware of it.

Among its advanced features, the car is fitted with a system called "Road Sign Assist" (RSA) which can detect and interpret a limited set of road signs, including speed limits. This system informs the driver via an icon on the Driver Support area of the information cluster (on the bottom towards the left-hand side on Figure 3.2) what the current speed limit is.[10] This system however, only operates – or at least, the driver information changes – at the point, or shortly after, that the vehicle passes the speed limit sign. So in the situation described here, we were travelling at 70mph approaching a tight bend with road signs indicating a 40mph speed limit. I could see both the bend and the speed restriction about 10 seconds before we reached it. As far as I knew however, the RSA would only detect the speed limit sign[11] as we passed it. I had learned that the car would reduce its speed when it approaches tight corners, though I had no way of knowing what it would do in this especially tight bend. While I was inclined to assume the vehicle would take the corner safely on its own, I had no way of being sure. And I certainly was not going to wait and see – so I took control.

As another example, in low temperature conditions the car displays a text warning of the risk of ice, without disengaging the ACC/LTA functions. A human driver is expected to reduce their speed and drive in a way that takes account of this threat. However, while the message was intended to be read by me, the driver – although in the role of supervisory controller of the autonomous vehicle at that time – I had no way of knowing if the vehicle would also behave in a way that reflects the risk from an icy road.

In situations such as these, other than simply trusting the system, I have no information that gives me confidence the vehicle is aware of the threats ahead that I know about or that it is going to take the necessary action in sufficient time. This discrepancy - I can see threats in the road ahead, but don't know if the vehicle is aware of them - creates a sense of unease about whether I need to take control. That unease – which could easily become anxiety in some people - is amplified by the time-critical nature of the decisions involved: how long should I wait before acting? If I leave it too long, it may be too late.

The natural tendency in such situations, of course – at least my natural tendency – is to minimize the feeling of risk and take control even if it was not necessary. And to do it early. Even after I had gained confidence in the system's ability to slow down to maintain a minimum distance behind the vehicle in front, I frequently find myself taking control and manually slowing the vehicle when travelling at speed and I can see slow moving traffic ahead. The point is, that I have no way of knowing that the system is aware of threats ahead.

Of course, I was exploring the cars capabilities carefully with a high degree of concentration and attentiveness. It is possible that if I was less attentive in my role, I would be less aware of the situation ahead, and therefore less inclined to be concerned about whether or not to take control. Though that is precisely what is *not* wanted or expected of the drivers of autonomous vehicles. They are assumed to be attentive and ready and able to take control at any time: what SAE J3106 terminology refers to as being "receptive". To quote from the warning given in the Lexus NX450h+ User Guide *"...The driver is solely responsible for paying attention to the vehicles surroundings and driving safely"* [1].

This kind of experience led me to realise that what I, in the role of human supervisory controller of the autonomous vehicle, really need is information not only about the state the automation is in right now, but what it "thinks" lies in the immediate future and what action it is intending to take. To put this in psychological terms, users need to know that the system's situation awareness includes awareness of the same threats that they themselves are aware of. And they need to know that the system both has the ability and is actively preparing to take the action needed to manage those threats.

The driver needs to be aware of what the systems situation awareness (SA) is at all three levels. That the system has detected the state of the world ahead (for example a particularly tight bend – this is Level I SA[12]); that it understands what that state means in terms of control of the vehicle (the bend is too tight for the vehicle to manoeuvre safely at the current speed – Level 2 SA); and that the system has planned and is preparing to initiate action in time to deal with the event (that it is intending to reduce speed in sufficient time to manoeuvre the bend safely – Level 3 SA). That means giving the user some form of forward projection, or a look ahead in time.

To build confidence and trust in the system, the user needs to know that the system is aware of threats they know about themselves that may threaten the safe operation in the immediate future. Providing some kind of forward projection would significantly ease the need for the driver to be continuously making judgements and deciding whether to intervene. Critically, if the driver decides they do need to take manual control, providing a forward projection would provide additional time and space to take control and perform whatever action is needed.

Using screen layout to support salience

I spent a lot of time reflecting on how easy it was to move my eyes from the road ahead, to locate and then understand the various pieces of information I needed from the driver's user interface. For the purpose of illustration, this section is going to focus on the use of screen position as a means of attracting attention and conveying the importance of information – making it salient – for me as the driver transitioning to understanding and using the automated features of my car.

Figure 3.7 illustrates how, using spatial location only (i.e. without using information design features such as size, colour, luminance or the use of flashing symbols, etc), I would prioritise the main information areas in the driver's user interface

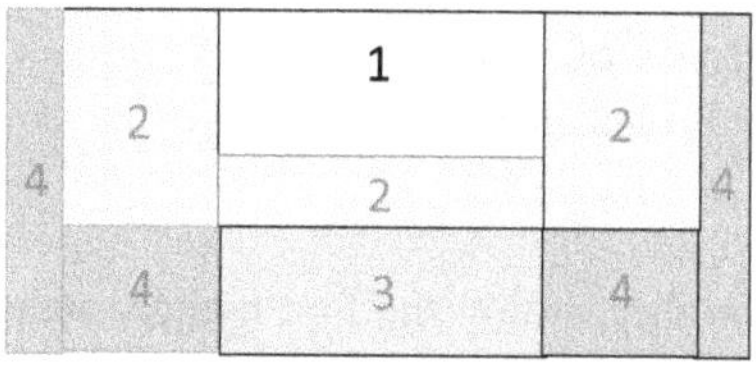

FIGURE 3.7 My prioritisation of information areas on the Lexus NX450h+ driver user interface

of my Lexus in terms of the ease of moving my eyes and focusing my vision to acquire information from that area while driving.

- when taking the eyes off the road to use the interface, it is quickest and takes least effort to read information in the top middle area of the screen. This should be Priority 1 for presenting information;
- areas to the right and left and slightly below the primary area demand slightly more effort to bring the areas into focus. I would consider these as Priority 2;
- the area in the middle bottom feels like Priority 3;
- areas to the bottom, and then towards the outside edges are the least easy to bring into focus – Priority 4.

Based on my experience with my new Lexus, Table 3.1 gives examples of the kind of information I felt I needed to support the various tasks involved in driving my new car in different modes. For comparison, the table also includes two tasks – planning and monitoring a journey and managing the cars fuel efficiency – that are of lower priority in terms of their demand for information from the user interface compared with the core driving tasks. Other tasks that might be performed in real-time while also driving either manually or under automatic control, such as controlling music or making or receiving phone calls through the car's audio system, would be of even lower priority.

Table 3.1 indicates that when both the ACC and LTA modes are simultaneously engaged and the driver is acting in the role of a supervisory controller, at least nine pieces of information can be considered as Priority 1;

A. the vehicle's current speed;
B. the speed to be maintained as set by the driver ("Commanded speed");
C. the minimum separation as set by the driver from the vehicle immediately ahead ("Commanded separation")[13];
D. the vehicle's awareness of other vehicles in the same lane immediately ahead;
E. the vehicle's awareness of lane boundaries;
F. the vehicle's awareness of how lane boundaries change in the road ahead;
G. the vehicle's awareness of current position relative to lane boundaries;
H. the vehicle's awareness of the behaviour of other road users with the potential to require a control action (such as vehicles entering from slip roads that could potentially require avoiding action);
I. the vehicle's awareness of potential hazards, threats or changes in speed limits or road conditions in the road ahead that could require a control action.

Of these nine information items, only the first five are included in the driver's user interface on my Lexus. Figure 3.8 shows where each of these five Priority 1 information items are located on the interface. In the case of the other four, the vehicle gives the driver no information.

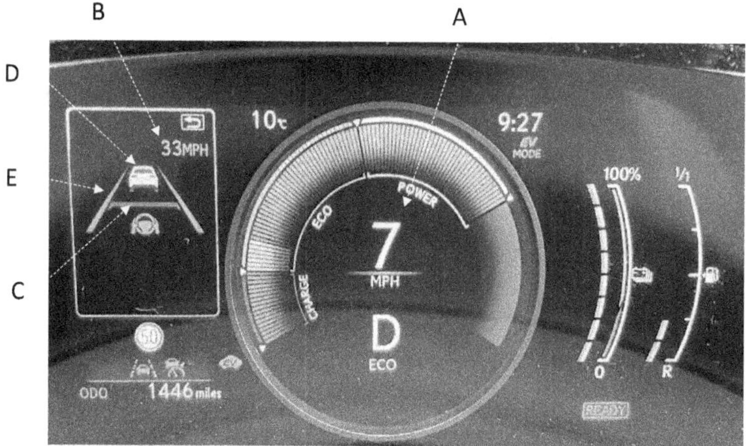

Priority 1 information items F,G,H and I are not shown on the driver's user interface

FIGURE 3.8 The location of priority 1 information needed to support driving in ACC/LTA more

A comparison of Figures 3.7 and 3.8 shows that of the nine priority 1 information items needed in the ACC/LTA mode;

- only one (current speed) is in the primary viewing area;
- four are in a priority two zone (colour is used in this area to indicate whether the vehicle is in autonomous driving mode);
- four pieces of priority 1 information are not available through the driver interface.

The most striking thing to notice about Figure 3.8 is that, when driving in the fully autonomous mode, the most salient features of the driver interface remain the same as in manual driving mode: power usage indicator, current gear and battery and fuel status. None of these are especially important to the core supervisory control task in these moments: which is to proactively monitor for internal and external threats and to decide, in real-time, whether there is a need to intervene. Indeed, from the perspective of the key task of ensuring the safety of the vehicle and other road users, the power usage, current selected gear and fuel and battery status are irrelevant. They support tasks that, while ACC/LTA mode is engaged, are much less important.

At the time of writing, I am 67, slightly short-sighted and wear glasses for driving.[15] I have no reason to believe my vision is not typical, worse than some though better than others, of the owners of Lexus cars around the world: I would guess tens of thousands of drivers. The Lexus NX450h+ is extremely capable, fitted with many features and functions and capable of displaying a great deal of information. The layout and design of the driver's user interface, as shown on Figures 3.1 and 3.2, is stylish and conveys an impression of high quality. The downside of such an impressive range of capabilities, however, is that it demands a great deal from the designers

TABLE 3.1
Example of information needs associated with key driving tasks: motorway driving during daytime (Automatic car)

		Priority 1 information		Priority 2 information	
	Driver's task	**Driver's user interface**	**Other sources**	**Driver's user interface**	**Other sources**
1	Manual driving; • *Manually control:* position of the vehicle on the road while remaining within current speed limits. • *Monitor:* movement of other road users and road conditions and state of vehicle.	Current speed.	Lane markings; position and relative movement of other road users in immediate vicinity; position and relative movement of objects near the road; current speed limit[14]; Imminent changes in current speed limit; potential hazards in the road ahead.	Current gear. Warnings (fuel, oil, ice, etc.)	Location and behaviour of other road users not in immediate vicinity; road signage; environmental conditions.
2	Drive in Lane Tracing Assist mode: • *Manually control:* speed and position relative to vehicle immediately ahead; • *Monitor*: As manual driving plus horizontal lane position.	As manual driving plus; Vehicle's awareness of; (i) lane boundaries; (ii) how lane boundaries change in the road immediately ahead; (iii) its position relative to lane boundaries.	As manual driving	As manual driving	As manual driving
3	Drive in Adaptive Cruise Control mode: • *Manually control:* horizontal lane position; • *Monitor:* As manual driving plus current speed and distance to vehicle ahead.	Current speed; commanded speed; commanded distance to vehicle ahead; Vehicles awareness of; (i) position of vehicles immediately ahead in the same lane; (ii) potentially threatening movement of other vehicles not in own lane;	As manual driving	As manual driving	As manual driving

		(iii) vehicles awareness of potential hazards or changes in speed limits, or road conditions in the road ahead.			
4	Drive in ACC/LTA mode: • No active manual control. • *Monitor:* speed, position relative to vehicle ahead and lane position.	(A) current speed; (B) commanded speed; (C) commanded separation from vehicle ahead. Vehicle's awareness of; (D) other vehicles in the same lane immediately ahead; (E) lane boundaries; (F) how lane boundaries change in the road immediately ahead; (G) current position relative to lane boundaries; (H) the behaviour of other road users; (I) potential threats or changes in speed limits or road conditions in the road ahead.	As manual driving	As manual driving	As manual driving
5	Plan and monitor journey			Current fuel level; distance with current fuel; distance to next turning; distance remaining; estimated time of arrival.	Road signage and navigation features.
6	Drive fuel efficiently			Engine power usage.	Engine sound.

of the user interface to present the information the driver needs as and when they need it, in a way that is easy to read and understand.

To close this chapter, here is a summary of my main observations of how well the fonts, symbols and colours used in the driver interface to my Lexus NX450h+ support me in the role of the vehicle's human supervisory controller;

1. The prioritization of information across the available space is unbalanced. What the design assumes as being the most important information the driver needs (power usage, current speed, current shift position and battery and fuel status) are displayed in the centre and to the right-hand side of the space. None of these are of top priority when the human is filling the supervisory role. Nearly all the key information needed in the supervisory role (the Multi Information display, Control Display and Driver Support System), some of which is complex and non-intuitive, is presented in a secondary area that amounts to less than about 20% of the available screen space on the left-hand side of the screen.
2. Much of the text is essentially unusable under driving conditions. It may be legible while the vehicle is stationary. However, it is not effective when the vehicle is moving, and the driver must locate and acquire information quickly, often in not much more than a brief glance if the eyes are not to be removed from the road for a meaningful time.
3. The interface uses many acronyms that are meaningless to a new driver (such as ACC, LTA, EV, HV, etc).
4. Many of the key symbols associated with the use of the automatic driver modes (see Figure 3.4), while being relatively complex visually, are displayed in a location (bottom left of the screen) and at a size that makes them virtually unusable.

NOTES

1 Note that this description is based on my experience of the vehicle. There is no directly equivalent description of this in the User Guide. (In fact, the Speed Limiter mode and Adaptive Cruise Control modes described here are not even mentioned in the User Guide). See the discussion in Chapter 4.

2 In theory, though not (yet) in legal practice, at least on Scottish roads. Though see the numerous media reports of drivers of self-driving vehicles being found asleep at the wheel. I discuss what happens if the driver does take their hands off the wheel in Chapter 4.

3 The difficulty of knowing precisely which type of roads these refers to is discussed in Chapter 4.

4 It is not possible to activate the Adaptive Cruise Control mode using voice control.

5 In the vehicle, the Driving Assist icon shown on Figure 3.3 is shown here rather than this text.

6 In Chapter 5, I reflect on where I am now after more than two years experience with the vehicle.

7 Note that I am only concerned here with how the various features available to the designer (screen space, fonts, shapes, colour, luminance as well as sound and touch) are used to give the driver the information they need in the different driving states. The design of controls (i.e. what facilities the driver uses, and what actions they take to make changes to the interface or exert control over the vehicle) are not of concern here.
8 Other things being equal, brightness is the most important static design factors in attracting a user's attention. The single brightest item on the cluster is the "READY" indicator (far left). Although I have not measured the luminance, my impression is that, after these, the brightest items in the cluster are probably those with bright green colouring. Though the luminance balance is comfortable, at least to my eyes.
9 Up to four of these lines can be shown, depending on the set vehicle-to-vehicle distance. They are enabled, and the set distance is controlled, by successive presses of the control at the right-hand side of the steering wheel. See Figure 3.1.
10 If you exceed the current limit, the colours on the icon change and it may issue an audible prompt. Though the driver has the option to turn the audible prompts off.
11 If indeed it did detect it. The system seems not to be 100% reliable. The user guide lists numerous conditions when it may not work. And see the discussion on the use of the speed limit icons in Chapter 4.
12 I am using Mica Endsley's three level of model of Situation Awareness. Chapter 8 includes a brief introduction to the psychology of Situation Awareness.
13 Note that if the vehicle detects slower moving vehicles in the lane immediately ahead, it will automatically reduce the actual speed to maintain the commanded distance.
14 Better would be showing the difference between current speed (and/or commanded speed) and the current speed limit.
15 For those interested to know, at the time of writing, my eye-sight prescription is: Right - Sphere -0.50, Cylinder -0.75, Axis 55; Left - Sphere -1.25, Cylinder -0.75, Axis, 85. My distance acuity is 6/6 and near acuity is n.5 in both eyes.

REFERENCES

1. Lexus. NX 450h+ User Guide. Toyota Motor Corporation. 2021.
2. Endsley, M.R. Autonomous Driving Systems: A Preliminary Naturalistic Study of the Tesla Model S. *Journal of Cognitive Engineering and Decision Making.* 2017;11(3):225–238.

4 A journey of discovery

Given my initial experiences described in Chapter 3, it was with a sense of wariness that, over the weeks after taking ownership of my new Lexus, I cautiously began to explore the car's driver assist features.[1] I experimented with the control options on the right-hand side of the steering wheel (shown on Figure 3.1), monitoring how the car reacted, and what changed on the information cluster. With cautious testing on quiet roads, I began to build my mental model of what the system could, and was, doing, and what I needed to do to control it. I also began to discover some of the limitations to what the automated features were capable of.

THE DIFFERENCE BETWEEN SELECTING AND ACTIVATING FUNCTIONS

The first piece of my mental model was the difference between selecting and activating each of the driver assist functions: I eventually realised the difference between what the two controls on the righthand side of the steering wheel did. Through a process of trial and error, I learned that each press of the 'Driving assist mode select' switch (see Figure 3.1 in Chapter 3) toggles between the three functions - Cruise Control, Speed Limiter and Adaptive Cruise Control - but does not actually activate any one of them. It is only when the 'Driving Assist' switch is pressed that the currently displayed function is activated. This was unlike my previous experience in previous cars, where simply pressing the Cruise Control button activated the function: as there was only one option, there was no need to select which one I wanted. Not realising or having been told when I took ownership of the vehicle that my new car had four driver-assist functions, I had not appreciated the need to select one before activating it.

Of the four functions, the one that most interested me was "adaptive cruise control" (ACC). The second piece of my mental model was an initially incorrect belief that, once ACC was activated, the car would both maintain the set speed and steer itself, keeping within the current lane. It was some weeks before I realized this piece of my model was wrong: ACC only controls speed and distance from the vehicle in front: it is the Lane Tracing Assist (LTA) function that controls lane keeping. It was only after careful study that I realised LTA was different. Although it was active at the same time, it was not part of the ACC function.[2]

WHEN IS IT SAFE TO ENGAGE THE AUTOMATION?

A few weeks after taking ownership of the car, when I was feeling reasonably comfortable using the adaptive cruise control function on motorways, I decided to see how the system worked on smaller, non-motorway roads. Having tried it for a short distance, I quickly decided I did not trust the system sufficiently to use

DOI: 10.1201/9781003679479-5

it on those roads: among other things, narrower roads, poor lane markings and frequent bends, with even less margin for error compared with motorways, all led me to lose confidence in the abilities of the automation. So I frequently took back manual control.

I concluded that I didn't feel I could trust the autonomous system in other than motorway or dual carriageway conditions - and, even then, only when I was happy that the road ahead was sufficiently well defined and free from unusual hazards. The question I asked myself was, how does a vehicle manufacturer expect drivers of their vehicles to know the limitations on the use of the driver assist features of their vehicles?

It was only after testing the car on smaller roads, when trying to find information in the User Guide, that I found the following, on page 130, twenty-two pages into the chapter describing the Driving Support System:

> Use the dynamic radar cruise control only on highways and expressways. [1]

This is written in a normal font in a light grey colour though at a size that, for my eyes, is not easy to read. And it forms part of the standard text, rather than being emphasized in a warning box, or emphasized in any way. There is no attempt to highlight it or otherwise draw the driver's attention. The warning does not make use of the guide's standard warning boxes, which normally appear with a border, a warning icon and label, and a pink background. These styling features are intended to make the content of a warning stand out, drawing the reader's attention to them. But they are not used to draw attention to this critical restriction on use of the Driving Support System. It seems to me unreasonable for any manufacturer to simply rely on their customers reading their user guide from cover to cover (even if it was well designed) to discover such a critical limitation of their system.

And what exactly are "highways and expressways"? In the UK at least, these terms are surprisingly difficult to define. Presumably they refer to what UK drivers would recognise as motorways. Do they include dual carriage ways? Or what in the UK are referred to as 'A' roads? Under common law in England and Wales a "highway" strictly occurs where there is always a public right of passage over land "without let or hindrance" that follows a particular route. Scottish Law does not appear to define the term at all. And neither term is defined in the UK Highway Code – something that all new drivers are formally tested on and are required to pass a driving theory test to be awarded a UK driving licence. So how is the owner of a vehicle having automated – never mind autonomous – features supposed to know which type of roads the manufacturer advises them that they can be used on?

We will return to this when we look at the support provided to assist the driver later in the chapter.

CAN I TRUST THE CAR TO STOP ITSELF?

Exploring further, I knew that autonomous vehicles were meant to maintain a minimum distance behind the vehicle in front, whatever the commanded cruising speed. I had also read the description of how this worked in the Lexus' User Guide. So the next step in my journey of discovery was to test this capability. Very tentatively, over a period of some days and with extreme caution, I gradually built confidence in the ability of the car to control its speed to maintain the minimum set distance behind the vehicle in front.

As my confidence in the vehicle continued to grow, I realised it was possible to give the vehicle control of the car and it would successfully drive itself in a truly autonomous mode. But I was not prepared to drive "hands-off" just yet: I would reduce the strength of my grip on the steering wheel just enough to allow the vehicle to turn it, while maintaining contact. My foot would hover over the brake, though without exerting pressure. And I remained very much "brain-on". Even if my hands and feet were "off" the controls, my concentration was very much ON, paying close attention to what the car was doing and the state of the road and other vehicles around me. I would frequently override the system to make manual adjustments to reduce my perception of risk to a level I was comfortable with.

With this sense of wariness, I was eventually willing to trust the system to the extent that I allowed it to slow down as it approached slow moving traffic, even to the point of stopping itself if the vehicle in front stopped, all with no manual input from me. For someone who has been driving manually for nearly 50 years, this was a very strange experience.

What happens if I don't respond?

My next question was, what would happen if I kept my hands completely off the steering wheel for any length of time? Very cautiously I experimented, at slow speeds, in back roads and when the roads were quiet. With growing confidence over a period of weeks, I discovered that if I completely released my grip on the wheel, three things would happen;

1. After around 12–15 seconds (as accurately as I've been able to estimate it in the circumstances), a warning text prompt appears in place of the ACC/LTA icon (i.e. mid left-hand side of the user interface, see Figure 3.2). Headed by a hands-on-steering-wheel icon drawn in red, the message says, "LTA Unavailable Soon Take Control of Steering Wheel".
2. If you don't respond to this prompt and keep your hands off the wheel, after around another five seconds, an audible prompt sounds. This is a soft, though clearly audible, 'beep' at a frequency of around one per second. At this point the on-screen message changes to read "No Driver's Operation Hold Steering Wheel".
3. If you continue to ignore the warnings, after around another 10 seconds, a number of things happen;
    a. both the ACC and LTA are disabled, and the car spontaneously returns to manual control mode;
    b. the textual warning is removed, and the ACC/LTA icon re-appears drawn in near white (meaning it is not engaged);
    c. the periodicity of the audible warning doubles to around two beeps per second;

d. the sound from the audio system is disabled;
e. there is a noticeable change in the sound of the engine, as well as a perceptible slowing of the vehicle.

At the moment control is passed back to the human, the combination of the change in the audible warning, turning off the sound from the audio system and the feeling of the vehicle slowing are all – for me at least - highly salient. It would seem difficult for an alert person, even if attending to something else at that moment, not to notice them.

Despite what seems to be an effective approach to catching the inattentive driver's attention at the end of this sequence, two issues in this strategy struck me;

i. the decision, in the event of a driver failing to respond ten seconds after two separate notifications from the vehicle to take control, to turn off the automation and pass control back to the (unresponsive) driver seemed odd;
ii. The presentation of information in the user interface had not been optimized to support the objective of capturing the driver's attention.[3]

It seems questionable whether simply turning off the driver assist functions in the event of the driver having not responded to two sets of text and audible prompts - in a vehicle that could be moving at 70 mph in the outside lane of a busy motorway - is the safest option. The driver's lack of response could be for many reasons: they may simply be distracted or otherwise choosing not to attend to the state of the vehicle. They could be seeking, as I was, to understand how the system works. Or they could be unconscious or otherwise incapable of responding due to a health episode. These cases may need a different approach both in the user interface design and the eventual response of the vehicle.

There is clearly the potential to use both the visual and audible prompts in a way that expresses more urgency and is more likely to gain the attention of a driver who is distracted or otherwise inattentive. Obvious alternative options other than simply turning off the automation and passing control to the driver, might, for example, be sounding a more salient audible alert inside the car, while simultaneously setting the distance to the vehicle in front to the maximum, slowing down in a controlled manner, and turning the hazard warning lights on as a warning to surrounding drivers. In the case of a driver who is unconscious or otherwise incapable of acting, bringing the car to safe stop while alerting other road users seems more likely to result in a safe outcome than simply passing manual control to an incapacitated driver.

In this situation, the design of the driver's user interface needs to be effective in performing two functions;

i. attracting the user's attention to the need to retain control, and;
ii. letting the driver know the system has reverted to manual mode.

A third would be to make drivers of surrounding vehicles aware that the car may not be under control.

I had learned from my tentative exploration that the design used a combination of audible and visual warnings to attract my attention: first a text message combined with a warning icon (in red); followed by a gentle audible alert. Although the warning icon was coloured red, hinting (and no more than hinting) at a potential need to act, its size and location, as well as the size and location of the warning text, are the same as all other text and icons used in the interface: there is nothing in the design intended to increase their salience and effectiveness in gaining the drivers attention to the need to act. They do not appear in the primary visual area (where, as Chapter 3 discussed, the vehicles current speed and gear is displayed), significantly larger, or with visual features enhanced for greater salience (such as flashing elements).

Similarly, the first audible warning, while it is clearly audible, lacks urgency or an attempt to convey the importance of the need for immediate action. It is possible for the audible warning to be masked if the driver is listening to loud music.[4] The clear conclusion to me, was that there was a lot of scope to improve the design of the driver's user interface in what clearly has the potential to be a critical situation.

Either way, I had learned that the car will drive itself in the autonomous mode, with the driver completely "hands-off" at up to the maximum cruising speed set by the driver, for up to 30 seconds before automatically disengaging and returning to manual control mode.

HOW DOES LANE TRACING WORK?

As my experience grew, it became clear that my initial mental model both of what the car was doing, and my role in it, were overly simplistic. The first time I experimented with the car's lane keeping abilities, I noted that the car stayed roughly in the middle of the lane: so, my mental model assumed that the vehicle automatically maintained its position in the centre of the lane. However, at other times the vehicle seemed to track one of the edges of the lane, moving from one edge to the other as the road curved. This was not a comfortable experience. For my risk tolerance, the car frequently approached too close to vehicles in adjacent lanes for comfort: even if it could be trusted to remain within its' lane, the margin of error in the event of movement of surrounding vehicles was too small for my comfort. I found myself paying a great deal of attention to the car's lateral position in the lane, frequently making manual adjustments and over-riding the car's steering system before handing control back once I felt more comfortable.

Based on this experience, I revised my mental model to one that assumed the lateral position was set at the point when I activated the system: it was down to me to initiate the function when the car was in the middle of a lane, rather than towards one side. In the more than 30 months that have now passed since I bought the car, I'm still not sure quite how this works: sometimes the car tracks the middle of the road, at other times it seems to gently oscillate from one edge to the other, according to the curvature of the road. My mental model of how this works remains hazy.

SPEED LIMIT ICONS

I accidentally stumbled across another feature of the design of the user interface that left me wondering about the reliability of some of the car's autonomous features, and the way they support the driver as its supervisory controller. This concerned the design and behaviour of the speed limit icons, located in the lower left-hand quadrant of the driver's display (see Figure 3.2 in Chapter 3).

The speed limit icon comprises a coloured circle containing a number showing the speed limit on the current section of road. Figure 4.1 illustrate the three states this icon can take. On the left (Figure 4.1a), the current speed limit is 50mph, and the vehicle is currently not exceeding the limit – the background is drawn in white. In the centre (Figure 4.1b), the vehicle is exceeding the speed limit of 50mph, so the background has turned red. Both these icons are used whether the car is in manual driving mode or if the full automation is engaged. I have no idea what the third icon means. I have noticed it several times while the automation is engaged but have not been able to work out what it is trying to tell me.

Unfortunately, the User Guide doesn't help. Section 5 of the guide uses four pages to illustrate a list of 59 warning symbols the driver interface can show. The icons on Figure 4.1 are not included. A section in Chapter 6 of the guide, describing the Driving Support System, shows another 26 icons that the vehicle's Road Sign Assist feature uses when it automatically detects different types of road signs. Again, those illustrated in Figure 4.1 are not included.

Apart from not being able to work out, even after more than 30 months experience with the car, what Figure 4.1(c) is telling me, after a period of careful checking, I realized there were surprising inconsistencies behind the behaviour of the icons shown as Figure 4.1a and b. Most of the time, they seem accurate (to within a second or two) and reliable. Here are two examples that illustrate the anomalous behaviour of these icons;

a. I was recently driving on the A91 between Cupar and St Andrews in Fife in the East of Scotland when I was stopped at temporary traffic lights due to road works. The prevailing speed limit on that section of road is 50mph. A temporary 20 mph speed limit sign had however been positioned about 400m before the lights. As I approached and passed the 20mph sign, I was impressed that the speed icon changed from showing a limit of 50mph, down to 20mph. And as I exited the area of road works, it increased again to

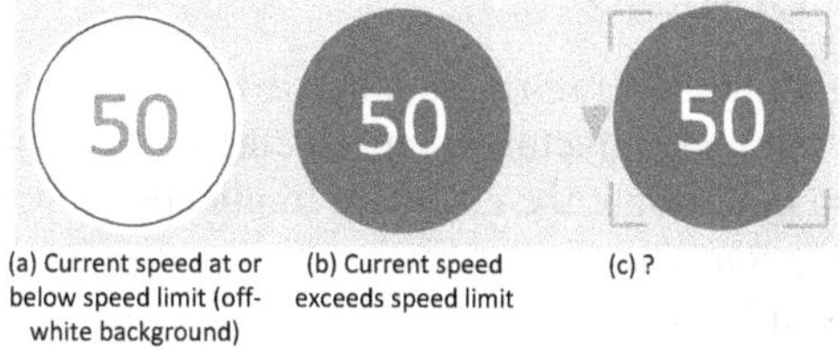

FIGURE 4.1 Speed limit icons used on Lexus NX450h+ driver interface

50. I assumed that there must be a process whereby the local council advises the highways authority when temporary speed limit signs are introduced, and the location is entered into some satellite-based system allowing GPS systems to detect the temporary limit. I was impressed.

b. On the other hand, there is a stretch of busy main 'A' road from close to my home heading north towards Glasgow. Over a period of about 3.5 miles, the speed limit had been 40mph for many years, until 2024 when that stretch of road was re-classified as having a limit of 30mph. Given what I had noticed about the car's ability to know the speed limit in Fife with temporary traffic lights, I assumed it would also know the new permanent speed limit on this stretch of busy A road in Glasgow. Alas not. My car showed the limit as being 40mph for the entire stretch.

I don't know the reason for the difference between these situations; perhaps local councils have the discretion whether to advise those who manage highway data about changes in the speed limit in their council areas. I have no idea. What I do know, is that I cannot trust the speed limit indicator on my car.

Another curiosity struck me as I reflected on this unreliable behaviour of the speed limit icon. Note first that there is nothing in the car's systems that prevent me from engaging the fully automated features at any speed and under any conditions I choose. Let's assume I choose to engage them at 20mph. Earlier in this chapter, I considered how, as a new owner of my Lexus, I was supposed to know the manufacturer's expectations about when it was safe to use the vehicle's self-driving capability;

> Use the dynamic radar cruise control only on highways and expressways.[5]

It seems reasonable to assume that, if the car is able to detect that the current speed limit of the road it is on is 20mph, which it can, it should be able to work out that, at the point I have tried to engage the autonomous driving functions, the car is not on a "highway or expressway".[6] And, therefore, according to their instruction, the features should not be used.[7] The fact that this does not happen is probably due to the relative immaturity of the technology. It also gives another impression of a lack of joined-up thinking in the design of the driver's user experience: it suggests one of many items of functionality that had been developed and tested and then implemented in the user interface. It does not suggest a user interface that has been developed based on a thorough analysis and understanding of the driver's needs.

THE COGNITIVE DEMANDS

As well as understanding their information needs in the new relationship between the users and the system, manufacturers introducing autonomous features to existing products need to understand the cognitive implications for the user. It is often assumed that automation and autonomous operations reduce the complexity of what the user needs to know about the system and simplifies the user's workload. The reality, as a consistent body of applied research has found (see for example [2]) and as I quickly realized as I became familiar with my Lexus, can be exactly the opposite.

The fourth lesson from my experience getting to know my new car, was that the cognitive effort I was required to put in to filling my role in supervising the autonomous capability, was different from any of my previous driving experience. It was both more and a different kind of mental effort, as well as demanding different knowledge about how the system worked, and how to control it. The more I gained experience with the autonomous system, the more I became aware that the mental state, or cognitive demands associated with driving with the ACC and LTA simultaneously engaged was very different from the experience of being engaged in driving the car manually.

For an experienced driver in a traditional car in most situations mental demands are relatively low. The driver frequently performs at a skilled level, with most of the mental activity occurring at a pre, or perhaps sub-conscious level. The process of transitioning to driving with both ACC and LTA engaged is very different.

Filling the role of a passive, though 'hands-on' supervisory controller of the ACC/LTA system, the driver is expected to be ready and able to intervene at short notice and potentially with little time to act. They are expected to be what the SAE/ISO described as "receptive": "…(a) fallback-ready user is considered to be receptive to a request to intervene and/or to an evident vehicle system failure, whether the ADS[8] issues a request to intervene as a result of such a vehicle system failure" [3]. The demands on real-time attention and the difficulty of staying alert when there is little to do are well documented and widely understood. What is not widely understood but became clear from my experience learning to use the autonomous capability of my new Lexus, was the extent of mental demands involved in continually monitoring whether I need to take control.

The core task is one of visually monitoring the vehicle's movement and the outside world. The mental demands in situations such as approaching other vehicles, approaching tight corners, or when near the edge of the lane with other vehicles in the adjacent lane, arises from the need to continuously make real-time judgements about whether to take over control or to continue to trust the system. In these and innumerable other situations, unless the driver simply puts complete trust in the vehicle, there is a need for continuous visual attention, reasoning in space and time, real-time risk assessment and decision making. Risk assessment and decision making that frequently draws on the ability to reason about how events may develop in space and over time. And in a context where the time available to take corrective manual action, if that is what the driver decides is needed, is limited and can be rapidly reducing.

This demand for real-time attention, reasoning and decision making when acting as a supervisory controller of a vehicle operating in an autonomous mode is very different from the mental demands involved in driving conventional vehicles. It raises many questions. For example, can it simply be assumed that all drivers who pass their driving test will be able to meet the cognitive demands in these situations? The UK Driving test now includes a hazard perception test. Should drivers proposing to use autonomous features also have to pass a cognitive test demonstrating their ability (assuming no degradation from fatigue, distractions, etc), to carry out the kind of reasoning and real-time decision making needed to be effective as a supervisory controller? What about fatigue and other states that affect a driver's cognitive abilities? Or individual or sex differences? There is now a strong psychological research base showing that male drivers are generally inclined to make riskier decisions than females.[9] There is also evidence that the visual scanning abilities that are an important part of supervisory control are affected by individual differences in spatial reasoning abilities.[10] And what about elderly drivers? Do the cognitive abilities needed to fill the role of a supervisory controller of a vehicle operating in an autonomous mode degrade with age?

USER SUPPORT[11]

My Lexus provides support to the driver in two ways:

- the vehicle's in-car information system, accessed through the touch screen display mounted (in the case of my right-hand drive vehicle), immediately to the left of the driving position.
- the User Guide [1] provided in hard copy format.

THE IN-CAR INFORMATION SYSTEM

The in-car information system provides access to the many driver-configurable options. Two sections cover the "Driver Support System". They contain, for example, information about options such as changing the type of driver alert (audio, text or both). They also support setting features of the driver assist system, such as the "curvature reducing limit".

Although the information system supports changing the settings of these features, it gives the driver no information about what the settings do. In many cases, this is unnecessary, as the feature is obvious (such as choosing the style of alerts – text, audio or both). In other cases, however, the driver is left to guess exactly what the control does, and what the options mean; such as, "curvature speed reduction".

Although the in-car information system is presumably intended to provide the specific details available in my version of the car, there are no cross-references between that system and the User Guide: indeed neither even mentions the existence of the other. There is nowhere the driver can go (other than, possibly, the dealership) to find out what many of these controls do, and what the options mean.

THE USER GUIDE

After I arrived home on that first, day (described in Chapter 3), I spent time studying the User Guide. I started, as I suspect most people do, with the important things: how to connect my mobile phone, how to connect to the Lexus network, and how to use the car's navigation system. Being my first electric vehicle, I paid special attention to charging and making the most of the cars electric power system. It was only after I had started experimenting with the driver assist features that I turned to Chapter 6, covering the Driver Support System. I was in for disappointment, not to mention some startling surprises.

To start with the positives, the guide is compact and visually looks good. Unfortunately, that's about it. This is not the place for a comprehensive critique of the design of the guide, but a few examples will illustrate just how little attention has been given to the needs or experience of the user, and how little support it provides to someone, like me, trying to understand what the automated features do, how to use them, or their limitations.

- first, the guide is difficult to read: the font used is small, and most of the text is in a light grey colour, offering limited contrast between the text and the background white pages. It is pleasing to look at, but hard to read;
- much of the language is complex with much of the content conditional on specific conditions and the equipment fitted to the particular vehicle. As just one example of how difficult this can make it to follow and understand, under the heading "Multi-information display", the guidance states that "When a menu icon other than [icon[12]] is selected, if the driving support system operates, the system operating state will be displayed" (p 102). This type of language is common throughout the guide.
- the figures are at times almost incomprehensible. Many of them are simply wrong: they do not show all the controls and features in my car;
- many of the functions that are available in my car, especially those associated with the driver assist systems, are missing from the guide. For example, the guide refers to "Dynamic radar cruise control", the user interface to my car refers to the "Adptv. (I assume this is Adaptive) cruise control". I have not been able to work out if these are the same features;
- the LTA control on the steering wheel in my car is not shown in the diagrams in my User Guide;
- the Index is extremely limited, and the content seems arbitrary: for example, there is an index item for "Using the dedicated floor mats correctly", but no entry for driver assist functions such as Adaptive Cruise Control, Lane Tracing Assist or Speed Limiter Mode. There is no index entry to help you work out how to change the set vehicle-to-vehicle distance (i.e. the minimum distance the driver assist system will allow the car to come before slowing down to the same speed as the car in front);
- the guide includes descriptions of many optional features that may or may not be fitted to my car. It is not always possible to tell if the features are in fact fitted. As an example, page 120 of the guide describes a "Lane Change Assist" (LCA) function. I have not been able to work out if this function is fitted to my vehicle or not. The guide also says the LCA system can be turned off "using the LCA switch". LCA is described as being activated when the signal lever is held partway: there is, however, no indication anywhere in the guide of where the LCA switch is to be found. The numerous descriptions of optional items, as well as the descriptions of variants (such as with or without a heads-up display) all add to difficulty and confusion trying to find information and understand the content that is relevant to my car.

In my journey trying to understand how to use the car's many features, and especially the meaning and use of the information displayed, I realised that much of the functionality only makes sense when the vehicle is moving. Many messages, and much of the displayed information is only visible on the driver's user interface in specific situations. An example is the use of green colour-coding on icons to indicate when automated features are engaged. This adds to the challenge of understanding the system by referring to the User Guide.

From the manufacturers point of view, the User Guide content seems to be intended to cover every option available in every model of the vehicle. It is a comprehensive listing of all the displays and controls for, I assume, all models of the NX450h+ vehicle, with whatever extras are fitted. For example, it covers, systematically and in a linear order, vehicles both with and without heads-up display. From this one user's user point of view at least, the guide was a long way from delivering what I needed from it.

It is notable that in her naturalistic study of her experiences driving a Tesla Model S in 2017 [4], Mica Endsley noted that she also needed to rely on self-discovery and trial and error to understand how the features of her Tesla Model S worked, due to a lack of support provided. In addition, and importantly, Endsley also noted that the frequent software updates to her car sometimes changed the car's behaviour in ways that were not clearly explained in the accompanying update notes. She noted the difficulty these frequent changes made in her ability to develop her mental model of how the features worked. In my case, my vehicle only received software updates when it was returned to the dealer for servicing, something that only happened once during the period covered by my reflections.

SUMMARY

The Human Factors challenges associated with self-driving vehicles may be among the most complex and difficult for any autonomous system. Both the number of variables constraining the design space (such as the quality of maintenance of the road infrastructure around the world, as well as the many cultural differences in the behaviour of road users), as well as the range of variability within those dimensions (not least the range of skills, experience, abilities and risk tolerance of drivers between the sexes and across national and racial groups) creates orders of complexity that probably few, if any, other domains share. Despite that, the core underlying argument set out in this chapter are true for those looking to develop many types of highly automated or autonomous real-time control system.

A competent developer or manufacturer intending to introduce autonomous capability, must be prepared to carry out a full re-analysis of the operator's information needs in all states where the automation could be in control. That means "normal" driving conditions within the intended design range of the system. But it must also include boundary conditions where the system is approaching the limits of its abilities, as well as transition states between manual and autonomous working.

To be clear, I am not arguing that the implementation of the driver's interface to the autonomous capabilities of my car are unsafe. I don't have any evidence to make such a claim, even if I could define what "safe" means. But it is clear that the implementation was not created with the aim of finding the most effective way of ensuring the driver always remains aware, not only of which mode the car is in, but of their role as a supervisory controller when the car is in autonomous mode.

To summarise the core lesson from the experiences set out in this chapter, when introducing autonomous capabilities, and most especially when they are being added to an existing product line, manufacturers need to re-analyse the user's information needs and priorities. Based on that analysis, they need to re-design their user interfaces to reflect the very different roles and information needs of the users when they are expected to support those systems operating in an autonomous mode.

NOTES

1 The User Guide was of little help, as will be discussed later in the chapter.
2 I assume I must have turned LTA on without realizing when I first drove the car. All the time I had been experimenting with ACC, as described in the chapter, I had not realized that LTA was also engaged.
3 The effectiveness of design features in a user interface to capture a user's attention is usually referred to as "salience".
4 Some experimentation suggests it would need to be heavy rock, or similar loud music with a heavy beat.
5 Although, as was discussed earlier, the terms "highways and expressways" seem to have no legal meaning in the UK.
6 Unless of course there are signs indicating temporary speed restrictions that the vehicle can detect. In which case I assume it would no longer meet the manufacturer's intentions behind the terms "highway or expressway" for the duration of the speed restriction. Note that this is different from the situation where traffic might naturally slow to 20-mph, although the prevailing speed limit is unchanged.
7 Of course, we need to allow for the possibility of using the systems at lower speeds, under strictly controlled conditions, to learn about the systems. But that is not how the features are presented to the driver. And there are ways around that.
8 ADS is the "Automated Driving System…the hardware and software that are collectively capable of performing the entire DDT on a sustained basis…".
9 I was one of the early contributors to this research base, at least in the context of driving. See [4].
10 See for example [1].
11 This chapter did not intend to offer specific criticism of the design of the Lexus NX450h+. Rather the intent had been to identify lessons likely to be of wider relevance, not only to automated vehicles, but to many other types of systems having autonomous capabilities. However, the design of the User Guide supplied with the Lexus +NX450 is so bad, from a HF perspective, that I find it difficult not to offer specific comment.
12 In the guide, there is a picture of the LTA icon at this position.

REFERENCES

1. Lexus. NX 450h+ User Guide. Toyota Motor Corporation. 2021.
2. Young, M. & Stanton, N. Driving Automation: A Human Factors Perspective. CRC Press. 2023.
3. SAE. Surface Vehicle Recommended Practice: Taxonomy and Definitions for Terms Related to Driving Automation Systems for On-Road Motor Vehicles. J3016. SAE International. 2021.
4. Endsley, M.R. Autonomous Driving Systems: A Preliminary Naturalistic Study of the Tesla Model S. *Journal of Cognitive Engineering and Decision Making.* 2017;11(3):225–238.

5 So what was learned?

This final chapter in Part 1 considers the main lessons I learned from the experiences set out in Chapter 3 and 4 about the process of transitioning to becoming the owner and driver of a car with autonomous capability. These lessons are relevant to any situation where automation is being introduced to control a real-time process or activity.

The experiences set out in this Part are the book, and the lessons drawn from them, were not intended as a critique of issues specific to the design and support of my Lexus. Rather, the intent was to identify lessons as a means of explaining general principles applicable to any industry and any type of system that is introducing autonomous capability to control activities previously controlled manually, whether in manufacturing industry, healthcare, mining, aerospace or elsewhere.

Before setting out those lessons, the chapter reflects on what must be among the most important issues that needs to be addressed in the introduction of any form of autonomous system, not just road vehicles. That is how the process of transitioning users from being the controller of a manual system, to being the supervisor of a system capable of performing autonomously is controlled and managed. The first part reflects on the process I had to go through to transition to being comfortable using the autonomous features of my new car: in particular how little support I received in that transition either from the design of the vehicle and it's support systems, or the organisations that sold it to me and allowed me to drive it on the UK roads. The part concludes with some reflections on how I experience the supervisory role now, after the transition has been completed.

SUPPORTING THE PROCESS OF TRANSITIONING

In Chapter 3, I discussed two key facts about the use of the driver support features in my car;

1. By simultaneously engaging two functions (Adaptive Cruise Control and Lane Tracing Assist), my car is capable of autonomous operation: within the limits of its designed abilities ("highways and expressways", whatever they are), the car can drive itself with no manual control input from the driver.
2. The introduction of this autonomous capability changes the role of the driver from being a manual controller to that of being a supervisor of the vehicle. This is a fundamental and significant change in the role of the driver from that which has been the norm for over 100 years. The previous chapters have discussed some of the implications of this change in terms of the psychological demands on the driver task, and the design of the driver's user interface to support those demands.

 DOI: 10.1201/9781003679479-6

Despite these significant changes, there was nothing about the process of buying my new Lexus or taking ownership and using the automated features on the UKs roads, that reflected what I, as the driver, needed to know about the vehicle to use the automated and autonomous features safely. As long as I could afford to buy the vehicle, and hold a valid UK driving licence, I could take it out on the roads. The briefing provided by the dealer's sales staff as I took ownership of the car made no reference at all to any of the autonomous capabilities. It did nothing to ensure I knew what they did, how to use them or constraints on their use. And as has been discussed in Chapter 4, the support available in the printed User Guide and the in-car driver support system had given the issue no thought and were of little help.

It is not sufficient simply to rely on a drivers' experience of driving manually as the basis for supervising an autonomous system: the roles are fundamentally different. To prepare the world properly for the time when there is widespread use of autonomous vehicles, manufacturers, the dealers that sell the vehicles to the public, as well as regulators need to fundamentally re-think how they go about letting drivers loose with autonomous vehicles.

As well as designing user interfaces that are effective in supporting users in the new supervisory role, manufacturers of any type of system introducing autonomous capabilities need to re-think how they inform their customers and users about those features, as well as what is expected of the user in their new role. They need to pay attention to preparing the user to fill their new role in supporting the autonomous capabilities. That includes ensuring the user understands the conditions and limits the system has been designed to operate in. And they need to ensure users know how to detect when those limits are being threatened, and when they may need to step in to take control.

Where autonomous features take control over functions with safety-critical implication (such as the control of a vehicle moving at speed), users must be properly trained to understand both the systems abilities – most especially the boundary conditions when the automation will degrade or fail – as well as the role they are expected to fill. A role that includes both initiating and giving control to the automation as well as taking back manual control, whether intentionally or unexpectedly. There is a case for introducing regulatory changes to ensure those taking ownership of autonomously equipped vehicles possess the skills and cognitive abilities needed to be able to perform the supervisory control task effectively.

At the very least, global companies such as Lexus/Toyota should be offering customers the option of voluntarily learning how to use the features of the vehicles using multimedia and virtual training environments, rather than simply relying on a hard copy user guide. That is something that could be done easily with today's technologies, at low cost.

In a world of internet enabled multimedia services that can be globally distributed near instantaneously, there is no reason to have to rely on sales staff or a written user guide to ensure users have both the knowledge they need and the skills to operator systems safely. As autonomous self-driving vehicles become increasingly common, those involved in selling them should be offering some form of simulation-based training – something that could be done at home, on mobile devices, or, even better, sitting in the driver's seat using the screen of the car's information system - as part of the vehicle hand-over process. Perhaps regulators or insurers need to introduce requirements that drivers should not be permitted to drive such vehicles – or at least use their autonomous capabilities – until they have demonstrated some degree of understanding and competence in the role of it's supervisory controller. There may even be a case for an amendment to national driving tests to ensure drivers have the cognitive capacity to monitor and supervise autonomous vehicles.

At the time of writing, it is more than two years since I took ownership of my new car. In that time, I have learned a great deal about what the car can do and how to use it: though I do still occasionally stumble on new features. I made a discovery just today, stimulated by a something I read while I was checking some of the references during the final editing of the manuscript.[1] I knew that, if the car was in self-driving mode and I pressed the brake, the automatic system disengaged, and I took over manual control of the car. I had assumed that the same applied to the accelerator: if I pressed it, the automation would disengage, and I would be in manual control of the car. However, I today discovered that if I press the accelerator while in self-driving mode, the automation does not disengage as I had assumed it would. I can override the automation and temporarily increase the speed of the car without disengaging the self-driving functions so long as I am pressing the accelerator. And when I release it, the car reverts to self-driving mode at the previously set speed. Presumably this is to assist over-taking manoeuvres. I assume this has been a feature of cruise control systems for many years. But no-one had ever told me, and I had never discovered the feature for myself. I could find no mention of it in the user manual. Having made the discovery, all those occasions of confusion I had experienced pressing the ACC button to re-engage the automation after manually increasing the speed (assuming the automation had disengaged), only to find that the ACC function was not active (because I had just disengaged it) suddenly made sense.

This chapter has focused on one model of a particular product (a car) manufactured by one company, sold by one dealership and used by one individual on the roads in one country. Some of the discussion is limited to that specific combination. More generally, however the intent was to identify lessons about the way previously manually controlled systems and operations are transitioning through partially automated to fully autonomous capabilities. Lessons that could apply to any type of highly automated system in any industry.

Where does responsibility lie?

The key to improvement must lie in responsibility. While there will be nuances according to the commercial and legal situation in different industries and different countries, the core argument is the same whichever industry and whatever type of system is being developed: those responsible for the transition to autonomous control systems must hold some responsibility, at the very least, to ensure they comply with established principles of best practice in Human Factors related to the design and implementation of their systems.

How do the four cornerstone concepts (ability, authority, control and responsibility) discussed in Chapter 2 apply to the situation I found myself in while exploring the autonomous capabilities of my new Lexus NX450h+[2]?

- ability: having the means and resources to execute control;
- authority: what each actor (people or technology) is or is not allowed to do;
- control: acting in real-time so the situation develops in the preferred way;
- responsibility: being accountable for the consequences of control.

Consider the situation where the vehicle is driving on a motorway ("highway or expressway"), with both Adaptive Cruise Control (ACC) and Lane Tracing Assist (LTA) functions selected and enabled: the car is effectively driving itself, although I remain alert, attentive, and prepared to take control if I need to. In this situation, and as long as the road ahead remains within the limits the car is designed for, the automation undoubtedly has the ability to fully control the vehicle for as long as the road remains within it's design limits. I also have the ability to execute control if I need to.

Because I have actively engaged both functions, I have explicitly given the vehicle the authority to exert control over the vehicle's movement until such time as I take over control. In the moments we are considering here, I am fully aware that the system is controlling the vehicles speed, lane keeping as well as distance to the vehicle in front. I have relaxed the pressure from my hands and feet on the controls to allow the vehicle to drive itself. By doing so, I have positively allocated the authority to the system to control the vehicle. For as long as I do not override the system, the automation is actively in control.

The challenge comes where responsibility is concerned. Under UK Law, and as is mentioned (if you can find it) in the user guide, I am legally responsible for any adverse consequences of the vehicle's movement, whether an accident or breaking the speed limit. Responsibility cannot, however, stop there. As could any other person in the UK who holds a driving licence and can afford to pay, I turned up at a Lexus dealer, uninvited and unexpected, and they were perfectly happy to sell the vehicle to me. On the day I took ownership, I lacked any knowledge of how the vehicles automated systems worked or what they did, what the different icons and abbreviations used on the information clusters mean, what the driver assist functions were capable of, how to engage them, or their limitations. Though I did have expectations based on my history of experience of similar systems such as cruise control in other vehicles. Despite this lack of knowledge, neither UK law, the manufacturer, the dealer, or indeed my insurers, did anything to prevent me taking the car onto the UK's roads and engaging the autonomous functions. If I failed to pay adequate attention, misunderstood what the vehicle was doing or otherwise failed to take control when I should, thereby causing an accident, I would be held legally responsible.

Clearly, it has to be the case that I, as the driver, can have no doubt of my responsibilities as soon as I climb behind the wheel and drive any vehicle. Cars kill large numbers of people, every day. But at the same time, it must be the case, morally at least, that all of those in the chain that enabled me – more than that, they actively engaged in activities directed towards tempting and encouraging myself and other potential customers - to buy and take control of the vehicle should also hold responsibility.

That means the vehicle manufacturers and their shareholders who design and develop the vehicles and sell them to dealerships; the dealerships (and their owners and shareholders) that promote and sell the vehicles to customers; and the regulators tasked with ensuring our roads are safe. To my, non-legally trained mind, that responsibility must, at the very least, be to demonstrate that they have done everything they could reasonably be expected to do to ensure the vehicle is safe to be operated in all of the modes they have provided, by the customers they intend, or are prepared, to sell the vehicle to.

At the very least, each of the involved parties should be able to show, within their area of responsibility, that they have followed recognised principles of best practices throughout the process of developing, selling, supporting and regulating vehicles that have autonomous capability.

COMING OUT OF TRANSITION

The experiences, reflection and learnings discussed in this Part of the book were nearly all gained within the first three to four months after I first took ownership of my new car. A period when I was transitioning from being someone who had more than 40 years' experience driving cars manually, to someone who, when I chose to engage the combination of driver support features that give the vehicle the authority to perform autonomously and drive itself, was now required to adopt the role of the vehicle's supervisory controller. It is that period of transition for the user that is so important, and has the potential to be fraught with risk, when autonomous features are being introduced in any type of real-time system.

At the time of writing, it is more than 30 months since I took ownership of my car. During that time, I have learned a great deal, not only about the car's capability, but about my own role in driving and supervising it. It might be, given the combination of my professional background and experience, and the extent to which I have reflected, thought and written about the process of discovering and learning about the car set out in these chapters, that the extent of that learning is unusual, if not unique.

To round out this material, it is worth reflecting on how, with more than 30 months experience, I now view the self-driving capabilities of my car, and my role as its supervisor. There are five reflections worth discussing;

1. I have developed a high level of trust in the autonomous features.
2. I think in terms of a single driver support function.
3. I have become a "highly engaged supervisor".
4. I exercise choice and discretion.
5. I no longer get confused about who is in control.

A HIGH LEVEL OF TRUST

The first thing that strikes me reflecting on where I now am as the supervisory controller of my self-driving car, is that I have developed a great deal of trust in it. To the extent that I now habitually enable the fully automated features as soon as I enter a motorway, dual carriage, or even a (good quality) 'A' road. I don't immediately engage it, and pass control over to the vehicle, but I enable it: I get ready. And, typically, as soon as I am comfortable with the quality of the road and traffic and weather conditions, I turn it on. But I enter it in a highly engaged state, as I will describe later in the chapter.

Undoubtedly, achieving that level of trust is partly a result of having developed a good understanding of the car's abilities and limits, and being able to recognise when those limits are being reached. As well as the fact that I have developed confidence that, within those limits the car performs reliably and well. By "limits", I don't necessarily mean the limits of the vehicle's capabilities: I suspect that, mechanically, the car is able, for example, to take corners at much higher speeds, and to brake in the event of slow-moving traffic ahead much more aggressively, both while remaining within safety margins I am confident have been extensively tested, than I am comfortable with. But that is the point.

Over time, and with experience, I have come to impose limits on what I am prepared to allow the car to do autonomously. And I have done so without meaning to or even knowing I had done it. It was only when I reflected on the way I now use the automated features that I became aware of that change. The sense of unease I frequently felt as I experimented with allowing the car to take corners, or slowdown in the face of slow-moving traffic ahead has gone. That is because I no longer allow the car to remain in control if I see something in the road ahead that I know will require it to manoeuvre in a way that I would not be comfortable with if I left it to its own devices.

This change in what I am prepared to allow the automation to do is limited to my own experience and may not be generalizable. Possibly because of the amount of experimentation and reflection I have put into attempting to understand the vehicle and my role in supervising it, I have come to adopt a style of driving that reflects my personal tolerance of perceived risk. Had I been a different person (or perhaps had I been the same person but 30 years younger), it is possible I would have stuck with it and trusted the car's ability. I would gradually have got used to the forces involved when the car cornered and applied the brakes in a way that was closer to its designed ability based on its mechanical limits. I would have recognised and come to feel comfortable with the car's automated performance, rather than restricting the car to performing in a way that kept those forces within my comfort zone.

I mentioned above that I now routinely engage the automation when I drive on some 'A' roads.[3] Though in Chapter 4, I discussed how, in my early transition, when I experimented using the features on other than motorways, I quickly concluded that I was not prepared to trust it. Whether I actually engage the automation is partly dependent on the quality of the road surface and lane boundary markers. Though it is mainly to do with the nature of the road: if the road is reasonably straight, without frequent sharp bends, I'm inclined towards engaging. Though if, as is the case with many of the roads I drive on in the West of Scotland, the road is winding and hilly, I would never use it.

All of these, and other changes in my driving style, are a consequence of the trust I now have in the car's abilities.

A single driver support function

On reflection after more than 30 months experience, I realise that although the vehicle and the user interface distinguish between four available driver support functions

(Cruise Control, Speed Limiter, Adaptive Cruise Control, and Lane Tracing Assist),[4] I don't. I leave Lane Tracing Assist on all the time, engage Adaptive Cruise Control with a single button press using my right-hand index finger and take back control by braking or accelerating. I make no mental distinction between the four functions: in my head, the self-driving function is either on or off, with the speed set at the point I press the button to engage the function. And, if I want to, I adjust the speed using the buttons on the steering wheel (which changes the set speed in one mph increments), without taking control of the vehicle.

The main consequence of this is that the user interface, which provides controls to operate each of the four functions, as well as icons on the driver's display showing their state, is much more complicated – and cluttered – than I need. The only icons I use are the Multi Information Display and the speed limit indicator (see Figure 3.1 in Chapter 3). I don't use any of the others. Apart from being difficult to visually locate and understand, I don't need them.

I am not party to the process by which Lexus designs their driver interfaces. I have no knowledge of the concept behind their interface styling or the decision making behind how they choose to create, test and implement design solutions. I do however have many years of experience working on the design and development of complex systems, much of it concerned with the development and evaluation of user interfaces. Based on that experience, the impression I have of the content and layout of information and controls in the driver user interface to my car is that it has been created as a result of function-centred, rather than user-centred design. By that I mean that research, development and engineering effort has gone into creating and testing technology and algorithms capable of performing a range of functions, each focused on solving a specific problem required to control the movement of a road vehicle: keeping it within lane limits, maintaining speed while avoiding colliding with vehicles in front; monitoring what the human driver is doing, and detecting and letting the human driver know if other threats are detected. Once that technical solution has been proven to the company's satisfaction, the functionality, as well as the control and information associate with it has been integrated into the vehicle.

This function-cantered design is in contrast with a user-centred approach. User-centred design starts with a thorough understanding of the role, tasks and needs of the system's human users. It understands what goals and tasks the human is trying to achieve, and what information and controls they need to achieve them. Crucially, it understands the context in which the human is expected to perform. From those, it develops user interface design concepts that satisfy the human's needs, and which are compatible with how the human brain is likely to think about the goals they are trying to achieve.

The functional-centred approach to the design of the user interface to my Lexus is based on engineering of the four driver support functions and integrating the controls and information needed by them into the driver user interface. This has resulted in what, by comparison with any previous model of car I have owned, is a significantly more complex driver interface than I have either used before, or, more importantly, need.

A "HIGHLY ENGAGED SUPERVISOR"

The third, and perhaps the most interesting, thing that strikes me on reflection, is that once I engage the autonomous functions, I adopt a style of supervising the vehicle that could be described as "highly engaged". This is almost the opposite of the well documented issues when people expected to supervise highly reliable automation over sustained periods become *disengaged* from the task. They find it difficult to maintain attention on proactively monitoring the automation for threats, leading, over time, to out-of-the-loop issues including loss of situation awareness.

What I mean by being "highly engaged" is that while I may have given authority to the automation to control the vehicle, and I am not actively providing control inputs via the steering wheel or pedals, I remain alert and focused on the road ahead and the behaviour of surrounding vehicles. My hands remain in position lightly resting on the steering wheel, though not exerting control. I know that the vehicle would allow me to remove my hands from the wheel for up to about 20 seconds, any longer and the vehicle will give me a warning[5] - there are times when I might do that, if only to confirm my confidence in the car's systems. My right foot (I drive an automatic right-hand drive vehicle) usually remains poised in position over the accelerator and brake ready to act if needed: though I often rest it on the inside of the wheel-arch which is convenient and provides support.

The key question must be, why bother? Why bother letting the vehicle do the driving, if I am going to remain alert and poised to take control? The answer has two parts, one to do with effort, the other to do with discretion and choice.

Taking effort first, in steady-state driving conditions, when traffic is flowing easily and there is a reasonable distance to travel, I experience a distinct reduction in the mental effort involved in adopting the supervisory role compared with driving manually. It simply feels mentally easier and more relaxing than having to make continuous, though frequently minor, control adjustments using my hands and feet. Sufficiently easier that the benefits of relinquishing manual control outweigh the costs of remaining attentive. If it wasn't, I wouldn't do it.

I was surprised when I realised this difference in my perception of effort. I have been interested in the psychological basis of mental workload throughout my entire career since being an undergraduate Psychology student. I followed developments in research for at least three decades and have led a few studies investigating physical and mental workload in applied settings. Despite that background, realizing that such a small change in manual activity could produce such a notable and clear reduction in my experience of mental effort, surprised me. It shouldn't have, as I had felt the same experience more than 15 years ago when I first started to use Cruise Control to let my cars maintain the speed I set. But it did.

It was some time after taking ownership of my car that I realized I had developed this highly engaged behaviour. In Chapter 4, I noted that during the initial few months while I was getting to know how the automation worked and what it could do, I felt there was a distinct increase in the mental effort needed of me to fill the role of the supervisory controller of the car. It was both more and a different kind of mental effort. It also demanded different knowledge about how the system worked, and how to control it. I'm not sure how I would describe my mental state in those early days, though terms like "worried", "anxious" and "cautious" come to mind. Whereas an important characteristic of the highly engaged role I now adopt, is that I feel relaxed and confident: relaxed because it is undoubtedly less effort driving with the automation engaged, and confident both in the vehicle and in my own ability to intervene if I need to.

This experience of adopting the role of a "highly engaged supervisor" has an obvious and important caveat in considering whether my experience might generalize to other situations, whether in road vehicles or other types of highly automated systems. That is that, for the most part, I do not have to adopt the highly engaged role for particularly long periods.

I live in Glasgow, on the west coast of Scotland. Compared with countries like the US and Canada, Germany, Australia and many others, generally, in the UK, most drivers – at least non-professional drivers - do not often undertake long distance motorway driving. In a typical month, my average daily drive is probably no more than 50 miles, with a median somewhat less, perhaps 10–20 miles. Though frequently, because of where I happen to live, a lot of that will involve motorway driving for at least some part of the journey. Which is when I now routinely let the car help with the driving. So when I adopt the role of a "highly engaged supervisor" of my vehicle, I'm not doing it for periods of hours on end: it's probably no more than perhaps 30–45 minutes at a time. I have made at least two journeys from Glasgow to Cornwall on the south coast of England using the automated features, a distance in excess of 450 miles, and taking probably around 10 hours, including rest breaks. I certainly adopted a similar highly engaged supervisor role during those journeys: though whether I was quite as engaged throughout those journeys as on my more usual shorter trips, seems unlikely.

It's also the case that, in my situation, I have no other tasks I'm responsible for while the automation has authority to control the car's movement. So, there are no distractions from preventing me choosing to adopt a highly engaged role. Had I been describing an industrial situation, where the individual assigned responsibility for supervising a highly automated operation also had other responsibilities competing for their attention, it would be much more difficult for that person to be "highly engaged" in the supervisory task. The obvious counter to that argument would be that in the industrial setting, perhaps the user's job should be designed in such a way that they have both the opportunity, and the motivation, to adopt the role of being a highly engaged supervisor. This is a significant issue, which, unfortunately, is beyond the scope of this book.

Choice and discretion

The second part of the answer to the question posed above about why bother letting the vehicle do the driving if I'm going to remain alert and poised to take control, is to do with discretion and choice.

It's 403 miles from my house in Glasgow to Brent Cross in north London. The significance of Brent Cross is that, starting from my home, it's at the end of the M1 motorway: Britain's first "full-length" motorway.[6] There is a junction to join the relatively new M77 motorway less than five minutes from my house.

Once on the M77, it is theoretically possible to drive all the way to Brent Cross with only a single change of road. That occurs about six miles after joining the M77 when it's necessary to take the left-hand lane to merge onto the M74. But after that, no further change of road is necessary. A little over 100 miles later, just across the border into England, the M74 changes seamlessly to become the M6 which, after another approximately 230 miles itself merges onto the M1 just north of Rugby in the English Midlands. A mere 100 miles or so, on a continuous single lane, to Brent Cross.

We need to make a few assumptions, but other things being equal, it should theoretically be possible for me to join the M77 near my home, engage the autonomous driving features of my Lexus, then sit back and watch it take me all the way, non-stop to London. Those assumptions are;

I. that the car would have sufficient fuel for the journey;
II. there is no need for any human comfort breaks;
III. that at no point does the traffic in my lane come to a stop for long enough for the automation to disengage itself;
IV. there are no road works on any of the motorways involved that require a change of lane.

But allowing those assumptions for the sake of argument, the car should, theoretically, be able to get me to Brent Cross with virtually no intervention on my part. Though, as the car has no autonomous capability to overtake, it will travel at the speed of the slowest car it encounters ahead of it.[7]

But of course, that is not how human beings want to travel. Until road vehicles are capable of autonomous over-taking, they will always be stuck behind the car in front unless the human chooses to intervene to conduct an overtaking manoeuvre. As the supervisory controller (highly engaged or otherwise), of my car that I have put into self-driving mode, there are frequently situations where I will be travelling on the inside lane (the left, slow, lane in the UK), moving, due to the speed of traffic ahead, significantly below the prevailing speed limit while I can see the next lane to my right is clear of traffic both ahead and behind. So, I choose to overtake: I indicate, move into the free lane, and accelerate up to the speed I want to travel at. The point is that I have chosen to intervene to satisfy a personal desire or goal – to travel faster. There was no need for me to intervene to complete the journey safely. The car would have carried on staying behind the vehicle in front perfectly well on its own without any intervention from me up to the point where I needed to exit the motorway. But I had the discretion to travel faster and chose to do so.

There are many situations where I exert my discretion and choose to intervene when there is no threat to the performance of the vehicle. As another example, there are many occasions when the car will slow itself down more than I feel is necessary to take a tight corner. I often find myself choosing to intervene to accelerate the car to the speed I want to travel at. Use of the windscreen wipers is another example of discretionary intervention by the driver: the autonomous systems in the vehicle don't care if the windows are covered in water or dirt: it doesn't use them. But to fill my role as a supervisory controller, or just for the pleasure of being able to see the outside world, I choose to activate the wipers to give me a clear view out.

The use of headlights isn't the same type of discretionary interaction: they aid me in my role as supervisor at night but are also important for the sake of other road users who need to see me. Indicators do very little for my task, but they are also important for other road users.

The purpose of these examples is to illustrate the general point I came to appreciate when reflecting on the way I now, after more than 30 months experience, use the autonomous features of my vehicle: many of my interactions with the vehicle and its' autonomous systems are discretionary. They are to do with how I choose to manage the journey, rather than being a response to objects or events that represent a threat to the vehicle's ability to control the car and complete the current section of the journey. I don't need to do them to complete my journey: but I do need to do them for that journey to be conducted and completed in a way that meets my wider objectives, including short-term goals as well as my personal preferences, and perhaps needs and values.

These discretionary interactions with the autonomous systems in my car are analogous to many industrial settings. Operators and other users will choose to intervene with a system to change goals, set points or other features to meet requirements other than simply performing a control activity: they are to do with ensuring the quality and efficiency with which the activities goals are reached. Setting these quality and efficiency goals generally lie in the domain of human discretion. An autonomous system may be capable of defining sub-goals, at a level of detail appropriate to the sophistication of the system. But, at the highest level, the overall goals of carrying out a real-time activity lie in the domain of human discretion, decision making. and choice.

To summarise, the question I asked earlier was, why bother letting the vehicle do the driving if I'm going to remain "highly engaged" - alert and poised to take control? The first part of the answer was because it feels mentally easier and more relaxing than driving manually. The second part is because I want to ensure the journey is carried out in a way that meets my personal goals, preferences, needs and values.

Who is in control?

When I was describing my experiences in the early days of getting to know my new car in Chapter 3, the issue that seemed most important to me in my new role as the car's supervisory controller, was being in no doubt who was in control at any moment - the car, or me. I wrote at the time; "The single most important thing the driver needs to know when the vehicle is in autonomous mode is that the vehicle is driving itself: everything else is secondary". I described how on at least one occasion I came close to a minor collision as I didn't realise I was in control as the car approached slow moving vehicles. And I noted that, having had no training, and with no support from the User Guide, I wasn't sure what information on the user interface told me where control lay.

Now, this no longer seems much of an issue. I now know that control is indicated by the colour coding used on the Multi Information Display located in the Control Display Area to the right of the user interface screen (see Figure 3.2). When the car is in self-driving mode (i.e. both the Adaptive Cruise Control and the Lane Tracing

Assist functions are engaged) the symbology in that display is drawn in a high luminance green colour. When I'm in control, the symbology is drawn in a lower luminance off-white colour. I have learned to glance at this symbol both to confirm the automation is working after I engage it, and as part of my regular monitoring of the vehicle's state.

I've also realised that, now I have experience with the car, it's obvious from the sensation if the car stops driving itself: there is a distinct feeling of the car suddenly slowing at the same time (if the petrol engine is running) as a change in the engine noise. Those sensations make it clear that I need to take control. As a novice with the car while I was in the transition process, I either wasn't aware of the significance of these sensations or wasn't as attentive to them. So it probably took me a little longer to realise when I needed to take control.

I have also learned that, if the Lane Tracing Assist function cannot detect the lane limit markers on the road, their colour changes to a kind of off pink/white. But, curiously, I have also discovered that even if the LTA fails, the car will remain in Active Cruise Control mode: even though the car cannot detect the lane limits, it will still retain control. Though, I assume without being able to steer the vehicle. I find this disconcerting at the very least. The 2009 crash of Air France Airbus A330-230 with the loss of all 228 passengers and crew on-board occurred when the aircraft's computers lost their air speed data because of icing of the pitot tubes on the outside of the airframe. Without airspeed data, the automation could not control the airframe, so the automation immediately stopped working, passing control back to the pilots.

In the design of the Airbus control system, the designers knew that without airspeed data, their algorithms could not compute the flight laws necessary to control the aircraft. The automation couldn't do its job, so it stopped working. The Lexus engineers however seem to have reasoned that, even though the car cannot detect the lane limits, and therefore cannot steer the car automatically, that is not a sufficient reason to stop automatic control of the vehicle's speed. Their reasoning seems to assume that the driver thinks about and understands the different functions in the same way they do. And that the driver will;

I. notice the change in colour and brightness of the lane limit icon – and do so soon after it happens;
II. understand that it means the vehicle cannot detect lane limits;
III. realise that the automation is still controlling the car's speed;
IV. take over control of the car's steering, though not necessarily it's speed.

Based on my reflections after more than 30 months experience driving my Lexus, these assumptions seem to me fundamentally flawed. As I described above, I make no mental distinction between the four driver assist functions available: in my head, the self-driving function is either on or off. Indeed, I probably never did, as I never really appreciated the differences between the four functions. They didn't matter to me: all I want to do is turn the automation on or off. I find this distinction between the automated functions, and the fact that one of them can fail, but the other remains engaged, unsettling.

Generalizing from this observation a little, the only situation I have come across during my time driving and exploring the automated features of the car where the self-driving functions will all stop and hand control back to the driver is when I keep my hands off the steering wheel for longer than about 30 seconds (this was discussed in Chapter 4).

Of course, as a now experienced user of the car's automated features, the bottom line in the issue of being clear who is in control is that, if I am in any doubt at all, I take control.

Minor observations

To conclude these reflections on my experience of owning and using the self-driving capability on my Lexus NX 450h+, two minor observations are worth making;

1. There are some things about the automated features, I still don't understand; an example being where the car positions itself in a lane when Lane Tracing Assist function is engaged.[8] And I certainly don't know what the limits of the car's performance is, in terms of cornering and braking under automatic control. Though I am comfortable that I know the limits and conditions under which I am prepared to give it authority to control the car's progress.
2. Over the course of documenting and reflecting on my experience using the car's self-driving functions, I have spoken with numerous people, family members, friends and acquaintances about the experiences I was documenting. Many of them also own or have owned, cars with similar features. I have however, not yet come across a single person that uses the self-driving features. Clearly an extremely small sample, but the consistent response surprised me.

No doubt, the specific vehicle automation functionality and design details on which my experiences discussed in these chapters are based will change over time: indeed, they may already have done so. They will improve in capability and reliability as the technologies and supporting infrastructure develop and as experience from previous models is fed back into new concepts and products. I assume they will increasingly become integrated until the driver is eventually faced with a single option to give the vehicle control or not. However, the point of the lessons from these experiences is not restricted to the specific functionality and design solutions offered in my car. They are intended to be generic, to do with how those conceiving, designing, developing, implementing, and selling or using highly automated real-time control systems think about the role of the humans they rely on to support the autonomous capabilities.

The future for autonomous vehicles is much more positive and exciting than I had appreciated before becoming an owner. However, my concern, which I hope is reflected in this chapter, is that the world has embarked on a period of transition from traditional manually controlled cars, to this new era of autonomous, self-driving vehicles. Without a change of mindset, and significant improvement in the way Human Factors are dealt with in the processes of developing, validating, regulating and supporting those systems, there is going to be a great deal of suffering and misery before the new safer world anticipated is realized.

ADDING AUTONOMOUS FEATURES TO AN EXISTING BRAND

Having concluded that the design of the user interface to the autonomous features of my new Lexus had not taken adequate account of the driver's information needs, I wondered, why not? Lexus is a leading luxury brand, part of Toyota, one of the world's most respected vehicle manufacturers. Lexus and Toyota undoubtedly understand Human Factors and Ergonomics: that is clear from many aspects of the design of their vehicles. So why have they not applied that understanding in this area of one of their newest products: a product introducing new and advanced features widely expected to play such an important part in the future of road transport? And one advertised as being the "All-New Lexus NX" (Figure 5.1).

I stumbled across the answer to this question by chance. During the period I was familiarizing myself with the automated features of my car, I had reason to return it to the dealer's workshop. They provided me with a replacement car for the day – one of the Lexus 250 models. The model I was loaned was a petrol-driven hybrid, that did not have the ACC/LTA capability. As soon as I looked at the controls and information cluster. the answer to my question became clear. Figure 5.2 shows the driver interface on the 250 model:

Comparing the interface shown on Figure 5.2 with that on Figures 3.1 and 3.2 in Chapter 3 (my vehicle) the answer to my question is obvious: while the vehicle may be advertised as being "All-New", the interface to the new NX450h+ is far from a new design. Rather, it is a variant of the existing brand styling, modified to support the additional information and controls associated with the new features. All the information and controls the driver needs are provided: but the design has not been re-thought to reflect the change in the role of the driver introduced by the autonomous capability available in the new model.

Neither the impact of these autonomous capabilities on the role of the driver, the impact on the driver's information needs when using the automation, nor the way

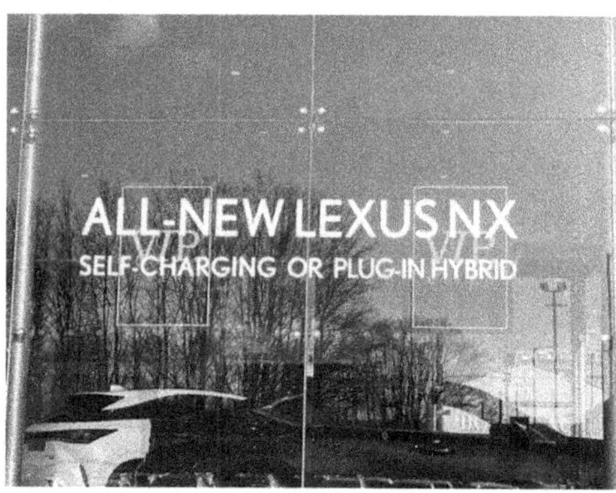

FIGURE 5.1 Marketing the "All-new" Lexus

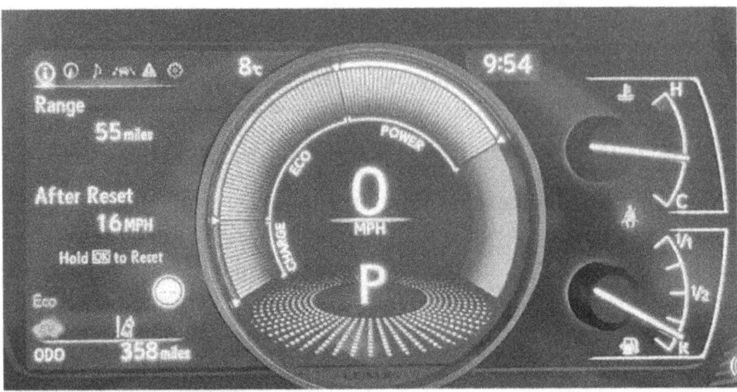

a) Information cluster on a Lexus 250

b) Steering wheel controls on a Lexus 250

FIGURE 5.2 Information cluster and steering wheel controls on a Lexus 250

those needs are supported by the vehicle, are reflected in the design of the driver interface to the "all new" NX450h+.

The assumption underlying the development of the NX450h+ must have been that the driver support features needed, including the potential for the driver to put the car into a fully autonomous state, are simply variants of the existing vehicles in the range. And that, therefore, the conceptual and design model of the relationship between the driver and the car that has been in place for over 100 years is unchanged.

The reality however is that the introduction of the autonomous capabilities fundamentally changes the role of the driver and their relationship with the vehicle when those features are engaged. It changes from one where the driver is an active manual controller of the vehicle, to one where they are a *supervisory controller* of an autonomous system. But where the driver is still expected, in fact legally required, to always be fully alert and attentive, able to step in and take active manual control of the vehicle at short notice, potentially with little time or space to think and act.

The lesson from these reflections is that manufacturers who are intending to add autonomous capability to an existing design, cannot simply treat the new version as a variant of a previous system. They need to recognise the new role of the user when those autonomous functions are engaged, and re-design the user interface accordingly. Manufacturers need to re-assess users' information needs and priorities and re-design their user interfaces to autonomous functions accordingly;

Simply adding the features associated with autonomous functions to a user interface designed for manual control is not sufficient: indeed, it may be highly risky. Users of highly automated systems have different information priorities than users of more conventional systems. That includes the information content as well as the way it is presented to the user: the most important information the user needs when acting as a supervisory controller of a system operating in an autonomous mode is not the same as a conventional system, even one that uses automation to support part of the users' tasks but where the user always remains in control.

SUMMARY OF LESSONS LEARNED FROM PART I

Issues around the challenges humans face in maintaining alertness and situation awareness in roles that require us to act as supervisory controllers and monitors of automated control systems are widely understood. They have been investigated and widely reported not only in the scientific and technical communities, but, increasingly, in the media. In the case of self-driving vehicles, they regularly feature in the mainstream media. There was no need for them to feature in these chapters.[9]

The chapters in this Part have explored other lessons that can be generalised from my experience as a new owner of a vehicle with self-driving capability. At least three main lessons can be identified that go beyond issues around maintaining alertness and situation awareness;

1. Users need to know what the system believes the state of the world it is operating in is going to be, and what action it is intending to take. They need information that lets them know that the system is aware of threats they themselves can detect in the immediate future and that it is preparing to take the necessary action.
2. Manufacturers introducing autonomous features to existing products, as well as regulators responsible for controlling the context in which they are introduced and used, need to fully understand the cognitive demands imposed by those systems, and the implications for users.
3. The transition of users from manual to autonomous operations needs to be taken much more seriously: that means not only re-thinking how users need to be trained, what knowledge they need and what cognitive abilities and other skills they need to have, but also improving user support in the form of on-screen instructions and user guidance.

Some of those lessons are recognised in the scientific and research communities. I am not however aware of them having been discussed as a consequence of a direct user experience in the way I have discovered them for myself in reflecting on my experience. Although the details of the user interface to my vehicle are specific to Lexus vehicles, the issues arising are generalisable to many situations where real-time control is transitioning to include some form of autonomous capability, whether in the world of road vehicles, energy, healthcare, manufacturing, mining, defence or anywhere elsewhere.[10]

HUMAN FACTORS PRINCIPLES

In 2022, the Chartered Institute of Ergonomics & Human Factors (CIEHF) published a White Paper based around nine principles for the design of highly automated systems [4]. For ease of reference, Table 5.1 lists these nine principles.

The automated driving features explored in this Part of the book are not new: vehicle manufacturers have been offering variants of these systems, and others, for some years. What is new is my personal experience as a driver transitioning for the first time from a vehicle where I am in manual control of the vehicle at all times, to one where, in some circumstances, the vehicle can drive itself. A transition that a great many of the world's drivers will need to go through over the coming decades if the vision of a future of self-driving vehicles is to become a reality. The nature of the cognitive demands placed on the driver in the role of the car's supervisory controller, as well as the information needed to be effective in performing the proactive

TABLE 5.1
Chartered Institute of Ergonomics & Human Factors principles for developing and implementing highly automated systems [2]

1	Understand the potential influence of other elements of the system on the automated components, as well as how the introduction of automation can affect those components. Automation must be seen in the context of the overall socio- technical system it exists in.
2	Recognise that automation nearly always changes, rather than removes, the role of people in a system. Those changes are often unintended and unanticipated. They can make the tasks people need to perform more difficult and can disrupt established relationships, lines of communication and the ability to exert authority.
3	Be clear about which of the four core functions (acquiring information, extracting meaning from it, making decisions and taking action) automation will have the ability to perform for each system task, and under what conditions it will be given the authority to control those functions without human oversight.
4	Be realistic in acknowledging that people, at some level, are going to have to monitor, supervise, and hold responsibility for, the performance of the automation. Design, introduce and support the automation such that those people can maintain awareness of the state of both the automation and the world it operates in.
5	Ensure effective, transparent and unambiguous communication between the automation and the human elements of the system, such that the human is able to remain in-the-loop and situationally aware at all times.
6	For each task or function an automated system has the ability to perform be as explicit as possible where the balance between authority, responsibility and control lies. Be clear about what the expectations about responsibility imply for the different stakeholders in the system.
7	Ensure the people relied on to support the automation understand what the system is doing and why. There should be no automation surprises.
8	Avoid making unrealistic assumptions about the ability of people to monitor and effectively intervene in any system where there is little for them to do over sustained periods.
9	Recognise that automated systems can increase the levels of task difficulty and workload imposed on the human elements in the system as well as the level of human reliability needed in the inspection, calibration, maintenance and testing of system components.

monitoring task that is central to the role of a supervisor are fundamentally different to when the human drives the car manually.[11] Those differences need to be acknowledged, understood and reflected both in the design of the user-interface and user support material for systems intended to automate the control of real-time processes and activities, as well as in regulations and other controls that enable and support their use.

Part 2 of the book explores the nature of the supervisory control task in some detail – whether it is the driver of a self-driving car, or any other real-time control system with autonomous capability - including the information needs arising from the task demands.

NOTES

1 It was the 2011 paper by Thomas Sheridan referred to in Chapter 2.
2 The A2CR model was developed by Flemisch et al [1] and is used in the CIEHF White Paper on Human Factors in Highly Automated Systems [2]. It is summarised in Chapter 2.
3 Chapter 3 notes that the car's User Guide states that the automated features should only be engaged on "highways or expressways", though those terms have no legal definition in the UK.
4 See Chapter 3 for a description of how these functions operate and how they are reflected on the driver interface.
5 What happens if I take my hands off the steering wheel for longer is explored in some detail in Chapter 4.
6 And, some say, also its most iconic, whatever that means.
7 It would be an intriguing exercise to know, if – or when - such a journey should be possible, what the difference in the time for the journey would be in an autonomous vehicle compared with a car driven manually.
8 This was discussed in Chapter 4.
9 For a recent review of the research literature relating to Human Factors aspects of driving automation, see [3].
10 It is possible that some of these issues, at least, may not apply to vehicles that are originally conceived and designed from the bottom-up to be "self-driving". Though I have not, in my research for this book, found this to be the case.
11 As a postscript to this chapter, Chapter 12 suggests a basis for estimating the quality of support provided for someone filling the role of a supervisory controller of an automated system using what is referred to as a "PMQsc score". For interest, I have retrospectively applied the method to my experience during and after the transition to owning my new Lexus. On day one of owning the car, I estimated the PMQsc score (with hindsight) at –14%. By the time I had become a "highly engaged supervisor", certainly more than four months later, the score had increased to +64% - not perfect, but a lot better. To understand what those numbers mean, you need to understand how they are derived. Which means reading Chapter 12.

REFERENCES

1. Flemisch, F., Heesen, M., Hesse, T., Kelsch, J., Schieben, A. & Beller, J. Towards a Dynamic Balance between Humans and Automation: Authority, Ability, Responsibility and Control in Shared and Cooperative Situations. *Cognition, Technology & Work.* 2012;14:3–18.

2. CIEHF. Human Factors in Highly Automated Systems. Chartered Institute of Ergonomics & Human Factors. 2022. Available from: https://ergonomics.org.uk/resource/human-factors-in-highly-automated-systems-white-paper.html
3. Young, M. & Stanton, N. Driving Automation: A Human Factors Perspective. CRC Press. 2023.

Part 2

Modelling human supervisory control

The first part of the book discussed and reflected on what was learned through the experience of transitioning to being the driver of a self-driving car. The discussion concluded by arguing that when introducing autonomous capabilities, and most especially when they are being added to an existing product line, manufacturers need to re-analyse their user's information needs and priorities based on an understanding of their new role as supervisory controllers. This is especially important during the process when users are transitioning from long established roles as manual controllers. Based on that analysis, developers need to re-think interaction concepts and re-design user interfaces and support systems in a way that provides effective support to the roles, cognitive demands and information needs of people when they are expected to support systems operating in an autonomous mode.

Although Part 1 was based on the context of taking ownership of a new car, the lessons learned – though clearly not the specific experiences those lessons were based on - are as relevant to those engaged in designing and developing, deploying, and licencing automated real-time control systems in any industry as they are to future autonomous road vehicles. Some industries seem already to be well up the curve in deploying systems that provide good support to human supervisory control. Certainly, based at least on the kind of research they support and the kinds of work they publish in the open literature, the likes of nuclear power, some aspects of space and aerospace engineering and (perhaps) some industrial process control systems can be expected to be well advanced. Others, however, are not.

DOI: 10.1201/9781003679479-7

Anyone who takes an interest in the technical press or even the general media, will realise that the world is being bombarded by organisations, including governments, seeking to take maximum and early advantage of the opportunities offered by new, frequently "AI-based", technologies. A moments reflection however makes clear that the great majority of these developments and ideas are based on technology-centred thinking. There is still little awareness, never mind recognition and understanding, of the complexities involved in ensuring the people that will be needed to monitor and support those systems will be capable of doing what is expected of them.

The purpose of this second part of the book is to provide a framework to help stakeholder organisations improve. It seeks to do so by introducing a comprehensive model of human supervisory control behaviour. A model that identifies and summarises some, at least, of the key challenges that need to be taken seriously if future systems are to enable humans to perform the role of supervisory controllers reliably and effectively. That means avoiding – or perhaps more realistically minimising – the risks associated with what are often referred to as "out-of-the-loop" (OOTL) issues.[1] OOTL issues have frequently been identified as sources of failure in previous generations of automation, both in the industrial world as well as in research settings. The comprehensive model introduced in this Part draws on learning and experience gained over many decades from scientific research as well as from many incident investigations. Learning and experience that has sometimes been gained at the cost of many fatalities and innumerable devastated lives.

The purpose of the comprehensive model is to draw attention to issues developers and operators of highly automated systems need to pay attention to, especially during system design, but also while preparing to deploy a new system or product. Issues fall into three general types;

1. Those to do with the design concept, or automation design paradigm. Specifically, the balance between what the automation is intended to do and what is expected of the human. That includes consideration of the wider context of what else the human is likely to be responsible for. In a work context, someone may have several assigned responsibilities that need to be met simultaneously with their role in supporting the automation. In a social, non-work context (such as the driver of an autonomous vehicle), the individual may have simultaneous interests and responsibilities that a system developer cannot control or perhaps even anticipate.
2. Issues to do with the design of the user interface, job design and the wider working environment. What kinds of information does someone need to fill the role of a supervisory controller, and how does that differ from when there are manually controlling the system? And is it possible to reconcile both sets of needs in a single user interface style or design? How to ensure information that is needed to be successful in the role of a supervisor is sufficiently easy to locate, attend to and understand. And is presented in a way that makes it easy for the supervisor to identify when support or intervention is needed. And what kind of distractions might have the potential to capture the supervisor's attention at times when they are expected to be monitoring for signs of trouble.

3. Issues to do with the characteristics and capabilities of the individuals expected to perform the role of the supervisory controller. These include different people's tolerance for risk as well as the perceptual and cognitive abilities needed to reason with and make judgements based on information that might be in different formats and displaced in space and time. They include issues such as an individual's value system and how they are motivated and incentivised. How clearly they understand what authority to act they have, and where responsibility for performance of the overall system lies: what are they likely to be blamed for if they get it wrong?

The impact of all this is summarised and encapsulated in a comprehensive model of supervisory control. To be clear, the model is comprehensive in the sense of being "more comprehensive than previous models", as opposed to being "definitive". Given the nature of human perception, cognition and thought, it will always be possible to build more detailed and richer models. That is the nature of the challenge. Undoubtedly, as experience of highly automated and autonomous systems develops over the years, and as we learn more about how people succeed or struggle in filling the role of human supervisory controllers, a more precise, psychologically accurate and richer picture of the essential features of supervisory control behaviour will emerge. But at today's state of maturity, the model presented is suggested as being sufficiently comprehensive.

A brief reminder is needed about the scope of the comprehensive model presented in this Part. This book is concerned with understanding the psychology of the role of the supervisory controller. At the point where action is needed, whether because the automation has failed, or because the human has decided to intervene, the human is no longer a supervisor: they have (hopefully) reverted to being a manual controller. The psychology of what is needed to ensure the human has a good chance of responding successfully in these situations is beyond the scope of this book. The model is limited to understanding the behaviour of the supervisory controller: from the point the individual steps in and acts, the model no longer applies.

STRUCTURE

The Part comprises four chapters;

- building on what was learned in Part 1, Chapter 6 considers what can be learned about human supervisory control from eight incidents involving automated systems. Three of these involve crashes of commercial airliners, one is from the world of nuclear power, one from fuel storage, one from shipping, one from the railways, and one from the field of professional sport;
- Chapter 7 reviews models of human supervisory control published over the past 50 years. Three models are considered: the first published in 1983 by Thomas Sheridan and his colleagues, the second published in 1986 by Neville Moray, and the third published by Mica Endsley in 2017. All highly respected authorities in the field;

- Chapter 8 explains why, drawing on my own lifetime's learning and experience as a Human Factors professional, as well as additional knowledge from the science base, I concluded that even the most recent of the models discussed in Chapter 7 is some way short of providing a satisfactory account of the psychology involved when a human is expected to fill the role of a supervisory controller. The chapter reviews what is missing and concludes that a more comprehensive model is needed;
- Chapter 9 presents and explains a suggested comprehensive model of supervisory control behaviour that builds on the models discussed in Chapter 7, adding necessary features that Chapter 8 argued were missing.

The comprehensive model is represented as a structured and logical flow of information, judgements, decisions and influences. Viewing the figure superficially, an uninformed reader might conclude that the thought processes behind supervisory control behaviour involve a similarly logical flow of considerations. As the chapter explains, such a conclusion would not only be superficial: it would be wrong. That is not how the human brain works, most of the time.

The material in this Part is intended to be relevant across a range of industries, different types of systems and applications. Many of the examples are taken from the world of autonomous road vehicles. This is largely because examples from this domain will be easily understood by a wide readership. The examples were also prominent in my thinking during the period I was writing the book. The points behind these road vehicle examples should however always be seen as instances of generic issues. Readers should use these to stimulate thinking of relevant examples from domains and systems they are familiar with.[2]

NOTES

1 The nature and scope of OOTL issues are discussed in Chapter 8.

2 I am always interested to learn of examples that illustrate the points discussed. Please send any to me at ronmcleod10@gmail.com.

6 Learning from incidents

Chapter 2 included a discussion based on a paper published by Sidney Dekker and David Woods in 2024 [1] that used the crashes of the two Boeing 737 MAX aircraft in 2018 and 2019 as a means of illustrating problems when systems that have both high levels of autonomy (are able to perform with little or no human involvement) and high levels of authority (are able to override human inputs) "misbehave".

This chapter will pick up on the arguments made by Dekker and Woods and look in some more detail at what can be learned about human supervisory control from the situation the pilots of the two aircraft were facing in the 737 MAX crashes. Crashes where, for reasons that stretch the limits of what would have seemed credible if they had not actually happened, the pilots were unable to recognise the nature of the internal threats they were faced with arising from the design of the highly automated and autonomous systems intended to help them fly the aircraft.

The Boeing 737 MAX crashes are powerful examples of the critical importance of ensuring human supervisory controllers have ready access to the information they need to be able to monitor automation. And especially to detect and understand internal and external threats. Many other incidents illustrate the same point, while also providing additional insight and opportunities to learn about other issues that need to be considered in developing systems to support human supervisory control. As well as the Boeing 737 Max crashes, the chapter briefly considers what can be learned from seven other incidents. In historical order;

- the incident at the Three Mile Island nuclear facility in the United States in 1979;
- the shooting down of Korean Airlines flight in 1983;
- the fire and explosion at the Buncefield fuel storage site in England in 2005;
- the crash of Air France flight AF447 into the North Atlantic in 2009;
- failure of a door-closing system on a passenger train in England in 2019;
- the loss of the Royal New Zealand Navy's ship HMNZS Manawanui off the coast of Samoa in 2024;
- the introduction of electronic line calling at the Wimbledon tennis championships in 2025.

THE BOEING 737 MAX CRASHES (2018/2019)

An enormous amount has been published about the events leading up to and following the losses of Lion Air flight 610 on October 29, 2018, and Ethiopia Airlines flight 302 on March 10, 2019.[1] Between them, 346 passengers and crew, as well as a rescue diver, lost their lives. The lives of countless families and friends were devastated. At the time of writing, both crashes are the subject of legal proceedings.

DOI: 10.1201/9781003679479-8

As is always the case when major tragedies occur in modern commercial aviation, the contributing factors were complex and diverse. Modern commercial aircraft only crash when a sequence of events that are normally, in themselves, extremely unlikely converge in the same time and space. Three factors converged in the case of the 737 MAX crashes;

1. The unprecedented way in which the 737 MAX aircraft fitted with a new Manoeuvring Characteristics Augmentation System (MCAS) that had both high autonomy and high authority to exert control over the aircraft was certified for commercial flight without any requirement for pilots to be trained, or even told the system existed.
2. The fact that the MCAS system had no built in redundancy but relied entirely on a single source of sensor data.
3. The fact that the pilots had no information directly available telling them there was a problem with an Angle-of-Attack (AOA) sensor providing critical flight data.

In the case of the Lion Air flight, two other events also converged to create the conditions for the crash. Although neither could be called "extremely unlikely", they were certainly not things that would be anticipated in an industry as safety conscious as commercial aviation;

- an engineer who installed a faulty AOA sensor to the aircraft the day before the crash did not carry out the required checks to confirm it was working correctly[2];
- a pilot who flew the aircraft with the faulty sensor the previous day, and who experienced the same difficulties as the pilots on the crash flight but managed to successfully control the aircraft, did not fully document the problem or how he solved it in the aircraft's maintenance log.

In both crashes, the immediate cause of the MCAS system misbehaving – as well as other automated systems not functioning as they were expected to - was failure of the left-hand AOA sensor. Sensors failing or going out of calibration is not in itself unusual in commercial aviation (or indeed, any other industry that relies on advanced technology for real-time control). Flight technicians regular change, re-calibrate and test numerous sensors in response to data recorded by the aircraft's health monitoring systems.

The purpose here is not to offer observations on what might be learned about the general Human Factors issues or to make other comment about the crashes. There are more than enough readily accessible sources of quality information prepared by those far more qualified to comment than I am. The interest here is specifically on what can be learned about the challenges and the ability of people to fill the role of supervisory controllers in complex systems capable of high levels of autonomous behaviour. In the case of the MCAS system fitted to the 737 MAX, a system that had not only complete autonomy to function, but had the authority to override pilot's control actions: in fact, a system that had been designed to act with sufficient force, regularity and speed to overcome the pilot's actions. More than that, a system that the pilots involved in the first crash did not even know existed.

A brief recap

For readers who are not familiar with the 737 MAX crashes, I need to provide a summary of the key facts behind the two crashes.[3] I am only concerned here with the details as far as they concern the challenge faced by the pilots as supervisory controllers tasked with identifying the threat to the safety of their flights.

The Boeing 737 MAX, which first entered passenger service in May 2017, was the 4th generation of the company's hugely successful 737 aircraft which had first entered service in 1967. The purpose behind the re-design of the existing 737 NG family was to provide Boeing's customers with levels of economic efficiency and operational performance that would make it attractive against competition from the new AIRBUS A320-family of aircraft.

To achieve this, the 737 MAX included larger, more fuel-efficient engines. To overcome technical problems, the engines needed to be mounted further forwards and higher on the wings compared with the previous model. This change created issues with the aircraft's flight stability that led the company to introduce an automated flight control system known as the Manoeuvring Characteristics Augmentation System (MCAS). MCAS received sensor data about the aircraft's Angle-of-Attack (AOA) that, in some situations, would indicate if the aircraft was entering an excessively high nose-up orientation[4], with the potential for an aerodynamic stall (i.e. a loss of lift due to lack of airflow over the aircraft's wings). The system used that AOA input as the basis for applying forces to the aircraft's stabilisers mounted on the tail, in order to induce a nose-down movement of the airframe, countering the potential stall. The system operated completely autonomously, with no input from the pilots.

The role of the angle of attack sensor

All 737 aircraft are fitted with two AOA sensors, one on the left and one on the right-hand side of the airframe, just outside the cockpit. The root of the problems faced by the flight crew of the Lion Air and Ethiopian Airlines flights was inaccurate real-time data on the aircraft's Angle-of-Attack (AOA) from the sensor mounted on the left-hand side of the airframe.

Without going into unnecessary detail here,[5] for the purpose of the discussion in this book, it is necessary to know a few things about the AOA sensors fitted to the two aircraft;

1. Data from the AOA sensors, along with signals indicating pressure, temperature and other parameters, are used by flight computers to calculate the aircraft's speed and altitude. The calculated data are sent to all aircraft systems that need them.
2. The MCAS system used data from only a single AOA sensor - the one on the left of the airframe. Data from the sensor on the right of the airframe was used by flight computers, together with that from the left-hand sensor, to

provide redundancy in the calculation of critical flight parameters including air speed, altitude and others.[6] But data from the right-hand sensor was not used as an input to the MCAS system.

3. In certain circumstances, such as some system failures, airspeed and/or altitude data will not be shown on the pilot's flight displays. Instead, a "flag" is shown in their place to alert pilots to the fact that the flight computers do not have valid data. These "flags" are displayed as vertically aligned "SPD" or "ALT" text drawn in amber within an amber box.
4. In the case of discrepancy between calculations based on data from sensors mounted on the left and right-hand side of the airframe, various alert messages are capable of being displayed on both pilots' flight displays. These include; "IAS Disagree" – meaning there is disagreement between the two airspeed calculations of more than 5 knots for more than 5 seconds; "ALT Disagree" – disagreement between the two altitude calculations of more than 200 feet for more than 5 seconds; and "AOA Disagree" – disagreement in angle of attack data of more than 10 degrees for more than 10 seconds.
5. For reasons that seem to be due to the sub-system delivered by one of its suppliers, Boeing's policy at the time of the crashes was that the "AOA Disagree" alert would only be displayed to the pilots if the airline had chosen to pay for an Angle-of-Attach Indicator on the pilot's flight displays. At the time of the crashes, the AOA Indicator was available as an optional extra but was not fitted as standard. Neither Lion Air nor Ethiopian Airlines had taken up the option.[7] Their aircraft therefore did not give the pilots any direct indication in the case of disagreement between the two AOA sensors.
6. Despite this, the Flight Crew Operations Manual (FCOM) provided to Lion Air by Boeing for its 737 MAX aircraft included details of the AOA Disagree alert. According to the US Final Report;

> The FCOM explained how the AOA Disagree alert was intended to work and provided absolutely no indication that Boeing was fully aware that the AOA Disagree alert on the Lion Air 737 MAX aircraft was not operational. As a result, Lion Air pilots who referenced Boeing's FCOM would have falsely believed that the AOA Disagree alert was functioning properly and would reliably warn them of a malfunctioning AOA sensor. [2]

The Pilots of the Two Aircraft Were in Different Situations

In terms of the roles they were in as supervisory controllers the pilots of the two aircraft were in very different situations. In common with every 737 pilot in the world (other than some employed by Boeing), the pilots of the Lion Air flight were not aware that the MCAS system existed. They were faced with trying to work out what was happening to the aircraft while both not knowing the MCAS - which was exerting control over the airframe - existed and having nothing to tell them that one of the aircraft's critical sensors was faulty.

By contrast, by the time of the Ethiopian Airways crash just over four months later, pilots around the world knew about the existence of MCAS, both from information provided by Boeing as well as from the publicity surrounding the investigation into the Lion Air crash. Boeing had provided airlines with details of the procedure pilots were expected to follow in the event they found themselves in circumstances similar to the Lion Air crash. Although the pilot in command of the Ethiopian Airlines flight had more than 5,500 hours experience flying 737s and over 100 hours flying the 737 MAX, the flight on March 10, 2019, seems to have been the first time he had flown the 737 MAX with an AOA sensor failure. The First Officer had 361 flying hours, of which 207 were on 737s (it is not clear how much was on the 737 MAX). For both pilots it was the first time they had experienced MCAS "misbehaving". Despite that, the pilots followed the procedure Boeing had advised to overcome erroneous MCAS activations. Though tragically, it was not sufficient to save them or their passengers.

Two phases

Both incidents evolved over two phases:

1. Pre-MCAS activation
2. Post-MCAS activation

Pre-MCAS activation

In the case of Lion Air, MCAS was first activated 4 minutes 52 seconds into the flight, while in the Ethiopian Airlines case, it first occurred after 1 minute 28 seconds.

With Lion Air, the AOA sensor had been installed on the aircraft with a -21deg offset bias.[8] Its data were therefore erroneous from the start of the flight. This caused an audible alert to be sounded in the cockpit, the stick shaker[9] to be activated and both the "Indicated Airspeed Disagree" and "Altitude Disagree" alerts to be displayed on the pilot's flight displays from the moment the aircraft was airborne. Lack of the AOA data also meant that the autopilot system, which would normally have been turned on very soon after take-off, was never available. The Lion Air pilots knew immediately that there was a problem, though they were never able to work out what was wrong.

In the case of the Ethiopian Airlines flight, data from the left-hand AOA sensor began drifting from the right-hand sensor from about 10 seconds into the flight,[10] leading to the same pilot indications as in the Lion Air flight. The pilot tried to engage autopilot twice without success, before being successful on the third attempt (51 seconds into the flight). Though the autopilot automatically disengaged again after only 37 seconds, at which point MCAS activated immediately.

The investigation into the Ethiopian Airlines crash [4] includes a compelling example of how automated systems with high autonomy can make critical decisions unknown to the human[11];

> While the autopilot was engaged, systems continued to be supplied by the erroneous LH AOA values. As a result, [...a left-hand flight computer...] computed erroneous ... minimum operational speed values which were higher than the current ... selected airspeed. As the ... minimum operational speed was greater than the ... selected speed at that time, speed reversion occurred (selection of the erroneous minimum operational

> speed as target speed) and autopilot commanded a pitch down to accelerate towards the erroneous minimum operating speed.

During this phase, in both aircraft, automated flight systems were misbehaving because of incorrect data from the left-hand AOA sensor, causing flight computers to disagree about key flight parameters, including airspeed and altitude. These disagreements led to flight computers, including autopilot and others, behaving in ways the pilots did not understand.

Post-MCAS activation

In the Lion Air flight, MCAS was first activated to move the aircraft's stabilisers and push the nose-down 4 minutes 52 seconds into the flight. It then activated another 23 times until the end of the flight 26 minutes later. Each time MCAS activated to move the nose down, the Captain or First Officer responded immediately - within an average of 4 seconds - to pull the nose up. The fact that the first two responses to the MCAS activation were the two slowest (6 seconds each), as well as the speed of the subsequent repeated responses over about the next 26 minutes, suggests that one of the pilots started anticipating and reacting to the nose-down activations. Ultimately however, the pilots lost the struggle, never having understood what was happening.

In the case of the Ethiopian Airlines flight (by which time the pilots knew about the existence of MCAS) although they did not know the cause of the flight control problems (loss of accurate AOA sensor data), they seem to have recognised that MCAS was at least part of the problem after the second time it activated (1 minute 48 seconds into the flight). Following the procedure contained in the FCOM Bulletin released by Boeing, they cut the electrical power to the tail stabilisers 14 seconds later, thereby preventing MCAS from moving them. However, because of the faulty AOA data, other flight systems continued misbehaving, including causing the aircraft to exceed its maximum safe operating speed (see the quote above). Being unable to sustain the force needed to move the tail stabilisers manually, they turned the stabilisers electrical power back on again 4 minutes 31 seconds into the flight. MCAS immediately acted to push the nose-down. The flight ended 39 seconds later.

The Ethiopian investigation report [4] noted that, between 1 minute 22 secs and 2 minutes 2 seconds into the flight, after the autopilot had disengaged and, in the period when the pilots decided to cut power to the tail stabilisers, preventing the MCAS activations, other automated flight systems were behaving in ways that would have been challenging for the pilots to understand:

> [Flight Management Computer] detected a significant difference between the RH and LH True Airspeed (...due to erroneous LH AOA value). From this time, FMC did not send any valid command to [Auto Thrust]. ... The loss of valid FMC command did not trigger an explicit alert but did result in the FMA continuing to display "ARM" instead of changing to "N1" as would normally be expected.
>
> As a result of the erroneous LH AOA value and the increasing airspeed ... computed LH minimum operational speed and LH stick shaker speed greater than [Maximum Operating Velocity] (340Kt) without any alert or invalidity detection.

Would the crashes have happened if MCAS had not existed?

Following the two crashes, attention has focused on the role of the MCAS system, and the failure of Boeing to inform pilots about it or train them on it. As the discussion in the previous section made clear however, MCAS activations did not enter the crash scenarios until sometime after the original problems occurred (4 minutes 52 seconds into the flight in the case of the Lion Air flight, and 1 minute 26 seconds in the case of Ethiopian Airlines). For the benefit of those who lack specialist knowledge of cockpit procedures,[12] but who want to better understand what is needed to properly support human supervisory control, it is worth briefly reflecting on whether, in the circumstances prevailing at the time, the pilots might have been able to control the aircraft and guide them to safe landings had erroneous activation of the MCAS system not complicated the situation.

I assume the answer is that the pilots would have identified the correct actions to take using the aircraft's Quick Reference Handbook (QRH). The QRH provides checklists that advise pilots how to manage non-normal flight conditions. The report into the Ethiopian Airlines crash includes details of how QRHs are used and includes examples of checklists for Runaway Stabiliser, IAS Disagree, Unreliable Airspeed and ALT Disagree events (as well as AOA Disagree, although that alert was not enabled on their aircraft). On the day before the Lion Air crash, a different pilot flying the same plane experienced the same situation as those of the plane that crashed.[13] The pilot on that flight managed the situation successfully using the "Runaway Stabiliser" procedure.

Prior to MCAS activating, the pilots of both aircraft knew they were experiencing flight control problems, though they did not know the cause was failure of an Angle-of-Attack sensor. Because the airlines had not taken up the option to purchase the "AOA Disagree" alert, the pilots had no direct indication of the source of the problem. They did have several indications of a problem, including a "MASTER CAUTION" alert, activation of the left stick shaker, as well as "IAS Disagree" and "ALT Disagree" alerts.

The captain of the Lion Air flight asked the First Officer (FO) to perform the checklist for "Airspeed Unreliable" 3 mins 30 secs after take-off. Twenty-six seconds later the FO advised he couldn't find the checklist, though he began reading it after another 46 seconds - 9 seconds before the first MCAS activation. The transcript of the Ethiopian Airlines flight includes no indication of either pilot making use of the QRH checklists.

It might be assumed that, had MCAS not complicated the picture, the pilots of both aircraft might at least have had a chance of successfully dealing with the consequences of the erroneous AOA sensor data. Though the Ethiopian Airlines investigation indicates that the issues that led to the loss of that flight were compounded by the way erroneous AOA data led to automatic flight systems calculating and acting to cause the aircraft to reach a speed that exceeded the safe maximum speed of the aircraft. It seems at least possible that the pilots would have been able to overcome the automated system's use of erroneous minimum speed data had MCAS not interfered with their efforts.

Safety and hazard analysis

According to the Ethiopian Airlines investigation, Boeing had identified the potential for "runaway" MCAS inputs as part of their safety analyses while the 737 MAX was under development. They specifically identified sensor failure as a potential cause of the worst-case potential failure. The analysis also made assumptions about the pilot's ability to detect and diagnose such a failure.

These assumptions however were made some years before the aircraft went into service, and before final decisions were taken about what pilots would know about the MCAS system. They were predicated on pilots knowing that the system existed and how and when it might be activated.

Boeing's safety analysis also considered the impact if MCAS made multiple, repeated stabiliser inputs due to receiving erroneous AOA data.[14] According to the report into the Ethiopian Airlines crash [4];

> Their assessment was that each MCAS input could be controlled with column alone and subsequently re-trimmed to zero column force while maintaining the flight path. Five seconds after cessation of the pilot trim command, the subsequent MCAS command could be controlled in the same manner as the previous instance. Eventually, use of the stabilizer cutout switches would be an option to stop the uncommanded stabilizer motion per the runaway stabilizer procedure (which is a trained flight crew memory item).[15]

Despite these confident assumptions, it is not what happened in the reality of both flights: the Lion Air pilots – who did not realise the effect cutting the stabiliser power would have - made 24 overrides in the space of 25 minutes but were eventually defeated. Note that Boeing's assessment quoted above must have been made in the knowledge that 737 MAX pilots would receive no simulator training in the MCAS system, which was a core element of the business case and design concept for new aircraft: something can only be a "trained memory item" if pilots are trained on it.

What can be learned from the 737 MAX crashes about human supervisory control?

Determining what can be learned from the 737 MAX crashes for the purpose here is not easy, given the charges levelled against Boeing for the way it is reported as having behaved during the development and certification of the new aircraft, and especially the MCAS system. If, as the evidence suggests, their behaviour bore no relation to how systems are normally developed, what can we learn from it that is useful to the mainstream of those engaged in developing systems with autonomous capabilities?

To summarise the situation in the cockpit of the Lion Air flight; we have two pilots engaged in controlling a modern airliner where most aspects of the aircraft's flight control – thrust, heading, trim, and so on - are normally controlled automatically. That is standard practice in modern aviation: pilots are taught to use automatic systems from shortly after take-off. Under normal conditions, most aspects of the aircraft's motion are controlled automatically under the pilot's supervision.

The pilot's time is normally spent monitoring and supervising the flight systems, perhaps making occasional minor adjustments, engaged in tactical planning and decision making (such as whether to divert around a thunderstorm) or communicating with ground controllers. These are tasks that have been studied and researched in enormous detail.

The situation the pilots of the Lion Air flight found themselves in however, was that after a period of nearly 5 minutes where they knew they had a flight control problem but had not yet diagnosed the cause or how to deal with it, a completely autonomous system they didn't know existed started exerting control over the airframe. An autonomous system that was using erroneous data about the aircraft's angle of attack: erroneous data that the aircraft's flight systems knew about, but that was not displayed to the pilots. And data that was also causing errors in airspeed and altitude calculations the pilots were being told about, though they did not know the cause. The investigation into the Lion Air crash [2] noted that;

> ...the effect of erroneous MCAS function was startling to the flight crews.

Very few autonomous systems to be developed in the future will having anything like the sophistication, complexity or safety implications of systems installed on modern commercial airliners. While that is undoubtedly true, and whatever shortcomings Boeing are eventually held to account for, it is also true that very few future autonomous systems will receive anything remotely approaching the level of supporting research, engineering analysis, simulation and testing, or quality control as part of their development that flight systems to be used on a commercial airliner do. Equally, few will be subject to the level of scrutiny, inspection, review and demands for evidence before receiving approval to be put into service that Boeing's products are normally subjected to.

MCAS is not, in the scale of things, an enormously complex or sophisticated system: the overall system it was part of certainly was, though in comparison with other pieces of software used in flight control, it was relatively simple. Certainly, no more complex than systems many enterprises around the world are easily capable of producing – and indeed, exceeding with advanced AI technologies becoming increasingly widespread. And doing so with very little, if any, of the scrutiny, testing and inspection that flight systems must pass to be certified for use. In that context, the lessons about the need to ensure effective human supervisory control over MCAS are applicable much more broadly than it may seem.

Many future autonomous systems are going to present their human supervisors with challenges that are similar to those faced by the Lion Air and Ethiopian pilots. Similar not in the specifics of the industries, technologies or tasks involved, but in the general characteristics of what the people expected to supervise those systems are faced with: autonomous features they don't know exist; systems capable of acting in ways and making decisions that the humans are not aware of or can't understand; systems that are reliant on real-time data with the potential to be erroneous; and systems that have interactions and dependencies on each other that are hidden from the human but that can have significant implications for the performance of the system the humans are expected to monitor and supervise.

To summarise, despite the implications of the accusations faced by the Boeing Corporation about the way they developed and put the 737 MAX aircraft into commercial service, there are several points those involved in developing systems with autonomous capability in other industries can learn about human supervisory control from these tragedies. These include

1. The pilot's role as a supervisory controller over automated sub-systems formed only a part of a more complex role. Pilots of modern commercial aircraft are faced with supervising and interacting with multiple automated systems simultaneously. Their role is a complex and dynamic balance between overseeing sub-systems, detecting, and reacting to information from the system, while simultaneously meeting other responsibilities (such as understanding information about the state of aircraft systems, identifying, and following procedures dependent on events, making tactical decisions and plans about the future flight, and, if necessary, stepping in to exert manual control over some functions). The same is likely to be true of many of those relied on to act as human supervisors of other types of systems.
2. The pilots had no information that directly told them flight systems were receiving faulty data from a key sensor.[16]
3. While there was little difficulty in detecting something was wrong, the pilots had no awareness of the existence of the MCAS and were therefore unable to understand the nature of the problem they were facing.
4. Even if they had worked it out and realised there was a problem with the sensor data, for it to allow them to work out what the threat they were facing was they would have had to; (i) know which sensor had failed; (ii) know about the existence and purpose of the MCAS system, and (iii) know that MCAS drew exclusively on data from the sensor that had failed.
5. Some of the aircraft's automated flight systems continued making decisions based on faulty sensor data, including giving the pilot data that was incorrect.
6. Safety analysis conducted during the development and certification of the autonomous system relied on incorrect assumptions about what the pilots would know and what they would be capable of doing in the circumstances when the anticipated failure occurred that were incorrect. Specifically, safety analysis assumed knowledge and training that the pilots did not possess and were not given.

There are however three specific learnings more than any other from these tragic and completely avoidable crashes, that those involved in designing, developing, implementing, and using highly automated and autonomous systems to exert control over real-time activities in the physical world should take to heart. Those are the importance of ensuring;

1. That the real-time sensor data those systems rely on are as reliable and accurate as they can be. That means;
   a. the sensors are properly tested during installation and confirmed to be working properly;

   b. sensors are properly inspected and maintained;
   c. faults are reported and dealt with quickly.
2. That the information available to the human is as direct an indicator of the state of the item in the world the human supervisor needs to monitor as it can be.
3. That the people tasked with monitoring and supervising those systems are left in no doubt if sensor data used by the automation becomes unavailable, erroneous or unreliable. And that, if sensor data does become erroneous, the people responsible for reporting and maintaining it understand the implications for the performance of the automation.

The two 737 MAX crashes seem especially awful in the light of what has subsequently been revealed about corporate behaviour and decision making and the lack of effective regulatory oversight of the flight safety certification of the new aircraft. In their 2024 paper that used the crashes to illustrate the dangers when automated systems that have high autonomy and high authority rely on data that turns out to be wrong, Sidney Dekker and David Woods [1] concluded that;

> Fundamentally, the design for supervisory management of MCAS as part of a suite of automated systems for different aspects of flight was virtually non-existent. In other words, the twin 737 Max accidents are a compelling exemplar of the risks from poor design of supervisory management when high authority/high autonomy automation misbehaves.

A supervisory control success story: Lion Air flight LNI043

Before moving on, there is one more feature of the story of the Boeing 737 MAX crashes worth being aware of. The crash of Lion Air flight 610 not only reveals how difficult it can be to provide human supervisory control over complex autonomous systems: it also provides an example of human supervisory control doing what is expected of it – at least in monitoring and intervening to control a real-time problem, if not in successfully following-up that intervention to ensure the continuing safe operation of the aircraft.

The day before the fatal Lion Air flight (28th October 2018), the same aircraft (designated as flight LNI043) had experienced virtually identical problems to those that were to occur the next day on flight 610. Though on this occasion, the crew were able to intervene effectively, despite not being aware of the existence of MCAS. It is worth exploring why that might have been.

Prior to flight LNI043, a ground engineer explained to the captain that the left-hand Angle-of-Attack sensor had just been replaced and tested. Unfortunately, the newly fitted sensor (which was itself coming back into service having been re-built in the US the previous year) was faulty: it had a 21-degree offset bias which was not detected when it was installed on the 28th.[17] The captain mentioned the replacement AOA sensor during his briefing to the crew prior to flight LNI043.

Shortly after take-off, the pilot's stick shaker activated, and the IAS alert appeared on the pilot's display. While the captain was engaged following non-normal checklists to determine how to respond to the problem, a third pilot - not part of the flight crew, but who happened to be "dead heading" in the cockpit's jump seat - intervened to advise the captain that the aircraft was pitching nose down.[18],[19]

The captain realised, shortly after the First Officer completed trimming the aircraft's nose up, that the trim system was being activated automatically to push the nose down again. After three cycles of this, and the First Officer commenting that the control column was becoming "too heavy" to hold back,[20] the captain concluded that they were experiencing what was referred to as a "Runaway stabiliser" – a problem which had been identified through safety analysis as a potential flight risk and for which a non-normal procedure had been included in the aircraft's Quick Reference Handbook.

The captain carried out the required non-normal procedure, which included cutting the power to the tail stabiliser motors, thereby (though unknown to the crew) preventing MCAS from acting on them. He decided, unusually, to carry on with the flight under emergency procedures, and eventually landed safely in Jakarta.

After landing, the crew returned the stabiliser trim cut-out switches to the normal position, which is where they were when the crew of flight 610 began their flight preparation the following day. The captain told the ground engineer about the problems they had experienced and completed a maintenance log reporting the airspeed, altitude and differential pressure alerts. Though he did not record either; (i) that they had used the trim cut-out switches to prevent the aircraft's nose being automatically pushed down; or, (ii) that they had flown the flight manually. The reasons for those omissions are not clear. One can only speculate on how different subsequent events might have been had he reported the full facts.

We are never going to know what was different between the two flights. That, faced with the same situation and the same set of alerts and other information, one captain successfully intervened to overcome an autonomous system that was misbehaving, while the other one did not. Both were highly trained and experienced. The fact that the captain of flight 043 had specifically been told about the changed AOA sensor before the flight seems to be important. His awareness of the new sensor was clear from the fact that he repeated it to his crew during the pre-flight briefing. So, when the alerts began shortly after take-off, the potential for an AOA issue may have already been easily accessible to his reasoning.[21] In the case of flight 610 the following day, so far as is known, no mention was made to the captain or crew about previous issues with the AOA sensor. As the investigation report stated; "The risk of the problems that occurred on the flight LNI043 were not assessed to be considered as a hazard on the subsequent flight". They had no reason to be aware of or anticipate issues arising from that source.

It may be as simple as the fact that the captain of flight 043 was fortunate in deciding to apply the non-normal procedure for "Runaway Stabiliser", while the captain of flight 302 decided to use the procedure for "Air Speed Unreliable". We will never know.

The presence of the "dead heading" crew on flight LNI043 may also have been a factor. All we know about the actions of this individual from the investigation report is that he drew the captain's attention to the fact that the aircraft's nose was pitching down. It is possible that the presence of a third highly experienced pilot in the cockpit was of some help to the captain in his reasoning. At least it seems to have made more resources available to him to help deal with the situation.

We are never going to know why the two situations led to such tragically different outcomes.[22] What we do know, despite whatever criticism might be levelled at the captain on 28th October for his subsequent decisions not to fully document the difficulties he had faced and overcome on the flight, was that he was successful in filling the role of a human supervisory controller under what must have been extremely testing and challenging conditions.

THREE MILE ISLAND (1979)

The design of the type of Pressurised Water Reactor (PWR) at Unit 2 of the Three Mile Island nuclear power facility in the United States required the reactor core to be permanently covered with water. As the level of water covering the core was not intended to vary, there was seen to be no need to provide operators in the control room with a dedicated indicator of it. There was an instrument showing the total level of pressurised water, though not specifically the level covering the core. The lack of such an indicator was behind the operators misunderstanding what was happening when a fault developed in the system. The decisions made and actions taken by the operators in the control room, based on their misunderstanding arising from the lack of any direct indicator of the level of water covering the core, led to a major incident. Had the operators taken no action, the automated plant control system would have safely managed the incident.

No one was hurt, though small amounts of radioactivity were released, and there was extensive damage to the reactor core. Despite this, Three Mile Island in March 1979 remains the most serious incident affecting the commercial nuclear power industry in United States history. It led to major change both in public awareness of the risks of nuclear power, as well as all aspects of the design, operation and regulation of nuclear facilities that remain at the heart of the industry today. All for the lack of an indicator of the level of water covering the core.[23]

WHAT HAPPENED?

For our purposes, we don't need to go into a lot of detail of what happened. Though a summary is needed for the benefit of those not familiar with the story.[24] The incident was what is referred to as a "Loss-of-Coolant-Accident" (LOCA). Pressurised hot water covering the reactor core was one of two water circuits used in the design. Its function was to move heat generated by the nuclear reaction to a steam generator located inside the reactor building, A separate system circulated cold water through the steam-generator, where it was heated by the hot water system and turned into steam to drive the plant's turbine, generating electricity for the grid.

Figure 6.1 illustrates the elements of the design involved in the incident. The incident, which lasted for around 150 minutes, began when the major pump circulating cold water around the steam generator failed. This was compounded by several other systems subsequently not doing what it was expected they would.

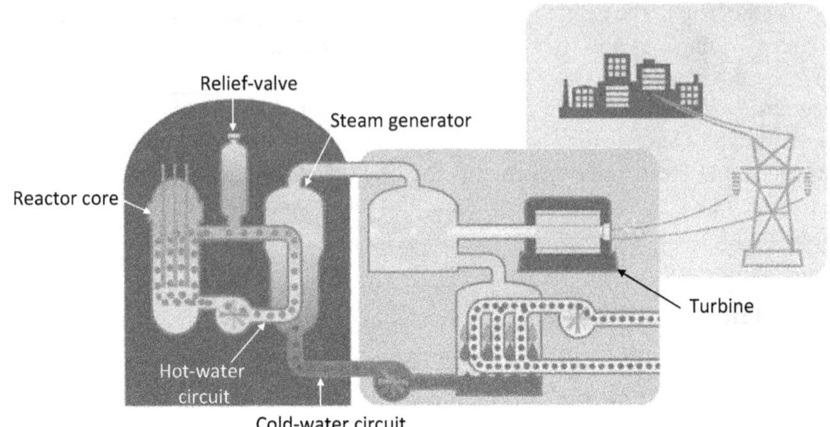

FIGURE 6.1 Schematic of Three Mile Island Unit 2 at the time of the incident in 1979 (Courtesy U.S. Nuclear Regulatory Commission)

That included a valve, that should have been open, having been closed sometime prior to the incident.

Failure of the steam generator caused the reactor's turbine, and then the reactor itself, to automatically shut-down. A relief-valve located at the top of the pressurised unit opened to release steam to relieve the growing pressure: though the pressure continued to increase sufficiently to cause the reactor to automatically trip, as it was designed to do. The release of steam from the relief valve gradually reduced the amount of water available to cover the core.

The relief-valve was expected to close again once the pressure returned to normal operating levels. However, not only did the valve become stuck open – causing a progressive reduction in the amount of water available to cover the core - but the indicator in the control room showing the valve's state told the operators that it was closed. It was not until 138 minutes into the incident that the operators realised the relief-valve was actually open.

Lacking a direct indicator of the level of water over the core, the operators reasoned that, as long as the total amount of water indicated on the hot-water level instrument was high enough, the core was properly covered. Faced with an unexpected, fast moving and increasingly dangerous situation they didn't fully understand, the need to reason about the behaviour of a sophisticated automated system controlling a process involving complex physics, and against a background of a flood of both visual and audible alarms, to avoid what they saw as the serious risk of avoid flooding the reactor building, they mistakenly turned off the emergency water pumps.

What can we learn about human supervisory control from TMI?

There is no need here to go into further details of the incident. Given the complexities of some of the physics involved, understanding the sequence of events properly requires a degree of study and effort beyond what is needed for our purposes here.[25] There are many excellent and easily accessed summaries of what happened.[26]

Three Mile Island was the first event to gain international attention that recognised the complexities of human supervisory control, especially in the task of diagnosing what had gone wrong and what action operators needed to take. The control room operators at TMI were in the role of human supervisory controllers of a highly automated system capable (had other systems worked as intended) of safely controlling the incident. Having the right information is fundamental to performing that role successfully. The mistake the operator's made (believing the reactor's core was covered with water, when in fact it had been exposed) was made in a challenging and highly demanding situation.

In their detailed study of the Human Factors issues associated with the incident published nine months after the incident, a team of experts from The Essex Corporation [5] concluded that:

> …the human errors experienced during the TMI incident were not due to operator deficiencies but rather to inadequacies in equipment design, information presentation, emergency procedures and training.

More specifically;

> In the absence of a detailed analysis of information requirements by operator tasks, some critical parameters were not displayed, some were not immediately available to the operator because of location, and the operators were burdened with unnecessary information.

Following Three Mile Island, a great deal of high-quality thinking and research was conducted, principally in the United States, to better understand the complexities of the human supervisory control role, as well as how to design systems and user-interfaces to fully support operators in the role. The discussion of models of supervisory control in Chapter 7 includes two of the models that came out of that research.

KOREAN AIRLINES FLIGHT 007 (1983)

Perhaps one of the saddest and most tragic stories that illustrates issues surrounding human supervisory control occurred during the loss of Korean airlines flight KAL 007 in 1983. Events surrounding the shooting down of the Boeing 747 continue to be shrouded in mystery. Occurring at the height of the Cold War, the unwillingness of both the US and the Soviet governments, both at the time and since, to reveal full details of what happened and what they knew on that tragic night leave a story that, from what is known, seems too full of unlikely coincidences and missed opportunities to be real. But what is real are the facts that 240 passengers and 29 crew members lost their lives in what must have been terrifying circumstances.

In his 2001 book "Taming HAL" [8], NASA research scientist Asaf Degani put together a narrative of the events surrounding the loss of the flight.[27] His description is based on both established facts as well as a degree of informed speculation about what the pilots are likely to have done, and how the automated flight systems are likely to have responded to the pilot's commands. Degani's purpose in telling the story is to illustrate important issues about the design of user interfaces to highly automated systems; non-determinism (when it is not possible for the user to determine what effect an action will have); issues of different kinds of transitions between states of automation; and the role of software "guards" that check whether pre-conditions are met before automation initiates events or make transitions between states.

Whether or not Degani's narrative is correct, the story of KAL007 illustrates important lessons about the role of human supervisory controllers, and what is needed to fully support them. For the purposes of this book, I have briefly summarised Degani's narrative. The interested reader will need to refer to Degani's original to fully appreciate the implications for the design of automated systems.[28]

What happened?

On the morning of 31 August 1983, Flight KAL 007 took off from Anchorage airport in Alaska en-route to Seoul in South Korea. Shortly after take-off the crew were instructed by air traffic control to follow transoceanic track R-20, passing a series of ten waypoints. The pilots are thought to have engaged the autopilot and rotated the MODE select knob to the Inertial Navigation position ("INS" on Figure 6.2). From then on, the aircraft's inertial navigation system should have automatically followed track R-20 all the way to Seoul.

In fact, the aircraft's autopilot had not transitioned from HEADING to INS mode but remained in HEADING mode. The aircraft maintained a constant heading of 245 degrees, progressively diverging further from route R-20, and taking the plane into Soviet airspace. And, as it turns out, directly on track not only to the town of Petropavlovsk, the home of a nuclear submarine base, but onwards to Vladivostok, the home of the Soviet Pacific fleet. Degani describes the complex sequence of events, some pushing the limits of credibility, that seem to have led to the flight eventually being shot down by a Soviet fighter and crashing into the sea near Sakhalin Island, north of Japan.

Degani's analysis is that the pilots did not understand what was necessary for the aircraft's autopilot system to engage INS mode. After the pilot commanded the change of mode from HEADING to INS, rather than engaging INS, the system entered a state he refers to as "INS ARMED" – still in HEADING mode but preparing to enter INS mode. While in INS ARMED mode the autopilot's software

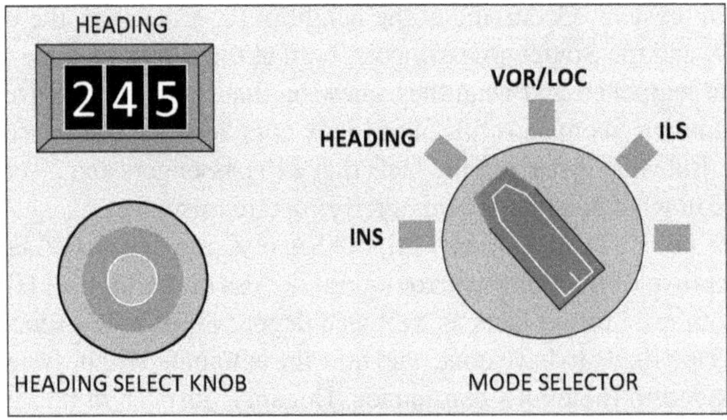

FIGURE 6.2 Representation of heading indication and autopilot mode select switch on KAL 007. Autopilot is in HEADING mode. Reproduced from Degani [8]

performed two logical checks every second; (i) was the aircraft heading towards the destination rather than away from it; and (ii) was the distance between the aircraft and the designated route no more than 7.5 miles? If at least one of these conditions was not satisfied, the system remained in "INS ARMED" mode. Degani refers to these conditions as examples of the kinds of "guards" that are widely relied on in automated and other software intensive systems.

While the aircraft was flying towards Seoul, it was never within 7.5 miles of route R-20. Because of this, and apparently unknown to the pilots, the aircraft remained in HEADING mode following the designated heading of 245 degrees, taking it towards Soviet airspace. Degani suggests that the pilots misunderstood the details of the second logical check, thinking the limit was perhaps 20 miles, rather than 7.5 miles. As far as the pilots were aware, from the moment they selected INS mode (see Figure 6.2), the aircraft should have automatically followed route R-20. But it didn't as the aircraft was more than 7.5 miles from the R-20 route at the time. From that moment, a combination of seemingly absolute trust in the autopilot system, a lack of any clear indication in the cockpit that the autopilot was still in HEADING mode, combined with no attempt to proactively check the aircraft's position caused it to relentlessly diverge from the required flight path.

The pilot's misunderstanding about the conditions necessary to enter INS mode were compounded by there being no indicator in the cockpit display directly telling the pilots which autopilot mode was engaged. There was an INS indicator light that turned green when INS mode was active. Though it was amber when the system was in INS ARMED mode – as it must have been from the point the pilot selected INS mode. There was nothing that unambiguously told the pilots the system was in HEADING while INS ARMED was waiting for its conditions to be met. As Degani discusses, there is no way of knowing what role this incomplete information display might have played. Though it cannot have helped the pilots recognise what was happening.

None of Degani's narrative begins to explain why the aircraft was ultimately shot down. But it is a compelling account of why it may have found itself in the wrong place.

There are many unanswered questions about the shooting down of Flight KAL 007. One of the key ones is why, despite not realising that the autopilot had not engaged the INS mode but remained in HEADING mode, the pilots seemingly did not detect any indications that would suggest they were not where they were meant to be. According to Degani's narrative, those indications included;

- although the cockpit was equipped with readouts of the aircraft's actual position, the pilots did notice that the aircraft did not reach any of the ten waypoints defined on transoceanic route R-20;
- the crew were not monitoring a radio frequency reserved for distress and emergency calls that the Soviets used to try to contact the Korean crew;
- although they had been outside civilian radar coverage since leaving Anchorage, the crew were in touch with air traffic controllers in Japan. If the aircraft's position was ever communicated, neither the crew nor the ground controllers questioned it;
- the crew did not notice a round of bullets passing them fired by a soviet fighter four miles behind the 747 or notice the fighter flashing its lights.

What can we learn about human supervisory control from KAL 007?

Based on Degani's narrative, three factors; along with a sequence of events and coincidences that are in themselves hard to believe combined to ensure the fate of flight KAL 007, its passengers and crew.

1. The pilots did not understand the conditions the system needed to satisfy to enter the inertial navigation mode.
2. The design of the user interface did not directly tell the pilots what mode the autopilot was actually in.
3. The pilots did not proactively monitor the aircraft's position throughout the remainder of the flight

Hindsight is, of course, a wonderful thing. And hindsight that draws on speculation to fill in for a lack of facts is fabulous. The design of cockpit displays and flight systems, pilot training and certification and flying procedures have all changed enormously since 1983. However, despite these changes and for the need for some speculation, the story we have of flight KAL 007 illustrates some important lessons relevant to ensuring effective human supervisory control;

i. the need to maintain a healthy balance between trusting automation to do what is expected of it, and maintaining a wariness for the presence of external or internal threats to its performance;
ii. the importance of being proactive in searching out information about the state of the system and potential threats, rather than simply relying on the system to draw the supervisor's attention to them;
iii. the importance of recognising and understanding the role of what Degani refers to as "guard" conditions: checks designed into the automation that need to be satisfied before it will change mode or act;
iv. the importance of automation providing clear feedback to the users about what state or mode it is actually in.

We can never know fully the situation and the context the pilots of KAL flight 007 were in that may have influenced their thinking during the flight. Nevertheless, it seems reasonable to speculate that, had the pilots been proactive in monitoring where they were, there was enough information available to them, and sufficient additional indicators that they would have noticed the divergence. But they cannot have been proactive. They must have been relying entirely on the autopilot, under the misapprehension that it was in INS mode.

BUNCEFIELD FUEL STORAGE SITE (2005)[29]

The explosion and fire at the Buncefield fuel storage depot on 11 December 2005 is a story of a relatively simple automated process being defeated by over-reliance on, and a lack of attention to, not just one, but two sets of critical instruments.

Fuel storage tanks are found all over the world, wherever refineries or other manufacturing processes produce fuel, in ports or convenient to airports, pipelines or other centres that store it prior to onward transport for processing or sale. The process of moving fuel between locations has been highly automated for many years. The Buncefield story comes close to breaking the bounds of credibility when we learn two facts about a site that was classed as a 'top-tier' site under the UK's Control of Major Accident Hazards Regulations, 1999 (COMAH)[30]: a site that was therefore expected to be subject to rigorous safety management procedures;

i. operators continued to rely on alarms as a control against over-filling a fuel storage tank despite knowing that the instrument driving the alarm was unreliable, and;
ii. the final control against over-filling the tank, an independent level switch designed to automatically stop the flow of fuel into the tank, failed to do it's job because a maintenance technician had omitted to attach a padlock.

What happened[31]?

At 05:37 on a Sunday morning the capacity of a fuel storage tank that had been receiving fuel since early the previous evening was exceeded. By just after 06:00, when the resulting vapour cloud ignited, more than 250,000 litres of fuel had escaped from the tank. The resulting fire burnt for five days: the largest peace-time fire in British history. It is fortunate that no-one was killed. Had the incident occurred during a working day, rather than on a Sunday morning, the likelihood of injury and fatalities would have been much greater.

The Automatic Tank Gauging (ATG) system, which is designed to automatically monitor the level of fuel in the tank, included three alarms set at progressively higher levels. Their purpose was to alert operators in the control room of the potential for a storage tank to over-fill. There are many Human Factors issues associated with alarms in process and other operations: from "flooding" due to too many alarms, many of which can be "nuisances", alarms that are confusing, not meaningful or are unhelpful in resolving a problem; to people who were not responsible for them acknowledging alarms without letting the responsible person know. At Buncefield none of these were the problem. It was more basic than that: there were no alarms. The Automatic Tank Gauging System failed:

> At 0305 hrs on Sunday 11 December the ATG display ... stopped registering the rising level of fuel in the tank although the tank continued to fill. Consequently the three ATG alarms, the 'user level', the 'high level' and the 'high-high level', could not operate as the tank reading was always below these alarm levels. [2, p. 10]

The design allowed for the possibility that the ATG system, including the operator alarms, might not be successful in preventing an overfill. The design therefore included, as a final control, a separate system, known as the Independent High-Level Switch (IHLS) designed to automatically stop the pumps if the level of fuel in the tank rose above the maximum safe level. Figure 6.3 illustrates the operation of the

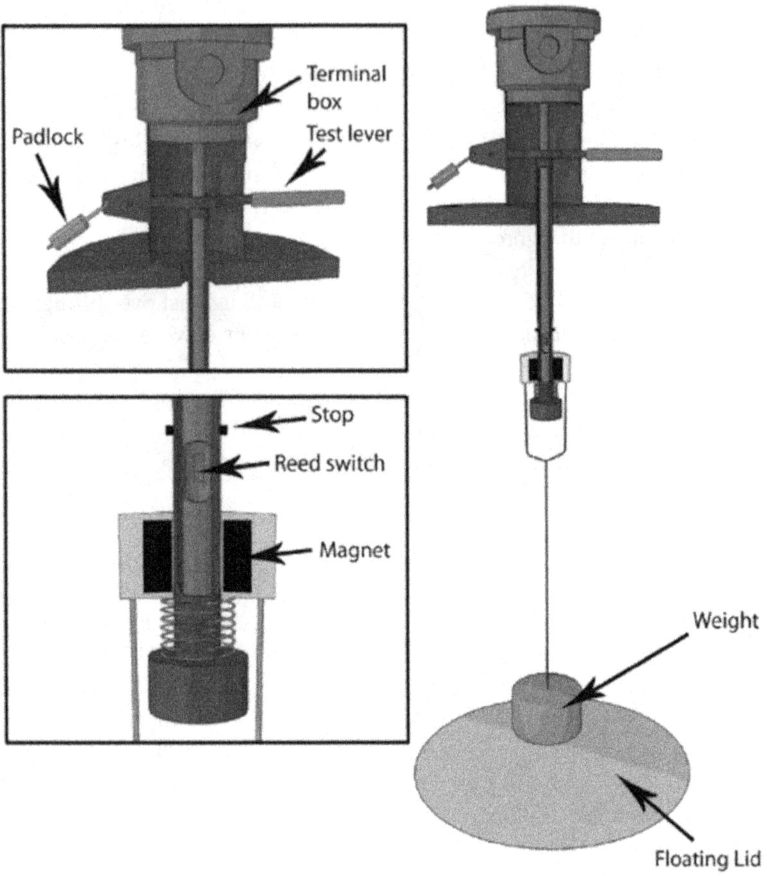

FIGURE 6.3 The working principles of the Independent High-Level Switch (IHLS) (From [11])

IHLS. The switch was based on movement of a weight suspended below a magnet located above the floating roof of the tank. When the tank was full, the fuel would make contact with the weight, causing it to lift the magnet and operate a switch located in the body of the device. Activation of the switch stopped the feed pumps.

As Figure 6.3 shows, a manual lever was fitted to allow the switch to be tested. For the switch to operate when the fuel level in the tank was high – the normal operating condition – the test lever was required to be in the horizontal position. If the test lever was moved to its lower position, the device would serve to detect low levels of fuel in the tank. A padlock was provided to secure the switch in the required horizontal position. The device would serve to detect low levels of fuel in the tank if the padlock was not applied, and the lever fell under gravity to its lower position.

Testing the switch involved unlocking the padlock and raising the switch to the vertical position, thereby activating the alarm circuit. Following the test, the test lever needed to be returned to the horizontal position, ensuring the device activated when the tank was full, and the padlock re-applied.

The technician who maintained the IHLS before the 11 December tank fill however was not aware of the critical role of the padlock in locking the test lever in the horizontal position. After removing the padlock and testing the switch, the padlock was not re-applied. The lever therefore dropped under gravity to the lowered position. Once that happened, the switch was incapable of initiating an emergency shutdown when the tank was full. In the words of the Competent Authority:

> Because those who installed and operated the switch did not fully understand the way it worked, or the crucial role played by a padlock, the switch was left effectively inoperable after the test. [9]

The switch continued to function as designed: it was not broken, but it was left in a state where it would detect a low level of fuel in the tank, rather, as was needed, than shutting off the pumps when the tank was full, preventing a fuel spill. Coming after the failure of the Automatic Tank Gauging System, this inability of the Independent High-Level Switch to do what it was expected it would do created the conditions for fuel to spill out of the tank into the surrounding area. The explosion and resulting fire occurred after the fuel found an ignition source.

What can be learned about human supervisory control from the Buncefield incident?

In my first book, I devoted several chapters to a detailed exploration of the Buncefield incident, focusing on the way various controls (or "Barriers"[32]) relied on to prevent tank overfills failed, as happened on that Sunday morning in 2005.[33] I was also sent by the Society of Petroleum Engineers (SPE) as one of their "Distinguished Lecturers" in 2016/2017 to deliver a lecture on learnings from the incident to membership groups across Europe and Asia. While there is no need to cover that material again here, it is worth drawing on some of that analysis in looking for lessons from Buncefield that are relevant to the role of humans as supervisors of highly automated systems.

What is of most significance in terms of human supervisory control is not that a tank level sensor and the associated alarms failed. The significance is:

a. that the control room operators had nothing to draw their attention to the fact that this critical sensor had failed;
b. that there was a history of repeated failure and unreliability of these alarms; and,
c. that the same control room operators who knew the alarms were unreliable continued to rely on them.

Here are some quotations from the investigation report [12] that summarise some of the issues associated with the failure of the tank sensor alarms;

> The servo-gauge had stuck … and not for the first time. In fact it had stuck 14 times between 31 August 2005, when the tank was returned to service after maintenance, and 11 December 2005.

> The Operations Coordinator had devised an electronic defect log but the supervisors did not use it properly. While the ATG gauge on Tank 912 had stuck 14 times during the three months before the incident, this was not recorded on the defects log and the Operations Manager was unaware of the frequency of failure.

Not only was the Automatic Tank Gauging system driving the alarms unreliable, but operators knew it was unreliable. They knew that the alarms could not be relied on; the ATG had failed fourteen times in the three months prior to the incident. It turns out there was a similar situation with the IHLS;

> ...by the first week of April 2004, it was known that the IHLS on Tank 912 was not working but the tank remained in use and a new switch was not fitted until 1 July 2004. Similarly, it was found that before this Tank 911, a very busy unleaded petrol tank, was operating without an IHLS for at least nine months.
>
> Faulty procedures and practices were not properly dealt with. The failings of the ATG system meant that there was greater dependence on the IHLS; as the IHLS was frequently left in an inoperable state, there was greater reliance on the ATG. The fact that both systems could not be relied upon meant that the overall control of the tank filling process was seriously weakened.

At the same time as the ATG alarms failed, operators were not proactively monitoring for signs to ensure themselves that the transfer was proceeding safely. They were monitoring reactively, relying entirely on the (failed) ATG alarms to draw their attention to the developing situation.

> Due to the practice of working to alarms in the control room, the control room supervisor was not alerted to the fact that the tank was at risk of overfilling. The level of petrol in the tank continued to rise unchecked.

How can this be? How could trained and experienced operators, whose own lives were potentially at risk, accept a situation where they would be responsible for a safety-critical operation knowing that not one, but both key systems they were relying on to avoid a potentially catastrophic event were unreliable? There is one thing that can be noted with certainty. That is just how often organisations (this is not limited to individuals on the front-line) seem willing to continue with highly hazardous operations far past the point when, at least with hindsight and the seeming rationality that comes with not being personally involved in events, the objective facts of the situation suggest the operation cannot possibly proceed safely. At least not without relying on luck and good fortune. The same seems to be true of the behaviour of the Boeing Corporation after the two crashes of its 737 MAX aircraft in October 2018 and March 2019. Despite knowing the role that faulty Angle-Of-Attack sensors had played in both crashes, they were prepared to allow the aircraft to fly with no changes until they were able to install a software change scheduled for more than a year later. According to the Final Report of the US Committee on Transportation and Infrastructure [1];

> Boeing continued to produce and deliver its MAX aircraft to customers before the Lion Air crash knowing the AOA Disagree alert was inoperable on most 737 MAX aircraft without informing them of this non-functioning component.

Research into the way pilots use automation has noted the same willingness to rely on automation even when they know the technology is not reliable.

Turning to the Independent High-Level Switch at Buncefield, there was no reliance on operator monitoring to enable the device to do its job: it functioned entirely autonomously if the floating roof of the storage tank hit the sensor. But, crucially, the IHLS did rely on human performance to ensure it was installed, maintained and configured correctly. Ultimately, it was human performance in putting the sensor back into service following routine testing that led to the failure of the automated emergency shut-down system at Buncefield. Superficially, not replacing a padlock correctly following maintenance of the IHLS device might seem a very simple human error, of a type that it should not be difficult to ensure could not recur. However, the Competent Authority's final report makes clear that there was a long trail of organizational failings that led to the failure of the technician to apply the padlock. Here are just a few quotations from the Competent Authority [12] that illustrate the scope of organizational issues behind this 'simple' human error:

> TAV was aware that its switches were used in high-hazard installations and therefore were likely to be safety-critical.
>
> ...the ordering process by both parties fell short of what would be expected for safety-critical equipment intended for such a high-hazard environment.
>
> They did not understand the vulnerabilities of the switch or the function of the padlock.
>
> At Buncefield the designers, manufacturers, installers and those involved in maintenance did not have an adequate knowledge of the environment in which the equipment was to be used.

The Competent Authority also made clear that issues around the failure of the IHLS were not limited to operational management, but that many began with failings in the design process;

> The design fault could have been eradicated at an early stage if the design changes had been subjected to a rigorous review process...
>
> ...the way the switch was designed, installed and maintained gave a false sense of security.
>
> ...the design, installation and maintenance of safety-critical equipment was just as important as the operational process controls.
>
> ...not only did the switch feature a potentially dangerous disabled position, which carried a risk that it would be inadvertently inoperable, but it was also a risk that was unnecessary to run.

To summarise, while there are many Human Factors lessons for process operations arising from the Buncefield incident,[34] the key learnings relevant to understanding and designing systems that properly support human supervisory control are;

1. The importance of proactive operator monitoring.
2. The importance of ensuring instrumentation providing data for the automation are not only fit-for-purpose, but are installed and maintained in a way that ensures their reliability.
3. Even in safety-critical situations, where the individuals own lives could potentially be at risk, supervisory controllers will sometimes continue to rely on automation, even when they know it should not be trusted.

AIR FRANCE FLIGHT AF447 (2009)

On 5 July 2012 the Bureau d'Enquetes et d'Analyses pour la securite de l'aviation civile (BEA) published its final report on the investigation into the crash of the Air France Airbus A330-230, flight AF 447 over the Atlantic whilst en-route from Rio-de Janeiro to Paris [13]. All 228 passengers and crew on-board perished. The sequence of events was initiated by a temporary failure in automatic flight systems.[35] But the crash only happened because of the actions taken by the crew after the system failure. The tragedy was essentially a failure of human supervisory control and of the controls that relied on pilot performance. The investigation report contains a significant analysis of the Human Factors issues that led to the crash.

The difficulties the crew had in diagnosing the situation, understanding what was happening and responding in accordance with their training has many of the characteristics of the challenges facing human supervisory controllers. For the purposes here, it is not necessary to go into technical detail of the causes of the incident or the precise actions the crew took. The discussion is therefore at a sufficiently general level to identify lessons that might be of value to understanding the psychology of human supervisory control.

WHAT HAPPENED?

The events leading to the loss of flight AF447 happened at a time when the crew were mentally preparing to transit an area of high turbulence (known as the Inter Tropical Convergence Zone (ITCZ). They anticipated a high demand on their skill, judgement and decision making during the transit. The aircraft's automatic flight systems suddenly, with no warning, stopped working due to a loss of air speed information when the 'pitot tubes', mounted on the exterior of the airframe, froze. The crew suddenly, and unexpectedly, needed to take manual control of the aircraft. Tragically, they failed to recognise the situation they were in: they mis-diagnosed what had happened and never recovered.

Loss of air speed information could have been safely managed if the crew had recognised, and applied, a pre-established Emergency Procedure for which they had all been trained a few months previously. A lot can be learned about why they failed to recognise the situation they were in, and therefore why they failed to apply the expected procedure.

The entire incident – from loss of automatic flight control to the crash - happened over a period of 4 minutes 23 seconds. Throughout that time, the crew would probably have been subject to at least four significant psychological stressors:

1. The flying pilot is described several times in the report as being 'surprised', 'startled' and experiencing 'emotional shock' from the moment he was required to take manual control.
2. Within 2 seconds of the auto-pilot disconnecting, the airframe roll angle increased significantly. The flying pilot's concentration was immediately fully absorbed regaining control of the airframe: he made a sequence of 'abrupt' and 'excessive' control inputs, described as being 'unsuitable and

incompatible with the recommended airplane handling practices for high altitude flight'. These lasted throughout most of the period when the crew should have been diagnosing the cause of the situation and planning a response.
3. The pilots' control inputs, together with the high level of atmospheric turbulence, would likely have led to a degree of spatial disorientation among even highly experienced flight crew. The actions of the crew should be seen in this context: it is very difficult to think clearly and rationally, whilst both stressed and subjected to severe motion.
4. The crew would probably have realised shortly after the abnormality occurred that they were in a situation where they were facing death if they did not recover very quickly. Flight crews regularly go through simulator training that can be highly realistic and psychologically compelling. Nonetheless, it is possible that the reality of potentially imminent death, and the fear and stress it would create, could have interfered with rational problem diagnosis and decision making.[36]

In addition, for thirty-four of the first forty-six seconds of the incident, a loud audible alarm was sounding in the cockpit. 'The C-chord alert therefore saturated the aural environment within the cockpit…' which '…certainly played a role in altering the crew's response to the situation.'

Similar, though probably not as extreme, psychological stressors could all exist in the moments immediately following significant unexpected upsets in many highly automated industrial processes. The actions and decisions of the crew also seem to have been confounded by a range of other factors, including apparent anxiety - or at least unease - on the part of the co-pilot who was flying the plane about the flight path through the Inter Tropical Convergence Zone they were about to enter, as well as failure of the captain to recognise or react to the flying co-pilot's unease with the chosen flight path, or to engage in discussion about possible alternatives. These are described in the investigation report as having created an environment in the cockpit that involved '…highly charged emotional factors…' There are repeated implications that this charged emotional atmosphere interfered with the ability of the crew to think clearly when they were taken by surprise by the loss of automatic control.

A range of other Human Factors issues

The investigation report covers many other Human Factors issues: to do with the design of the human interface to the aircraft systems and the design of procedures, as well as organizational issues to do with training, decision making, communication, and inter-personal relationships. Here are a few examples:[37]

- the design of the alarm annunciation system – especially the conflict between audible and visual alarms, and the duration and intensity of the audible alarms;
- the display of textual alert messages over the critical period while the crew were trying to diagnose the problem. In particular the way messages were

displayed and prioritised on the flight displays and the lack of any single message that showed the root-cause of the problem. ("...no explicit indication that could allow a rapid and accurate diagnosis was presented to the crew...", "...The successive display of different messages probably added to the confusion experienced by the crew...");

- breakdown in communication between the two pilots throughout the final minutes. "In general, the failure of both crew members to formalise and share their intentions made the identification and resolution of the problem more difficult";
- the fact that the flight system had different 'modes': the flight crew were apparently unaware which mode the systems were in at the time. The failure to recognise the mode the flight computers were in was fundamental to the pilots' reasoning about how to fly the aircraft at the time[38];
- differences between the way training for the Emergency Procedure was carried out, and the actual conditions (including flight conditions, the human-machine interface and alarm overload) that faced the crew at the time; "...the training scenarios may differ significantly from the reality of an in-flight failure";
- assumptions inherent in the design of the cockpit, procedures and training that pilots would unambiguously recognise a situation of approaching stall and take the necessary corrective actions were incorrect. This is despite the very rare experience of an approach to stall during the career of most pilots. "The safety model assumes that the abilities to identify the signals indicative of the approach to stall, and to recall the expected actions, remain sufficient over time, despite the low level of exposure." These assumptions were mistaken;
- various indications of a lack confidence among the flight crew in instrument readings, warnings and procedures. These were also found among flight crew in other incidents studied as part of the investigation;
- indications of cognitive tunnelling - loss of Situation Awareness - when the flying pilot appeared to focus on a single display indicator, rather than the bigger picture;
- design and implementation of Procedures: "...Procedures that are inappropriate to situations; A workload that makes it impossible to apply procedures; Procedures that are too numerous or too detailed";
- inconsistent human-machine interface behaviour in the actions needed of the crew in cases of disconnection and re-engagement of automatic systems.

What can we learn about human supervisory control from the AF447 crash?

While many Human Factors issues seem to have played a part in the loss of Air France flight AF447, there are at least five specific learnings that have wider relevance to designing automated systems that are effective in supporting the task of human supervisory control;

1. The difficulty of designing training strategies that are effective in ensuring human supervisory controllers will respond effectively to an unexpected loss of automated control should not be underestimated.
2. Similarly, the difficulty of designing procedures, including emergency procedures, that are effective in supporting human supervisory diagnosing threats and transitioning to manual control is challenging and should not be underestimated.
3. Assumptions inherent to a system design about the ability of human supervisory controllers to unambiguously recognise threats need to be challenged taking account of the realities of the situation and context they may be in at the time.
4. The pilots were captured by a mode error: they misunderstood the mode the aircraft was in at the moment they were handed control and consequently flew the aircraft into a stall. And, crucially, the design of the cockpit user interfaces was not effective in ensuring the pilots would be in no doubt which mode it was in at that moment.
5. The design of the user interface, and especially the various systems alerting the pilot to the situation they were in, did not support the pilots in diagnosing the problem. In particular, the pilots had no single clear indication of the reason why the automation had stopped working.

TRAIN PASSENGER DOORS (2019)

In recent years, trains in the UK (and, I assume, most countries), have been fitted with an automated system allowing the driver to close the train doors from the cab, replacing the need for a train guard to manually close the doors. An interlock prevents the train from moving if any of its doors are open. So, when the driver operates the 'door close' control inside the cab, the interlock system checks that all doors are closed. If the check is passed, yellow hazard lights on the outside of the coaches, as well as a visual indication in the driver's cab, are extinguished and the interlock is released allowing the train to move away under driver command.

WHAT HAPPENED[39]?

In 2019, a passenger train in England travelled for 16 miles (26 Km) at speeds of up to around 80 mph (128 km/h) with one half of a 'double-leaf' door to a coach occupied by passengers open. The partly open door was not visible to platform staff. The open door was reported by a passenger after the train had stopped at four stations.

It turned out that the door opening and closing mechanism on this class of train included a micro-switch attached to a drive mechanism driven by a pneumatic piston (see Figure 6.4a). The mechanism was designed to be physically attached to one door-leaf, with a drive belt connecting the other door-leaf, such that the mechanism and both door-leaves would move together. The interlock circuit monitored the status of the doors by detecting the position of the drive belt piston rod.

The incident occurred when two attaching screws worked loose and fell off the drive mechanism leaving the drive physically detached from the door (Figure 6.4b).

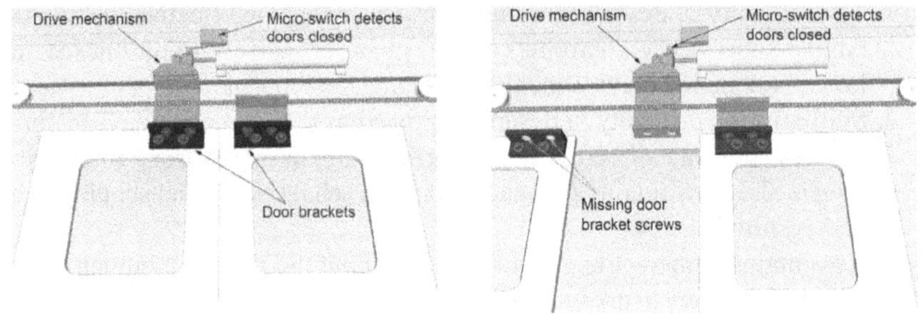

FIGURE 6.4 Illustration of door opening mechanism (a) as designed and (b) during the incident (Crown copyright. Reproduced from [14])

When the driver commanded the doors to close, the piston and drive mechanism moved to the closed position. Though, because the mechanism was no longer connected to the door, the door-leaf did not close.

The maintenance contractor required its staff to paint a yellow mark across the screw head, washer and bracket when they had tightened each screw to the specified torque. This was intended to help maintenance staff identify any loosening of the screw, as rotation results in misalignment of the marks. A second person was also expected to witness the tightening, make their own checks, and apply a blue dot to the screws which they had checked. The investigation found no yellow marks on the screws which fell from the connection between the bracket and the drive belt, and no yellow mark adjacent to the corresponding holes in the bracket. Though both screws carried a blue verification dot.

After the incident, the train company carried out checks on a further 480 doorways on refurbished trains. At least one screw was found to be loose on at least 60 doors.

What can be learned about human supervisory control?

At least two things can be learned from this incident relevant to human supervisory control;

1. The automated system did not tell the driver what they needed to know. The driver believed the system told them whether the doors were open or closed: in reality, what it told him was the position of the drive-belt piston rod;
2. Safe and reliable performance of the automated system in service depended on error-free maintenance. Replacing a human check with an automated system increased reliance on the reliability of maintenance. Maintenance that, as is often the case, had the potential for loss of reliable human performance.

HMNZS Manawanui (2024)

On 5 October 2024, the New Zealand Naval vessel HMNZS Manawanui grounded on a reef off the island of Upolu, Samoa in the Pacific Ocean. The ship sank the following day, fortunately with no loss of life or injury to any crew members. A report on the official inquiry into the loss of the vessel was published on 4th April 2025 [15].

What happened?

The accident happened when the crew were unable to use the ship's thrusters to turn the ship under manual control to avoid a reef. The ships autopilot was engaged at the time, overriding the crew's attempts to steer manually and keeping the ship on a heading towards the reef.

Two officers were directly involved in the events immediately preceding the grounding. I'll refer to them as W2 and W4. W2 was the Officer of the Watch (OOW) at the time of the incident. W4 was a senior officer in attendance in a supervisory capacity. The Ship's Commanding Officer (CO) was not on the bridge until shortly before the grounding.

The inquiry concluded that the direct cause of the incident was down to errors on the part of W2 and W4;

> The direct cause of the grounding has been determined as a series of human errors in that the Ship was put on a heading towards land and the autopilot mode was not disengaged to enable the Ship to turn in an easterly direction…
>
> The correct initial actions for an azimuth thruster failure were not initiated upon realising that the Ship was not responding to the planned starboard turn, the first action being to take the Ship in hand, which means to take the Ship out of autopilot mode. [15]

The inquiry identified twelve additional factors considered to have contributed to the loss of the vessel - ranging from a lack of training to what the Navy referred to as "hollowness" – as well as two further "aggravating factors" – incorrect procedures and inadequate preparedness. None of the three key players were eligible to hold a platform endorsement.[40] And neither W2 nor W4 had completed training on the use of the thrusters that they tried to use to manually steer the ship.

Prior to embarking on its final mission, the ship had experienced a problem with its thrusters. On entering the bridge around two and a half minutes after the ship failed to respond to the first attempt to turn manually - 38 seconds before grounding - the CO was apparently told that there was a problem with the thrusters. The inquiry identified three possible ways in which the CO could have recognised that the reason for the problem was that the ship was in autopilot;

1. The difference in the actual and ordered position of the ship's thrusters. The inquiry concluded it was not reasonable for this to have been effective, first because of the short time before the grounding and, second, because the CO had already been advised that the problem was due to a thruster failure.[41]
2. Confirming that the Bridge Cards[42] had been followed and therefore that the possibility of the autopilot being engaged would have already been considered. Again, the inquiry thought this was not reasonable based both on the short time before the grounding, and the fact that it was reasonable for the CO to have assumed that W4, who was the senior officer on the bridge at the time, would have already ensured the Bridge Cards had been followed.

3. Visual inspection of the autopilot switch. The switch was located under a Perspex cover mounted on the autopilot console. Again, the inquiry thought there was both insufficient time, and that the CO could reasonably have expected W2 or W4 to have checked the switch, either directly or by following the Bridge Cards.

Limitations of the investigation

The content of the report into the inquiry that was made available to the public was heavily redacted. Some of that redaction seems to have been to protect the rights and privacy of the individuals involved. From the point of view of learning about the Human Factors contribution to the incident, as well as the relationship between automation and the people that are expected to supervise and work with it, the findings of the investigation are disappointingly limited.

Despite concluding that "The Court was able to obtain a sound understanding of the direct causes of the grounding and the factors which contributed to this incident occurring" the inquiry showed no interest in issues to do with the *situation* the crew were in or the *context* in which they made decisions and acted. As far as it is possible to tell from the redacted account in the published report, the inquiry made no attempt to go beyond a finding of "human error" and to adequately understand the nature of the error, or how it could – perhaps should - have been avoided if more attention had been given to the design of the interaction between the ship's crew and the automated autopilot they relied on to steer the ship. The conclusion that the incident was caused by "human error" without attempting to adequately understand the nature of that error is frustrating.

The investigation took the form of a Court of Inquiry ordered by the Chief of the New Zealand Navy, comprising four military officers with the authority to summon witnesses in accordance with the country's Armed Forces Discipline Act.[43] The order for assembly of the Court defined the Terms of Reference for the Inquiry in the form of thirty-four specific questions that were to be answered, together with comments and recommendations. Unfortunately, the Inquiry missed the opportunity for deeper learning about the psychology of human interaction with automation.

Psychological and Human Factors Engineering issues

The importance of investigators putting themselves "inside the head" of people involved in incidents, trying to understand what those people might have been thinking and experiencing at the time they made decisions and acted – or omitted to act – has been understood by Human Factors professionals and professional incident investigators for many years. In its influential White Paper on learning about the human contribution to adverse events, the Chartered Institute of Ergonomics & Human Factors (CIEHF) [16] made clear that developing a picture of the situation and context people were in at the time they made decisions and acted is essential if we want to understand the contribution human performance may have made to incidents.

The Inquiry into the loss of HMNZS Manawanui did not pursue important psychological and Human Factors questions behind the incident. Most importantly, how was it possible for the individuals involved not to realise the reason the ship did not respond to their attempts to manually change course was because the ships autopilot was engaged? The same individuals who had manually engaged the autopilot only a few minutes prior to the ship hitting the reef. The inquiry simply put the cause down to "human error" against a background of lack of training and other factors, thereby missing the opportunity for deeper learning about the psychological basis for "errors" of this type and the role of design both in inducing human error as well as avoiding potential adverse events.

The inquiry also showed limited interest in whether the design and layout of equipment used on the ship's bridge might have contributed to, or prevented, the loss. It did not attempt to answer questions such as why the equipment on the bridge did not provide an indication to the crew of where control of the ships heading lay in the key moments that was salient enough to break into their thought processes. Or why, when the crew tried to exert manual control over the ship's heading, the system gave no indication that their attempts were being overridden by the automation. Or indeed, why the attempt by the crew to exert manual control over the ship's heading did not override the autonomous autopilot system.

These are important issues in understanding how to design highly automated systems that provide adequate support to the people that need to work with them. And they are issues that must have relevance across the rest of the New Zealand Navy's fleet.[44]

The Terms of Reference for the inquiry included requirements to consider both whether the charts and navigation aids being used were suitable and sufficient for the safe operation of the ship (ToR 26), as well as whether the "materiel State" of the ship (ToR 33) could have contributed to the loss. However, the inquiry concluded that, other than an earth fault with one system, which had no impact on the operation, "The materiel state of the Ship did not contribute to the grounding or subsequent loss of the Ship…". The term "materiel" is somewhat ambiguous. But it can be taken to refer to the physical design and layout of the ship and its equipment, its status and suitability to support the ship's mission. A deeper, Human Factors-based consideration of both ToR items could have led the inquiry to ask questions about whether the design of the human interface to the equipment and systems or the automation policy might have contributed. Unfortunately, the inquiry did not interpret either item in that way, but limited itself to the physical, mechanical and electrical characteristics of the ship and its equipment.

There is a lot of important detail that the inquiry report does not reveal. These include;

1. Which of the two officers involved (W2 and W4) engaged the autopilot system immediately before the incident. While there is an implication that it was probably W2, who was in command of the ship at the time with W4 supervising, the report does not confirm that assumption. Psychologically, this matters. The act of physically acting is associated with a level of mental engagement in a task that helps build situation awareness and retain knowledge in working memory, making that knowledge accessible to thinking and reasoning.[45] Lacking this detail in the report leads to speculation that perhaps the state of the autopilot may not have been activated in W2's working

memory at the time he tried to steer the ship with the thrusters because he had not turned it on. It also raises the question whether W2 could have assumed that, if W4 had been taking responsibility for operating the autopilot, he would have disabled the autopilot as W2 tried to take over manual steering. Indeed, it raises the question whether both could have thought the other had disengaged the autopilot.

2. The design of the autopilot control panel and specifically the prominence of the display that the autopilot was engaged. The inquiry notes that the switch was located under a Perspex cover, presumably mounted on the autopilot console. Though it gives no detail about things like it's location, visibility, prominence, luminance or any other detail that would indicate its ability to clearly indicate when it was engaged. The report simply assumes that a competent operator would know and would make the effort to look.
3. Why the autopilot was designed in such a way that an attempt by a human to take control over the function did not override the automation? Decisions such as this are a fundamental part of any automated systems design policy. In cars, for example, as soon as the human driver applies force to the brake, accelerator or steering wheel, the automated vehicle control system is overridden, and control is handed seamlessly to the human. The same is true in many (though not all) other automated applications. It appears that overriding a ship's autopilot by manual command is not standard practice in the shipping industry. However, an Inquiry into an incident into the loss of a national asset such as a naval vessel, would seem to justify at least asking the question whether the policy is the right one.

The inquiry notes that neither of the two officers referred to the "Bridge Card" that was intended to be used in the event of thruster failure. The card would have prompted them to confirm that autopilot was not engaged. There is nothing in the inquiry report to suggest why neither officer used the cards, other than a lack of training.

The most psychologically confusing issue is why, whatever their training, both crew members persevered with trying to steer the ship manually even when they knew it was not responding. The inquiry report documents at least five occasions when the thrusters were operated manually over a period just over two minutes, each time with an increased orientation and level of requested thrust. None of which were effective because the ship was in autopilot mode. A ship's autopilot controls heading, not speed. Although the crew were not able to manually change the ships heading to avoid the reef by changing the orientation of the ship's thrusters, by manually increasing the power applied to the thrusters, they were able to change the ship's speed: the actions the crew took to avoid the ship grounding on the reef actually caused the ship to accelerate from 6 knots to nearly 11 knots when it grounded.

We know that, prior to embarking on its last mission, the ship had experienced a problem with its thrusters. The repair had interfered with the ship's preparations and caused it to delay sailing by a day. We can assume both W2 and W4 were aware of this recent fault. And we might speculate that knowing about the previous fault, and if they were adopting a System 1 style of thinking, jumping to a conclusion based on the first thing that occurred to them (which is exactly what System 1 thinking does),[46] they might have concluded that the reason the ship was not responding to manual commands was due to a recurrence of the problem with the thrusters. That does not explain the repeated attempts, involving both W2 and W4, and with increasingly larger demanded turn angles as well as thrust force to use the thrusters to manually steer the ship. While there is simply not enough information available, they both seem to have been captured, in a kind of "groupthink", by a focus on getting the thrusters to respond, making neither able to question what they were thinking and doing. An attempt to understand how W2 and W4 thought about the autopilot and thruster control – i.e. what was their mental model? – could have been revealing about the why they continued with these actions.

The Inquiry referred to W2 having apparently been distracted by a question from W4 on a matter not related to the current activity around a minute before the first failed attempt to manually turn the ship. There is however no basis for knowing whether that distraction had any bearing on their subsequent thought processes.

What can we learn about human supervisory control from HMNZS Manawanui?

Despite the lack of detail available in the redacted inquiry report, and the evidence that the key individuals involved lacked the proper training and qualifications, against a background of major organisational failures, there are several lessons that can be learned from this incident that have wider implications for anyone developing or using highly automated and autonomous systems;

1. It is not reasonable to rely on training as the basis for ensuring that people will always know the status of a system with autonomous capability. It is reasonable, when systems are used in professional settings, to expect the training needed to be competent working with a system to have been specified and for those people assigned to use them to have developed that competence. But training – or any other organisational control against incidents such as procedures – should not be relied on as an alternative to ensuring systems are designed in such a way that provides the information and support people need.
2. Autonomous systems controlling real-time activities must be designed such that it is extremely clear to users, in the context in which the system will be used, whenever the system is in control. It is not sufficient, certainly in systems that have some safety-critical implications, to rely on people going to the effort to actively find out where control lies at any moment.
3. Similarly, it should not be assumed that people will always make rational decisions or behave in a way that seems clear and obvious with foresight or with hindsight. There is a substantial evidence base that people can find themselves in a set of circumstances (a situation), and a frame of mind (a context) where what might seem obvious to an outsider not engaged in the heat of the action is not obvious to those involved in real-time at the sharp end.

4. The automation policy used to guide the development of systems with autonomous capability needs to ensure it defines the appropriate balance in where priority lies between giving authority to the human or the system to exert control: specifically, under what conditions is it appropriate to ignore an attempt by a human to override the automation?
5. Investigations into incidents need to focus on learning, rather than finding someone to blame. And in incidents where both human and automated systems play a part in the events leading to the adverse outcome, investigations should consider how effectively the human was supported in the role of the system's supervisory controller, and what indicators could or should, have helped them identify and respond to the threat.

There is one final point to be made about the loss of HMNZS Manawanui. And it is a conjecture made based on no factual evidence whatever. Rather, it is made based on a lifetime of experience studying incidents where Human Factors have played a prominent part. That conjecture is that HMNZS Manawanui was not the only occasion when officers on the bridge of a ship have made the mistake of not realizing the ship was in autopilot mode and have tried to manoeuvre the ship manually. My experience, including many conversations with individuals who work at the front line in industries ranging from oil and gas and chemicals to defence, aviation, rail and even financial services has taught me that instances of "human error" that make their way into formal incident reports are rarely, if ever, unique. They have nearly always happened before. Sometimes they are even commonplace: everyone knows people regularly make a particular mistake, but nothing too serious happens, and nothing is done about it.

If I was a betting man (which I'm not), I would lay odds that if those serving or who have served as Officers of the Watch on naval vessels around the world were asked if they had ever not realized that an autopilot was in control of their ship, or had heard about others who had made that mistake, the answer from a significant number would be a resounding yes. It may not previously have happened in situations leading to significant loss as serious as that of HMNZS Manawanui that are investigated and made public. But situations have occurred where essentially the same error was made, but nothing untoward happened and/or the event was not documented, reported or otherwise made available for wider learning. That lifetime of experience adds to the sense of frustration when the rare opportunities that do occur to learn from the kinds "human error" that happen in real-world operations are not taken. So the world goes on, the errors continue to happen, users continue to be blamed, and systems that incorporate inherently bad user interface designs continue to be used around the world.

WIMBLEDON (2025)

The All England Lawn Tennis Club (AELTC) might not seem an obvious place to learn lessons about the role people play in overseeing autonomous systems. But events at the Wimbledon tennis Championships in 2025 provided just such an opportunity.

2025 was the first year The Championships had adopted a fully automated electronic line-calling (ELC) system to make decisions whether balls were in or out of play. Since 2006, line calls had been made by human line judges assisted by the chair umpire. Players were allowed three opportunities in each set to challenge questionable calls. The electronic Hawk Eye system was used as a means of reviewing challenged calls. Court-side crowds as well as millions of TV viewers around the world enjoyed the drama of big-screen replays of the track of the ball showing whether the player's challenge was to be successful. Until 2025, responsibility and authority to decide where to award each point lay with the chair umpire. Though while players could challenge calls made by the human line judges, neither they nor the chair umpires could challenge the electronic replays produced by Hawk Eye.

This imbalance in authority gained a very high profile – not to mention a degree of notoriety and even ridicule – during the 2025 Championships. In line with the world tennis authorities and every Grand Slam championship apart from the French Open, the decision was made to completely replace human line judges and rely exclusively on the "Hawk Eye Live" system to make line calls. Such was the confidence in the reliability and accuracy of the system that players were no longer allowed to challenge line calls, and chair umpires were not allowed to overrule calls made by the automation. This system had worked successfully across the professional tennis circuit for some time and was popular with the players.

The ELC technology works by tracking balls using up to 18 cameras around each court. The cameras capture the ball's movement, and a computer interprets its location in real time, producing an accurate three-dimensional representation of the court and the ball's trajectory within it.[47] In 2025, the system was accurate to at least 3.6mm and was continually improving. To provide human oversight, at least during the initial transition period, a centralised line-calling hub was set up inside the grounds of Wimbledon, where 50 operators monitored the ball-tracking footage from the on-court cameras. Chair umpires continued to sit at each court to enforce the rules of the game. During this transition period, although the automation had been given full autonomy to make line calls and the chair umpire had no authority to overrule the system on their own, there remained a level of human oversight via the operators in the hub.

On 6 July 2025, Anastasia Pavlyuchenkova was playing a fourth-round match against Sonay Kartal on Centre Court. With the score at four games all, and advantage to Pavlyuchenkova, TV pictures clearly showed a return from Kartal landing outside the court. But the ELC system did not call it as being out. Pavlyuchenkova complained to the chair umpire. Having also clearly seen the ball landing out, but having no authority to overrule the system, the umpire phoned the hub for assistance. He was told the system had been turned off during the game.[48] Rather than follow natural justice and award the point to Pavlyuchenkova on the grounds that the return by Kartal had clearly been out, the chair umpire ruled that because the Hawk-Eye technology had not tracked the ball, the point had to be replayed. Had the system been working and the ball been called out as it clearly was, or had the umpire had the authority to overrule the system, Pavlyuchenkova would have won the game. In fact, she went on to lose the game. Though she did subsequently win the set, and the match. It turned out that the system had been turned off by mistake for around six minutes.

There are at least three points relevant to the role of people when highly automated systems are being introduced worth reflecting on from the events at Wimbledon that day;

1. The authorities had sufficient confidence in the accuracy and reliability of the automation that they had given it complete authority to act autonomously. Chair umpires had no authority to overrule calls made by the automation, whatever the evidence of their own eyes or High-Definition TV cameras. The automaton continued to hold that authority – the chair umpire did not have the authority to step in to over-rule the lack of a call and award the point to Pavlyuchenkova - despite the fact the system had been turned off.
2. There was a centralised hub where humans were expected to monitor and oversee the performance of the automation. The people in the hub however were only able to advise the chair umpire that the system had not been working during the game. From the information available at the time of writing, it is not clear whether the people in the hub had access to the imagery from the cameras and could potentially have checked what had been happened. Or whether both the cameras and the decision algorithms that evaluated it were switched off. If they did have access to the imagery, they decided not to use it as a basis for overruling the authority given to the automation.
3. The decision eventually made by the chair umpire illustrates the kind of issues that, whether consciously or sub-consciously, can come into play to influence a human supervisor's thought processes when faced with the need to intervene in an automated system that is usually extremely reliable. Being forced to make a decision, and, under the protocol put in place by the AELTC authorities believing they did not have the authority to make line call decisions themselves, the umpire ruled that the point should be replayed on the grounds that the technology had not tracked the movement of the ball. As the BBC's tennis correspondent Russell Fuller put in on radio the following day;

   > ...if we'd had the same situation last year... if the line judge had decided that that ball was in and not said anything, chair umpire would almost certainly have over-ruled because it was very very clear, and should have been from his chair, that the ball was long. But I think the issue is that umpires are frightened to make decisions for themselves or nervous about making decisions for themselves, because there are so many replays they could look stupid when other people get a chance to see the replay. And it's preventing them making decisions which, really, they should have been able to do with ease.
   >
   > PM Programme, BBC Radio 4, July 7th

The response of the Wimbledon authorities to the clear mistake was also revealing. Repeated interviews with the media revealed their lack of understanding that the technology cannot be seen in isolation. They demonstrated a complete lack of awareness that they were reliant on a socio-technical system. A system that, despite the proven accuracy and reliability of the technology, contained elements that relied on human performance. Human performance that, on the one hand was necessary for the overall system to function, and therefore needed to be recognised and supported while, on the other, had the potential to introduce unreliability into the system.

Within 24 hours of these events, the Wimbledon authorities had advised the world's press that changes had been made to the system that made the same error impossible. They returned to being completely confident in the system. This was a kneejerk reaction to deflect further criticism. Chair umpires were apparently told that, in the event of further similar events, rather than being able to overrule the technology, they should declare that there had been a "malfunction" on a specific point, but that the system was now working properly again. Something that happened more than once in the remainder of that year's championships.

Another very public glitch with the system occurred the following day, when the system mistakenly called a "fault" in the middle of a rally, with the ball nowhere near the lines. In this case, the presence of a ball-boy on the court prevented the system from detecting the start of the point. The chair umpire ruled that the point should be replayed. Fortunately, the circumstances that time were not contentious.

The reason for these events will undoubtedly have been analysed after the 2025 championship was completed, no doubt taking into account the full scope of the system. But the response of the authorities at the time was revealing. Clearly, the Wimbledon authorities needed to act very quickly, with the eyes of the world media and tennis fans around the world watching and the reputation of the AELTC and the Wimbledon championships at stake. Though that is not unlike the pressures to make real-time decisions faced by leaders of other organisations and businesses introducing automation in many sectors. Lack of awareness at senior levels of leadership that, even with highly reliable automation, there nearly always remains a need for people to oversee and support the system at some level, and that those people need to be adequately supported if the system is to deliver what is expected of it, is not unusual.

What can we learn about human supervisory control from Wimbledon?

In summary, the issues experienced at the Wimbledon championships in 2025 can be considered as having arisen for two reasons;

1. A failure to recognise the full scope of the system involved in implementing the autonomous line calling system. Specifically, the fact that the system not only relied on human activity to set it up and monitor it, but that there was potential for people to do things that were not expected, and that could interfere with the reliability of the system.
2. Because of the failure to properly understand the system boundary, a failure to make allowance for the potential for chair umpires to ever need to have the authority to overrule the automation.

SUMMARY OF LESSONS IDENTIFIED

Table 6.1 contains a summary of the general lessons about designing systems to support effective human supervisory control that can be identified from the eight

TABLE 6.1

Summary of lessons learned about designing automated systems to support effective human supervisory control

Lessons	1 737	2 TMI	3 KAL	4 B'Field	5 AF447	6 Train	7 HMNZS	8 W'don
General								
The role of a supervisory controller often forms only a part of a more complex role. A supervisor often needs to divide their time and allocate their attention over a number of simultaneous activities.	√			√				
Supervisory controllers must be proactive in searching for potential threats to the system and the activity being controlled rather than simply relying on the system. They should not rely on reacting to prompts or alerts from the system or elsewhere.			√	√			√	
To be able to diagnose threats and, if necessary, intervene effectively, human supervisory controllers need to know automated functions exist, what they do and how they work.	√		√					
Supervisory controllers need to maintain a healthy balance between trusting automation to do what is expected of it and maintaining a healthy wariness for the presence of external or internal threats to its performance.			√					√
The automation policy used to guide the development of systems with autonomous capability needs to define an appropriate balance in where priority lies between giving authority to the human or the system: specifically, that includes being clear about under what conditions is it appropriate to ignore an attempt by a human to override the automation and exert manual control.							√	√
Those responsible for deploying highly automated and autonomous capabilities need to take a systems view, that recognises where there will continue to be a reliance on human behaviour and performance. They should recognise the role of the automation within the wider socio-technical system and ensure the automation policy and associated decision making reflects the continuing role of, and reliance on, people.								√

Thinking and behaviour in the real world								
It should not be assumed that people will always make rational decisions or behave in a way that seems clear and obvious with foresight or with hindsight. People often find themselves in a set of circumstances (a situation), and a frame of mind (a context) where what might seem obvious to an outsider not engaged in the heat of the action is not obvious to those involved in real-time at the sharp end.				√			√	
Supervisory controllers will sometimes continue to rely on automation, even when they know it should not be trusted.				√				√
Assumptions made during development and certification about what the controllers will know and what they will be capable of doing in threat situations should be challenged to ensure they are reasonable and realistic in the likely circumstances.	√	√		√	√			
Assumptions inherent to a system design about the ability of supervisory controllers to unambiguously recognise threats need to be challenged taking account of the realities of the situation and context they may be in at the time.						√		
Sensor data								
To avoid confusing or misleading supervisory controllers, real-time sensor data automation relies on must be reliable and accurate. That means; i. fit for purpose (i.e. designed and installed to detect what they are intended to detect); ii. provide a direct indication of the information the supervisor will use in their thinking; iii. properly tested during installation; iv. inspected and maintained; v. faults reported and dealt with quickly.	√			√		√		

(Continued)

TABLE 6.1
(Continued)

Lessons	1 737	2 TMI	3 KAL	4 B'Field	5 AF447	6 Train	7 HMNZS	8 W'don
Supervisory controllers should be in no doubt if sensor data used by the automation becomes unavailable, erroneous or unreliable.	√			√	√			√
Supervisory controllers need to understand the potential implications for the performance of the automation if sensor data does become unavailable, erroneous or unreliable.	√			√	√			
Safe and reliable performance of an automated system depends on error-free maintenance. Replacing human checks with automation increases reliance on maintenance. Maintenance tasks have the potential for human error.	√					√		
Information								
The information supervisory controllers need to be able to detect, diagnose and respond to threats needs to be systematically identified during system development Detailed analysis of information requirements by operator tasks should identify critical parameters and ensure supervisors are not burdened with unnecessary information.		√		√				
Information made available via a user interface should indicate the state of the automation or the item in the world the human supervisor needs to monitor directly.	√				√	√	√	
The design of information supervisory controllers need to be able to detect, diagnose and respond to threats needs to be easily accessible, salient, clear and meaningful.	√	√					√	

Training and Procedures					
The difficulty of designing training strategies that are effective in ensuring supervisory controllers will respond effectively to an unexpected loss of automated control should not be underestimated. Especially when the transition to manual control could occur unexpectedly in situations that have the potential for high levels of stress.				√	
Similarly, the difficulty of designing procedures, including emergency procedures, that are effective in supporting supervisory controllers diagnosing threats and transitioning to manual control is challenging and should not be underestimated.				√	
Interaction					
Avoid the potential for mode errors.		√	√		
Provide clear feedback to the users about what state or mode the automation is in.		√	√		
Supervisory controllers need to understand when there are checks or "guard conditions" designed into the automation that need to be satisfied before a user input to the automation will take effect.		√			
Control					
It must be extremely clear to users, in the context in which the system will be used, whenever the automation system is in control. It is not sufficient to rely on people going to the effort to actively find out, or to assume they will be able to work out, where control lies at any moment.	√				√

(Continued)

TABLE 6.1
(Continued)

Lessons	1 737	2 TMI	3 KAL	4 B'Field	5 AF447	6 Train	7 HMNZS	8 W'don
Learning								
Investigations into incidents involving highly automated and autonomous systems should consider how effectively the human was supported in the role of the system's supervisory controller, and what indicators could or should, have helped them identify the threat and respond to the threat.	√	√	√	√	√	√	√	√
In industrial and commercial settings, where work is organised on a shift or crew basis, many individuals may fill the role of human supervisory controller of the same system. Individuals who need to intervene to resolve abnormal, though critical, threat conditions, need to be rigorous in ensuring action is taken to resolve the problem and that others filling the supervisory role are aware of, and able to learn from, their experience.	√							

Key; 1. Boeing 737 MAX; 2. Three Mile Island; 3. KAL007; 4. Buncefield; 5. Air France AF447. 6. Train doors; 7. HMNZS Manawanui; 8. Wimbledon.

incidents considered in this chapter. The table is not intended as anything like a comprehensive summary of what has been learned about the psychology of human supervisory control from investigations into incidents involving highly automated systems. It is simply a summary of general lessons associated with the incidents reviewed in this chapter.

Among the most important issues in understanding the role of a human supervisory controller is knowing what information they need, and what is available to them about the state both of the world and of the system being supervised. To put it another way, what information are they going to need to be able to detect and understand both internal and external threats, and where are they going to get that information from? This is a theme we will return to, especially in Chapters 7–9 when we look at models of supervisory control.

Issues surrounding the information available to the human supervisor, it's reliability, where it is located, and what it means – that is, how directly it relates to threats and the state of the world the human is expected to think, reason about and monitor – have been associated with adverse events since the earliest days of automation. That includes systems ranging from very simple automation, through to some of the most sophisticated. Four issues in particular need to be given particular care and attention when designing interfaces to support human supervisory control;

i. knowing what parameters to instrument;
ii. assuring the reliability of data from instruments;
iii. knowing what information the human supervisor needs;
iv. how to convert the data from instrumentation into information the human supervisor can easily monitor and reason with.

NOTES

1 I have based the material in this section entirely on information available in the three government investigations, references [2–4].

2 The engineer told the investigation that he had completed the required calibration checks on the fitted sensor although, against required procedures, he did not record the test results.

3 At the time of writing, Boeing still disputes some of the findings. The reader who is unfamiliar with the incidents and wants to know more should read the Executive Summary of the Final Committee Report [2]. Rory Kennedy's documentary film "Downfall" includes simulation of the events on the flight decks leading up to the crashes and explains the context both before as well as after the crashes, up to 2022.

4 Angle of attack is the angle between the oncoming air or relative wind and a reference line on the airplane or wing.

5 The function of the AOA sensors, including the way the data are used by other flight systems to calculate airspeed and altitude, as well as the associated pilot displays and alert messages are explained in Section 1.6 of the Indonesian investigation report [3]. The report also contains details of the maintenance history of the AOA sensor fitted to the Lion Air aircraft.

6 Relying on a single source of data for a safety-critical function is unheard of in modern aviation. Boeing however justified the decision on the grounds that they did not consider the MCAS system to be safety-critical.

7 Boeing advised that, at the time of the crashes, only around 20% of the 737 MAX aircraft sold included the AOA indicator, and therefore the AOA Disagree alert.
8 For example, if the aircraft's actual AOA was 10 degrees up (nose pointing towards the sky), the sensor would be reporting 11 degrees down (nose pointing towards the ground).
9 The stick shaker is a mechanical device fitted to both pilots' control columns. Its role is to physically shake the column as a means of alerting the pilots that the aircraft's flight systems believe there is the potential for the aircraft to stall.
10 Actually, it was the anti-icing element on the sensor that failed.
11 For clarity for readers that may lack a detailed understanding of modern flight systems, I have removed several system acronyms from this quote. While this is a loss of detail, it does not affect the point of the quote.
12 Including myself. The details here are taken entirely from the investigation reports.
13 This is discussed later in this chapter.
14 According to the US Final Report, the assumption that pilots would be able to respond to repeated MCAS activations was questioned by some of Boeing's own engineers. Their own simulation trial had shown it took their pilot 10 seconds to respond.
15 From Ethiopian Investigation report [4], Section 1.17.14. p. 204.
16 In theory, it should have been possible to identify the source of the problem indirectly, by drawing on knowledge of how aircraft data are calculated by the system, and reasoning with that knowledge. Something of a mental challenge at the best of times.
17 The investigation report notes that, contrary to procedures, the engineer had not made a record of the results of the tests conducted on the AOA sensor when it was fitted to the aircraft on 28th October.
18 The dead heading pilot happened to be Lion Air's Group First Officer and was rated to fly the 737 MAX.
19 In the US House Final Report on the two crashes [2], the dead heading pilot is reported as having recognised what was happening and provided instruction to the two crew to turn off the stabiliser trim switches. The report concludes that this interaction enabled the captain to manage the problem. The Indonesian investigation report [3] however gives most of the credit for diagnosing and responding to the unknown MCAS activations to the captain.
20 MCAS was designed to be able to apply more force to the stabiliser than a pilot was capable of doing manually.
21 The Indonesian investigation report believes this was the case.
22 The Indonesian investigation report includes speculation about the reasons behind the different outcomes. See [3], Conclusions 68 to 72.
23 It is sobering to consider what might have happened somewhere else in the industry had what was, in the scale of what could have happened, a relatively minor incident not occurred, and the learnings and improvements that came from the numerous investigations and enquiries not been identified and implemented.
24 A report prepared by The Essex Corporation under contract to the US Nuclear Regulatory Commission and published in December 1979 provides a comprehensive and detailed treatment of the full range of Human Factors issues associated with the incident, including a timeline of events. See [5].
25 At least for a non-physicist.
26 As well as the Human Factors report by the Essex Corporation referred to earlier, the US Nuclear Regulatory Commission's web-site [6] includes an animated page providing a clear and authoritative background to the incident: Wikipedia also provides a comprehensive and well referenced description of the events preceding and following the incident [7].

27 The International Civil Aviation Organisation's (ICAO) 1993 report on the incident [9] sets out the facts as far as they could be established at the time.
28 I would commend Degani's book to anyone looking to understand how to design effective user interfaces to automated systems – or in fact any other kind of systems. Although many of the examples used (such as mobile phones) are dated having been published in 2001, the underlying principles remain sound and are well worth understanding.
29 Some of the material in this section is a summary of a more extended discussion in my first book [10].
30 The COMAH regulations, which first came into force on 1 April 1999 with amendments in 2005 and 2008, are the UK's implementation of the European Seveso II Directive introduced across Europe following the major accident at Seveso in Italy in 1976.
31 The report of the formal investigation into the Buncefield Incident is at reference [11]. Reference [12] contains further insight into the Human Factors associated with the incident.
32 See Chapter 12 for a discussion of the concept of "Barriers" in this sense.
33 See Chapters 16–20 in [10].
34 It is clear from the investigation by the Competent Authority into the incident [11], as well as other sources in the public domain (such as [12]), that the entire Buncefield incident occurred because of a near complete failure of the management of Human Factors across the entire operation.
35 Actually, they functioned as designed, given that the flight computers were receiving conflicting data from flight sensors.
36 There are numerous other aviation incidents where pilots have avoided a crash despite also being in a situation of potentially imminent death: perhaps most famously US Airways Flight 1479 which landed on the Hudson River in New York in 2009. The Air France incident appears unusual in terms of the immediately preceding context, emotional state of the crew, lack of any clear indication what had happened, and possible spatial disorientation.
37 There was a lot of focus in the press at the time about the fact that the use of side-arm control for flight commands, as opposed to the traditional large central 'yoke' column, means the non-flying pilot cannot see the flying pilot's control inputs.
38 I.e. The difference between 'normal' flight law – when it is technically impossible for an Airbus to stall – and 'alternate' law, when a stall is possible.
39 Details of the incident are taken from [14].
40 I.e. they had not completed the basic training expected by the New Zealand Navy to be assigned as a crew member of the HMNZS Manawanui.
41 Psychologically, this is known as "Priming", where thinking is captured by expectations based on an initial piece of information. For a discussion of the psychology of priming, see Kahneman [17].
42 Bridge Cards are the standard procedures expected to be followed in the event of various types of equipment breakdown or other abnormal events. Effectively the same as the non-normal procedures defined on Quick Reference Handbooks used by pilots. See the discussion of the Boeing 737 MAX crashes earlier in this chapter.
43 It is worth commenting that approaching the investigation into this incident as a Court of Inquiry in this way is not consistent with normal practice when investigating maritime incidents, as least in some countries. For example, Regulation 5 of the United Kingdom's Merchant Shipping Regulations (Accident reporting and investigation) Regulations 2012, states that

> The sole objective of a safety investigation into an accident under these Regulations shall be the prevention of future accidents through the ascertainment of its causes and circumstances. It shall not be the purpose of such an

investigation to determine liability nor, except so far as is necessary to achieve its objective, to apportion blame.

44 Not to mention every other naval operator in the world, whether military or civil.
45 A good example of this happened over the 1980s and 90s when Air Traffic Control operations around the world were trying to move towards digitizing their operations. Prior to digitization, controllers maintained physical flight strips to represent each aircraft they were controlling, The flight strips recorded the aircraft's call sign and other details. Controllers manually moved the strips around their console as the aircraft progressed and physically handed them over to controllers in the next sector when each aircraft left the airspace they were responsible for. Initial attempts to automate flight strips met great resistance from controllers who believed they maintained much great awareness and engagement with the progress of the aircraft through the process of physically manipulating the flight strips. See for example [18].
46 System 1 thinking, and its relevance to situations where people make seemingly unexplainable errors, is discussed in Chapter 8 See [17].
47 The Independent newspaper carried an explanation of how the Hawk Eye Live systems works on 6 July 2025. https://www.independent.co.uk/sport/tennis/wimbledon-electronic-line-calling-elc-tennis-hawkeye-live-b2783629.html.
48 Apparently because of "human error", though neither the nature of that error, or what led to its being made, has so far been explained.

REFERENCES

1. Dekker, S.A. & Woods, D.D. Wrong, Strong and Silent: What Happens When Automated Systems with High Autonomy and High Authority Misbehave? Journal of Cognitive Engineering and Decision Making. 2024 April. DOI:10.1177/15553434241240849
2. The House Committee on Transport and Infrastructure. Final Committee Report on the Design, Development and Certification of the Boeing 737 Max. 2020 September.
3. Final Aircraft Accident Investigation Report KNKT.18.10.35.04. Komite Nasional Keselamatan Transportasi Republic of Indonesia. 29 October 2018.
4. Investigation report on accident to the B737-MAX8 Reg ET-AVJ operated by Ethiopian Airlines 10 March 2019. Federal Democratic republic of Ethiopia, Ministry of Transport and Logistics. Aircraft Accident Investigation Bureau. Report AI-01/19. 2022 December.
5. Malone, T.B., Kirkpatrick, M., Mallory, K., Elery, D., Johnson, J.H., Walker, R.W. Human Factors Evaluation of Control Room Design and Operator Performance at Three Mile Island – 2: Final Report. The Essex Corporation. 1979.
6. US Nuclear Regulatory Commission Website. Accessed 2 April 2025. Available from: https://www.nrc.gov/reading-rm/doc-collections/fact-sheets/3mile-isle
7. Three Mile Island Accident. Accessed 2 April 2025. Available from: https://en.wikipedia.org/wiki/Three_Mile_Island_accident
8. Degani, A. Taming HAL: Designing Interfaces beyond 2001. Palgrave Macmillan. 2001
9. ICAO Destruction of Korean Airlines Boeing 747 on 31 August 1983. Report of the Completion of the ICAO Fact-Finding Investigation. ICAO. June 1993.
10. McLeod, R.W. Designing for Human Reliability: Human Factors Engineering for the Oil, Gas and Process Industries. Gulf Professional Publishing. 2015.
11. Buncefield Major Incident Investigation Board. The Buncefield Incident 11 December 2005: The Final Report of the Major Incident Investigation Board. 2008. Available from: https://www.fabig.com/media/tpuaseey/buncefield-incident-miib-final-report-volume-1-dec2008.pdf

12. Health and Safety Executive. Buncefield: Why Did it Happen? The Underlying Causes of the Explosion and Fire at the Buncefield oil Storage Depot, Hemel Hempstead, Hertfordshire on 11 December 2005. Available from: https://www.icheme.org/media/10706/buncefield-report.pdf
13. Bureau d'Enquetes et d'Analyses pour la securite de l'aviation civile. Final Report on the Investigation into the Crash of the Air France Airbus A330-230. BEA. July 2012. Available from: https://bea.aero/fileadmin/documents/docspa/2009/f-cp090601.en/pdf/f-cp090601.en.pdf
14. RAIB Safety Digest Door Open in Traffic between Shenfield and Hockley stations. D10/2019: Hockley. 22 August 2019.
15. Court of Inquiry for the Purpose of Collecting and Recording Evidence on the Circumstances that Resulted in the Loss of HMNZS Manawanui off Upolu, Samoa, on 6 October 2024. Accessed 18 April 2025. Available from: https://www.nzdf.mil.nz/assets/Uploads/DocumentLibrary/MAN-COI-ROP-FINAL-31-Mar-25_Redacted-v2.pdf
16. CIEHF. Learning from Adverse Events: A White Paper. Chartered Institute of Ergonomics & Human Factors. 2020. Available from: https://ergonomics.org.uk/resource/learning-from-adverse-events.html
17. Kahneman, D. Thinking, Fast and Slow. Allen Lane. 2010.
18. Mackay, W.E. Is Paper Safer: The Role of Paper Flight Strips in Air Traffic Control. *ACM Transactions on Computer-Human Interaction.* 1999 December;6(4):311–340.

7 Models of human supervisory control[1]

This chapter summarises models of human supervisory control behaviour published over recent decades by three of the most influential researchers and thinkers about the relationship between people and automation: Thomas Sheridan, Neville Moray, and Mica Endsley. The models selected were identified through a combination of searches of the research literature,[2] as well as manually following leads from papers identified in those searches and elsewhere. The searches were looking for models that attempted to capture and represent the nature of the behavioural and cognitive relationship between the human as a supervisory controller and automated control systems.[3] The only model identified that was not included in the discussion was one that sought to use statistical approaches to predicting the behaviour of a human supervisor [2]. Because this model focused on predicting outcomes, rather than identifying and explaining anything about the mechanisms behind those outcomes, it was considered as being outside the scope of the chapter.

SHERIDAN'S HIERARCHICAL MODEL

In 1983 the US National Research Council put together a workshop of experts to advise Congress on what, at that time, were the priorities for investment in Human Factors Research [3]. The review identified "Supervisory Control Systems" as a key area where research was needed. The NRC subsequently invited Thomas Sheridan[4] and Robert Hennessy to organise a workshop attended by nine leading experts to explore the concept of human supervisory control and identify the most effective approaches to conducting experimental research into the subject [4]. Despite the intervening 40+ years of extraordinary developments in computing, automation and Artificial Intelligence technologies the NRC report prepared by Sheridan and Hennessey still provides a valuable introduction to many of the human factors issues involved when the human is expected to fill the role of being a supervisory controller of a largely automated real-time activity. The report included a model of human supervisory control that emphasised the hierarchical nature of the component sub-systems.[5]

The model proposed by Sheridan and his colleagues considers the relationships across four hierarchical levels of an automated system;

I. a set of tasks to be performed in the real-world, typically performed by mechanical or electrical devices;
II. a set of "dumb" controllers using negative-feedback servo loops to control the state of those devices to maintain the tasks within predefined limits;

DOI: 10.1201/9781003679479-9

III. an "intelligent" computer providing an interface between the human and both the task and the "dumb" controllers, that may have some ability to manipulate the controllers independently of the human, and;
IV. the human operator in the role of a supervisory controller, setting and adjusting overall goals and limits for performance of the tasks, as well as monitoring the entire system and being available to intervene to take control if needed.

My variant of Sheridan's model has been modified to emphasise some of the complexities in the user interface, as well as to highlight the importance of sources of information and control that may fall outside the automated system. My variant is shown on Figure 7.1.

At the bottom of Figure 7.1 are a series of individual tasks which, taken together, are intended to achieve the goals of the overall process. For example, a vessel in a chemical plant might involve several individual tasks, each with control loops over, perhaps, temperature, pressure, and flow rates inside the vessel. Or in an automated vehicle, control of the vehicle's lane keeping and speed are identifiable tasks each having dedicated control loops. Each of these tasks is controlled by what is referred to as a "Task Interactive System" (TIS), comprising the sensors, actuators and processing needed to provide closed-loop control of the tasks.

Data from the TIS is fed to up to a "Human Interface System" (HIS) that provides the immediate user interface between the human user and the TIS. Of course, for a system of any complexity, the system architecture between the local task controllers and the user interface will likely be much more complex than the elements and the simple hierarchical relationship shown on the figure. Though the point here is not to try to represent any real system architecture, but to think about the relationship

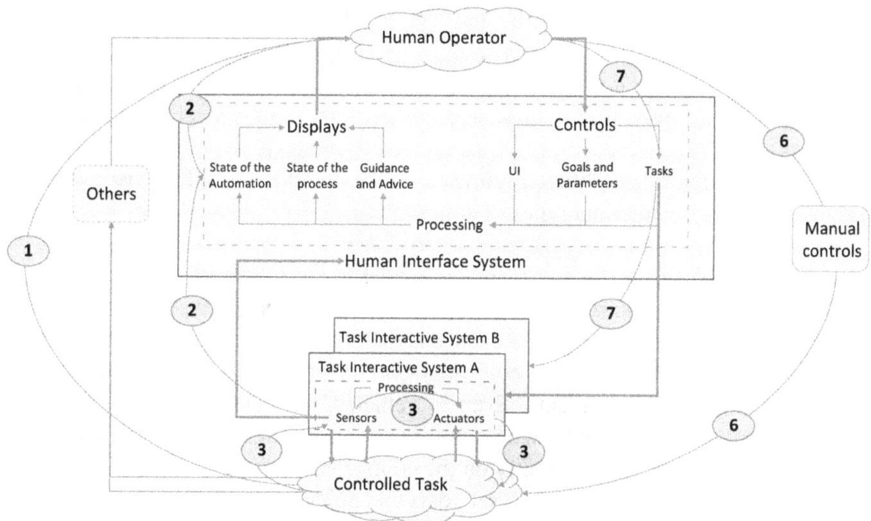

FIGURE 7.1 Modified version of Sheridan's Hierarchical model of human supervisory control (Modified from entry 7.309 in [1])

between the human supervisory controller and the system and tasks they are ultimately responsible for.

A great deal can happen in the HIS. My variant of the model emphasises the different types of information presented to the human, as well as the distinct types of control it supports. The layer provides the user with (at least) three types of information;

i. information about the state of the activity being controlled. That is, data derived from each of the TIS's. This might be represented in different formats, at different levels of abstraction, from "raw" data from each of the local sensors and comparisons with individual task goals, up to high level overviews of the state of the entire system under control in relation to the overall system goals. The data may be real-time, showing the most recent state of the variables, or non-real time, showing historical changes in the variables over time and/or projections about expected future states;
ii. guidance and advisory messages (such as alerts or warnings) intended to advise the user about issues they need to know about, suggest or recommend actions, or give explanations of why the system has done something;
iii. information about the state of the automation itself. Such as what underlying mode(s) are currently engaged, what the system knows about the local environment (for example, a vehicle with Automated Lane Keeping capability is likely to indicate whether it can currently detect the limits of the lane), as well as what it believes or expects is likely to happen in the immediate future. It will also show any known faults in the system.

Figure 7.1 shows that the supervisory controller will often have access to information about the performance of controlled tasks that come from sources other than the HIS: these might include direct line-of-sight, smell, or the sound of the process, information communicated by third parties, or other sources. The availability and ease of access to these information sources become especially important when we recognise how important it is that the supervisory controller monitors performance of the system in a way that is proactive, rather than simply reacting to alarms or system prompt.[6]

In terms of control facilities, Figure 7.1 illustrates that the HIS layer must also provide at least three distinct types of controls;

i. controls that allow the user to make changes to the TIS sub-systems controlling individual control loops. In a system of any complexity, these controls are unlikely to be "direct" in the sense that the user does not directly interact with the task interactive control system: control is mediated by software and other system elements. Even in vehicles, with electrically controlled power steering and fly-by-wire control systems in aircraft, it is increasingly rare for the user to act directly on the controlled element;
ii. controls that allow the user to set goals and adjust individual parameters. In an automated vehicle, for example, these might be setting or adjusting the cruising speed via manual controls on the steering column, or by adjusting the vehicles speed through the braking and acceleration systems;

iii. controls that allow the user to manipulate the user interface itself, such as choosing what information to display, how to display it, changing ranges or timescales, as well as acknowledging system messages. These types of controls include those that allow the user to make modifications to the system set up.

So far, the Sheridan model is really an elaboration in more detail of the basic models of supervisory control shown on Figure 2.2 in Chapter 2. However, its value as a model of human supervisory behaviour comes from the ten different types of human interactions the model identifies as occurring across the four hierarchical levels. By way of illustration, five of these interactions are illustrated on Figure 7.1 (the numbers on the list below refer to the numbered circles on the figure);

1. The task is observed by the operator directly, without mediation from either the TIS or the HIS.
2. Observation of the task by the human is mediated by the performance of the sensors in the TIS, and the way the information is represented in the HIS.
3. The task is controlled entirely within the TIS, with no involvement from the HIS or human.
6. The human interacts directly with the task by manipulating local controls, without mediation by the HIS or TIS.
7. Operator interaction with the task is mediated by both the HIS and the TIS.

Any or all of the ten types of interactions shown on Figure 7.1 can be going on, often simultaneously, at any time.

MORAY'S MODEL OF THE DISTRIBUTION OF INFORMATION AND CONTROL

Drawing on the hierarchical framework devised by Sheridan et al, Neville Moray suggested a means of thinking about how knowledge and control can be distributed between the different hierarchical layers in a supervisory control system (see entry 7.308 in [1], as well as [4]). This model is also worth exploring, as it can bring insight and aid understanding, at the time automation concepts and user interfaces are being thought about, of some of the difficulties humans might face in filling their role as a supervisory controller. It also suggests a form of analysis that could usefully be carried out as part of the process of developing or verifying the design of user interfaces to highly automated systems.

Moray used a form of Venn diagram to illustrate how the knowledge and control facilities needed to perform effectively as a supervisory controller can be distributed among the three system levels identified by Sheridan (the human, the Human Interface System, and the Task Interactive System as shown on Figure 7.1). Importantly, Moray's model makes a distinction between information and controls being *available* to the user, and the user being *aware* of or understanding the meaning of that information or of how to use the controls. This is an important distinction. System designers may have invested time and effort making information and controls available in a user interface. But if the users are not made aware of their existence, or do not understand how to use them correctly, that effort will be wasted. Examples of this are abundant in modern IT systems, software applications and the controls provided in our phones and tablets. I have been a regular user of Microsoft's Office suite of applications since they were first launched on the world back in the 1980s: until a few years ago I might even have considered myself something of a "power" user of some of them. Though even a brief browse round the options available in the menus and toolbars quickly makes it obvious that I am probably only comfortably familiar with no more than about 50% of the available options. Many are a complete mystery.

Figure 7.2a and b shows two variants of my modification of Moray's model. While retaining the key features of Moray's model, my modifications are intended to achieve three things;

i. to simplify the representation;
ii. to focus on information – which is something that has a chance of being translated directly into things that system designers can support - rather than the more abstract notion of "knowledge" – which is something that exists in the head of a user, and is neither directly accessible to nor under the control of, system designers and;
iii. (as with my modification of Sheridan's model) to explicitly represent information available from sources outside the strict three levels within the boundary of the automated system: including the environment, other people and other systems, as well as things like space and the passage of time.

Conceptually, the boundary to Figure 7.2a represents all the information that an individual in the role of a supervisory controller could, at least in theory, need to be able to deal with every possible situation where they might need to intervene. That means information about the controlled process and the world it is in, as well as information about the state and capability of the automation itself. Within this overall conceptual space, the figure illustrates how information might be shared between the human user, the user interface, and other sources.

Figure 7.2a illustrates at least eight distinct ways in which information might be distributed among the various system elements;

A. shows information available directly from the user interface;
B. information not reflected in the user interface, but directly available from other sources;
C. information available via the user interface, but that the human has only partial awareness or understanding of;
D. information available in the user interface, but that the user has no awareness of or does not understand;
E. information the user is not aware of, but that could be accessed from other sources;
F. information the user does not know about, and that is not fully represented in the user interface;
G. information that is not available to the user through any source;
H. information that is only available to the user from prior knowledge, training, or experience.

Figure 7.2b shows the same conceptual framework as Figure 7.2a. Though in this case, the content is concerned with how functions providing control over individual tasks are shared between the system elements.

In the same way that Figure 7.2a mapped how information was distributed and shared around the system elements, Figure 7.2b illustrates at least six distinct ways

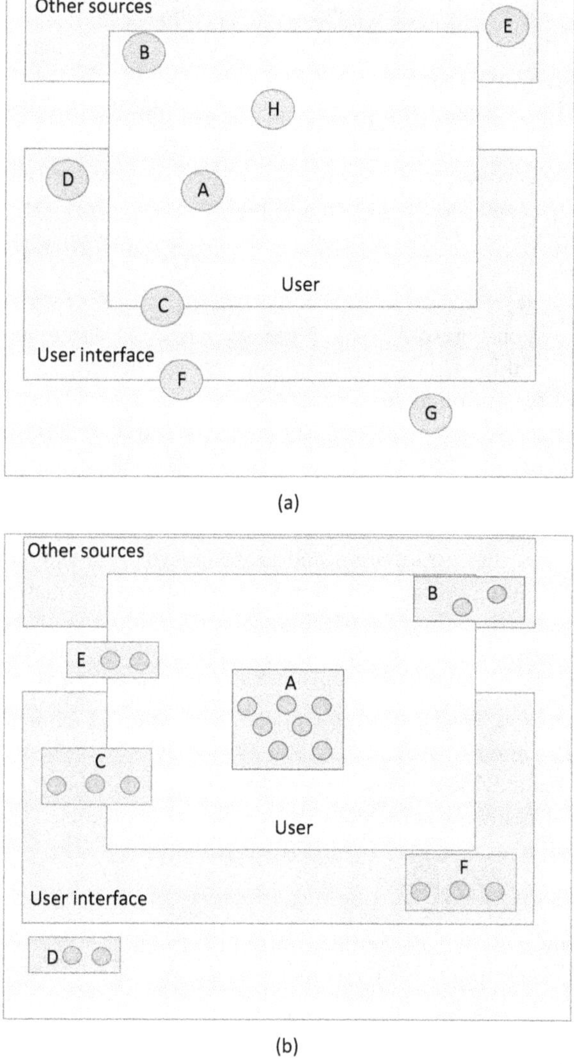

FIGURE 7.2(a) Modified version of Moray's (1986) Venn diagram showing how the information a human supervisory controller might need can be distributed around elements of a system (Based on entry 7.308 in [1])

FIGURE 7.2(b) Modified version of Moray's (1986) Venn diagram showing how the control functions a supervisory controller might need can be distributed around elements of a system (Based on entry 7.308 in [1])

in which control functions needed by the human supervisory controller might be distributed among the system elements;

A. functions that can be controlled through the user interface. These are fully understood by the user and fully supported by the user interface;
B. control functions known to the user but not supported in the user interface. They can only be performed outside the user interface;
C. functions that can be controlled via the user interface, but are not fully understood by the user;
D. automated functions performed by the system that the user has no knowledge of, and are not reflected in the user interface;
E. automated functions performed by the system that are not reflected in the user interface and that are only partly understood by the user;
F. automated functions reflected in the user interface that the human does not understand.

The models shown on Figure 7.2a and b are explicitly conceptual and generic: they suggest a way of thinking about the relationship between the information and controls human supervisory controllers need to fill their role, and how – or indeed whether - that information and those controls are supported in a system design. These models do not attempt to represent the reality of any particular system. They provide a basis for thinking about, assessing or challenging a system design that is focused on the information and controls the human will need to perform the role of a supervisory controller, and how that information and those controls are supported. They also highlight the critical importance not of only providing the information and controls, but of making sure the system users are aware of them and know what they mean and how to use them.

Although they are conceptual, Moray's models directly lend themselves to be used practically during the development or evaluation of complex automated systems or during the investigation of incidents. As the 1984 NRC report put it, the model: "… suggests problems may be expected to arise if the human supervisor only, the HIS computer only, or neither is sufficiently connected to the task itself" [4].

ENDSLEY'S HASO MODEL

As well as being one of the world's most eminent and respected Human Factors Engineers (and, so far, the only one to be appointed as Chief Scientists to the United States Air Force), Mica Endsley's body of published work is probably among the most highly referenced and respected of anyone working in applied cognitive sciences. Her work over many years has addressed a wide range of issues associated with human performance in complex real-time systems. Her work defining, understanding and applying the concept of Situation Awareness has to be among the most significant and influential contributions to the professional practice of Human Factors and related safety disciplines.

Perhaps the central theme of Dr Endsley's work has been seeking to understand and optimise the balance between technology and human performance in the development of highly automated systems. As well as leading many empirical research studies along with colleagues and collaborators, she has regularly published reviews and summaries of the state of thinking and knowledge on key topics concerned with human performance in complex systems.

In 2017, Dr Endsley first published a theoretical model known as the "Human-Autonomy Systems Oversight" model (HASO) [6,7]. The model sought to summarise what had been learned from several decades of research into a range of significant problems associated with human performance in highly automated systems: not least, difficulties people face in maintaining situation awareness when they are expected to support highly reliable automation, as well as what are known as "Out-Of-The-Loop" (OOTL) problems.[7] OOTL problems arise when, because of losing situation awareness, people are slow to detect that automation is having problems and are consequently slow to diagnose what is happening and what they need to do. She encapsulates the range of these problems in what she terms the "Automation Conundrum":

> The more automation is added to a system, and the more reliable and robust that automation is, the less likely that human operators overseeing the automation will be aware of critical information and able to take over manual control when needed. More automation refers to automation used for more functions, longer durations, higher levels of automation and automation that encompasses longer task sequences. [6]

The HASO model, illustrated on Figure 7.3, identifies the relationships between, on the one hand, characteristics of the design of a highly automated system and, on the other, the key cognitive functions and cognitive performance necessary to allow the human to provide the monitoring, decision-making, and intervention needed to

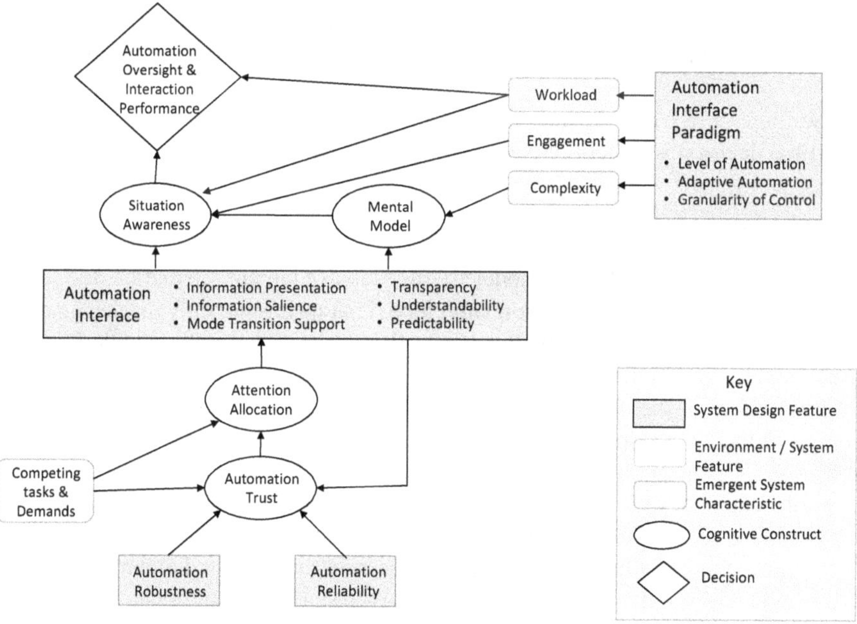

FIGURE 7.3 Endsley's human-autonomy systems oversight model (re-drawn from [6])

be successful in the role of a supervisory controller. The purpose of making these relationships clear is to help those responsible for the design and implementation of systems with highly automated elements recognise and appreciate how the decisions they make in systems design and engineering can impact directly on the ability of people to provide oversight. Achieving those two objectives (i - recognising the relationships, and ii - being aware of the impact that design and engineering decisions can have on them) are extremely important in developing and implementing automation successfully.

Inevitably the HASO model itself is a simplification. Such a simplification cannot go so far as to provide an understanding of the nature or dynamics of the relationships between components of the model, nor to explain the complexity of the interactions and influences involved. Achieving even a moderate understanding at that level requires a degree of knowledge and topic-specific expertise that very few engineers, managers or decision makers have or, indeed, need: that level of understanding lies in the domain of the genuine technical expert, and, as such, is beyond our needs here. The following material will therefore concentrate on explaining important features of the model without burdening the text with extensive references to the supporting research literature and other evidence. For a deeper understanding, the interested reader can refer to Endsley's original paper [6] or the description of the model in her book with Debra Jones [7], as well as the body of associated literature referenced in them as a starting point.

The HASO model is based on the relationships and interactions between three main elements that support – or interfere with - successful supervisory control;

- Features of the design of the technical system;
- Characteristics that emerge from decisions about the system design;
- Cognitive constructs assumed to be going on in the head of the controller.

These three core elements interact to influence the mental process supporting the real-time decision whether there is a need to intervene at any moment.

System design features

The first element in the HASO model are the system's human interaction design features embedded in the **Automation Interface**. That means not only the design of the user interface to the automation, but also to the conceptual paradigm underlying the use of automation in the system.

The quality of the design of the Automation Interface depends on factors under control of the team responsible for developing the system, principally;

- the extent to which the system provides the information the human needs to be able to monitor the system and make decisions;
- how easy it is (that is, how "salient" the cues are) for the human to detect and understand information that indicate the state of the automation, including things like which mode it is currently in and how close it is to its design limits;

- how well the design supports the human in making the transition from manual to automated control, as well as transitions back from automation to manual control – especially when those transitions back to manual control occur rarely, are unexpected and unanticipated and, when they do occur, usually need to be made under significant time and other sources of pressure;
- how transparent the interface is. That means how effectively it allows the human to understand what the system is currently doing, as well as predicting what it is likely to do in the immediate future.

The **Automation Interface Paradigm** goes beyond the information presented in the user interface and the controls and displays associated with its use. The term "interface paradigm" refers to all the decisions about how automation will be used that, implicitly or explicitly, have consequences both for the role and tasks expected to be carried out by the human, and for the physical and mental demands arising on the supervisory controller from the nature of their relationship with the technology. Not least, the paradigm includes expectations about the level (in terms of the extent of human support expected) at which the automation is expected to be able to perform its various functions, both when operating inside the intended operating conditions, as well as when conditions exceed those design limits.

The HASO model includes two other system design features, **Automation Reliability** and **Automation Robustness.** Reliability is defined as the "ability to operate accurately", while Robustness is the "ability to operate across a wide range of possible conditions". These features are clearly different from the system design features of the automation interface and the interface paradigm. They reflect the quality of performance arising from the system (including its design, validation, and implementation as well as ongoing support) rather than being features of the design *per se.* While engineers and designers will hope that the performance of the system they have developed will prove through experience to deliver high levels of robustness and reliability, these are not qualities they will be sure about until the system is in service.

These two features however need to be included in the HASO model to capture the important influence they have on the human's cognitive performance as a supervisory controller: specifically, they impact directly on the level of trust – or the lack of it – the human has in the system's capabilities. Via the mechanism of trust, they influence the time and effort the human will put into monitoring the performance of the system.

Emergent system characteristics

The Automation Interface Paradigm gives rise to characteristics of the system that emerge as a consequence of the system design. Referred to as **Emergent System Characteristics,** these form the second element in the HASO model. Three emergent characteristics are of particular importance to successful supervisory control;

- the *workload* imposed on the human;
- the extent to which the human is *engaged* in the overall system performance, and;
- the *complexity* of the human's remaining tasks.

Workload: It has long been assumed that, among other things, automating human tasks will both simplify what those humans that remain must do, and reduce the workload imposed on them – because the automation is expected to be doing most of the work. One of the ironies identified through applied research over many years, and a finding that has been found in many different contexts, is that, while automation usually reduces the number of people needed, and gives them fewer physical things to do, it frequently increases both the amount and the complexity of the work that remains. It increases the work on perceptual and cognitive resources – searching for information, understanding and making sense of it, making real-time judgements and decisions and working out what to do. All frequently under pressure both from lack of time to make decisions and act, and from the potential adverse consequences if the human gets it wrong. It also creates the burden of trying to maintain accurate real-time situation awareness, both of the performance of the system and of the state of the world it is acting in and on. That itself makes demands on attention and concentration that can be challenging for humans to maintain over more than short periods when there is little to do. As the prominent Human Factors specialist Peter Hancock wrote in connection with the accident with a self-driving car that killed Elaine Herzog in 2018, "…if you build vehicles where drivers are rarely required to respond, then they will rarely respond when required." *Peter Hancock.*[8]

Engagement: The second characteristic that emerges from the automation interface paradigm, is what is referred to as "cognitive engagement". Engagement, in this sense, refers to the extent to which the human's attention and mental resources are actively involved ("engaged") in the performance of a task. Endsley summaries the conclusions of a body of research by noting;

> It is inherently difficult for operators to fully understand what is going on when acting as a passive monitor of automation as compared with when they are actively processing information to perform the task manually due to lower levels of cognitive engagement. Actively engaging cognitively in a task improves ones understanding of that task and retention of critical information. [6]

Lack of engagement has frequently been identified as one of the key problems that emerges when automation is both highly reliable (able to operate accurately) and robust (able to operate across a wide range of conditions).

Complexity: Adding features and modes to a system increases the complexity of the human's task both in understanding what the system is doing, and in being able to unexpectedly transition to manual control if that should be necessary. It is much harder for the human to understand what a system is doing, and to predict what it is likely to do, when that system has many features, often structured into different modes of working. That complexity is compounded when the system is capable of complex logical situation-based reasoning which varies according to the information available to it and the current goals and constraints. And when the system's reasoning utilises the results of machine learning, perhaps extracted from many hundreds of thousands of pieces of source information rather than simply pre-defined logic, the complexity the human faces can be daunting.

As Endsley points out, the key issue here is that the more complex the design of the automation becomes, the harder it becomes for the human to be able to build an effective mental model of how the system works. Being able to build an effective and sufficiently accurate mental model is fundamental to the ability of humans to develop any level of skilled performance at any task.

Cognitive constructs – the 'Big 4'[9]

The third element in the HASO model comprises four **Cognitive Constructs** representing the key cognitive components most usually associated with human performance in complex systems; (i) *Situation Awareness*, (ii) the humans *Mental Model* of the system, (iii) *Attention Allocation*, and (iv) *Automation Trust.* These might be thought of as the "Big 4" psychological topics in terms of research into human performance in any complex system. All four are big and messy psychological constructs. The first three have substantial bodies of empirical research in the mainstream psychological sciences built over many decades: the body of research on trust in automation – as opposed to the nature of trust as a core psychological construct - is somewhat smaller though none the less compelling and reasonably consistent. It is hard to do justice to these four constructs without going into a lot of detail, but to give a flavour for the purposes of this book:

Situation Awareness refers to the extent to which the supervisory controller can build and maintain a real-time (or at least near real-time) understanding of the state of the system and relevant objects in the world. As a result of much previous work by Endsley and her colleagues, we recognise that the brain needs to be able to develop Situation Awareness at (at least) three levels: 1. Detection and perception of information and cues "given off" by the system and objects in the world; 2. Interpreting those cues and information to develop an understanding of what they means in terms of the current state and performance of the system and objects in the world and how those objects are changing, and; 3. Projecting into the future to predict what those current states and changes might mean in terms of future states, threats and events.[10]

The *Mental Model* construct in HASO refers to the ideas (or "model") our brains develop to help us understand how a system works: its components and how they work together, and how they behave and interact both with each other and with other elements in the world. Our mental model of a system can be complex and multi-dimensional, comprising visual, auditory, and other types of sensory-based imagery, facts and knowledge, as well as spatial and temporal relationships and what we know about the systems dynamics and how it performs over time. (The section below describes mental models in more detail).

Attention Allocation refers to the cognitive processes involved in controlling the allocation of our brain's limited ability to pay attention to things in the world in a way that allows the human to monitor and provide oversight of the automation. The fundamental requirement here is that, to extract information from objects in the world (Level 1 SA) we need to direct our attention towards that object. The challenge is that, while there can be a great many simultaneous sources of information available to us in the external world, our ability to pay attention to more than one source at any one moment is limited. Which is why attention allocation is such an important element of the HASO model. (The section below goes into some more details on the psychology of our ability to pay attention to information from multiple sources at the same time).

Automation Trust. Conclusions both from experimental research as well as investigations of numerous incidents are clear that the effort people put into actively monitoring the performance of an automated system depends largely on the extent to which they trust the system. Endsley cites a useful definition of what we mean by trust in this sense as being "the attitude that an agent will help achieve an individual's goals in a situation characterized by uncertainty and vulnerability".

Completing the HASO model

In addition to System Design Features, Emergent System Characteristic and Cognitive Constructs, the HASO model includes two other components necessary to appreciate how decisions made in systems design and engineering can impact on the performance of the human supervisor;

- a **Decision** that the represents the overall performance of the human as a supervisory controller, providing oversight of the performance of the automation and intervening as and when needed. This is represented as a decision because, at its heart, is the decision whether there is a need to intervene to support the automation. The decision is influenced by the quality of the user's situation awareness, as well as being influenced by the overall workload on the human at any time;
- other features of the environment and system. Specifically, the model recognises that **competing tasks and demands** can impact on the effectiveness of human supervisory control in two ways. First, competition for the user's time and attention can impair the ability to monitor what the automation is doing effectively. Second, when people are distracted or feel they need to spend their time on other tasks, they can be more inclined to put greater trust in the automation than they would if they did not have those competing demands.

SOME MORE ON MENTAL MODELS, ATTENTION, AND TRUST

Before moving on it is worth saying a little more about some aspects of the "Big 4" cognitive constructs.

It is of course a gross over-simplification to consider the Big 4 cognitive constructs as though they are in some way independent from each other.[11] It is certainly the case that the four constructs are highly inter-connected, drawing on the same neural structures and thought processes, and subject to similar limitations, constraints as other cognitive processes. They are also far from the only cognitive constructs that influence human performance in complex systems; issues such as emotional relationship and engagement, our morals and value systems, and even our sense of whether we are being treated fairly in comparison with our colleagues and peers to name just a few, have important roles to play. Generally, constructs such as these are somewhat less well researched than the "big 4" explicitly included in the HASO model. Though the distinction between the two styles of thinking brought to the attention of a global audience through Daniel Kahneman's 2011 book "Thinking, Fast and Slow" [8] and the characteristics of the two systems seem especially important.

The Big 4 constructs are convenient packages constructed by academics and researchers who have historically had interests in how the human brain performs specific types of tasks. Nevertheless, they play a useful and important role in allowing the HASO model to summarise how system design decisions and features influence the ability of people to perform as supervisory controllers in highly automated systems.

Some more on mental models

A mental model might include things like; where *we believe* different objects are spatially located relative to our current location or line of sight; how *we think* a system and its components work physically or mechanically; how *we understand* objects and events are related to each other and how they interact; how often *we expect* elements in the model to change – and therefore how long we can *be confident* that an observation of the state of an element can be assumed to be accurate; *beliefs* about what are normal (or safe) states; and what *we believe* the signs are that indicate when the systems performance is becoming abnormal (or unsafe). All of this is largely sub-conscious, put together by our brains based on experience of watching and interacting with a system and what we have been told about it. We might be capable of articulating some part of our mental model if we were asked to. Though our verbal description will be unlikely to capture all the details our brain uses.

In the paragraph above, note how often I have used italics to emphasise the subjective nature of a mental model: "we think", "we believe", "we expect", "we understand", and so on. In other than perhaps a few very special cases[12] a mental model does not need to be – indeed, for a system of any real complexity, it is unlikely to be - "correct" in the sense that all the elements, mechanisms, relationships and behaviours the brain uses to represent how a system works are technically accurate. The model we hold will be based on beliefs, expectations, our understanding and other subjective judgements about how we think a system works and behaves. But, for a skilled performer, the model will be sufficiently accurate to support a standard of human performance that is acceptable most of the time. We only really appreciate how far mental models differ from reality, even in critical operations and with highly trained personnel working in highly controlled operational environments, when things go badly wrong, and the subsequent investigation takes the time to explore what the operators involved thought and expected about the system. Unfortunately, such insightful investigations are rare.

Clearly, much of the experience and beliefs about the behaviour and performance of a system used to develop our mental models of a system will be derived from information through the automations, user interface. This is usually predominantly information displayed visually, though other senses, including smell as well as sound and touch, can also provide powerful cues. In many cases, that information from the user interface will be enhanced by cues arising from other sources, including direct observation and perception of the state and performance of the system itself as well as the behaviour and impact on objects in the external world. Feedback from other sources, including other people and other systems, can also provide cues about how the system we are engaged with behaves and performs.

Our brain develops mental models of our world, the things we engage with in it, and the tasks we perform whether or not we go to any conscious effort to create one. Mental models are the basis of all skilled and expert performance: indeed, they are the enablers of any form of non-trivial human performance. A mental model underlies the awareness of the current and predicted state of the world that comes from Situation Awareness. The mental model of a system we develop over repeated exposure and experience, as well as what we are formally told, is the basis by which the brain is able to interpret what is happening in the world in real-time (that is, build Level 2 SA) make predictions about what is likely to happen next (L3 SA), as well as determining what actions we would need to take, how and when, to bring about changes we think are needed. Importantly, our mental model of the behaviour of a system that operates in real-time is dynamic: it is only current for a short time and needs to be constantly re-informed with new information about the state of the system and the world.

Experience your own mental model

Try this simple exercise to give you a subjective experience of the nature and use of a mental model. It is best if you do this in a space you are familiar with. Assume, in this scenario, that there are no animate objects in the room, and that everything other than yourself is fixed in position and unable to move - we will call this Scenario A[13] . Take a good look around. Now close your eyes and try to leave the room without opening your eyes.

Initially, perhaps for a few seconds and especially before you make any movement, your confidence will be (relatively) high. Drawing on your mental model of the space, you will be confident you know where you are, where you need to get to, what objects you need to avoid, where they are, and how big they are. As time passes, and as your position changes, your confidence declines as your mental model becomes out of date: your confidence in where you are, where the door and objects you need to avoid are, and the movements you need to make next to get to the door all reduce. Simultaneously, your sense of risk that you may be about to hit something, and perhaps fall, increases.[14] That growing sense of risk creates an increasing desire to open your eyes and look about you: you are experiencing a desire to "sample your environment". It is not only your visual attention that is becoming out of date: you will certainly have been continuing to pay attention for any sense of touch and possibly sound as you moved around without visual guidance. Now open your eyes to check the room layout, then close them again and carry on moving. Repeat until you reach the door.

The mental model you drew on to perform this exercise is relatively simple, comprising not much more than the layout of the space, and your position in it relative to the door and to other objects. There are no dynamic elements: nothing other than your own position is changing as the task progress. The model in your head only needs to try to keep track of where it thinks you are relative to your goal and to objects that might represent a risk to you as you move.

This task will be far easier for some people than for others. Though however good your brain is at updating your mental model, however long your personal confidence in your model remains high, and however slowly your sense of risk grows, eventually[15] your confidence in what your mental model can tell you of where you are relative to the door and to other objects will reach the point where you feel impelled to open your eyes to check.

Let's now make a small change that will illustrate how our experience of trust interacts with our mental model to influence how we allocate attention. In scenario A, we assumed there were no animate objects in the room, and that everything other than yourself was fixed in position. Let's now introduce a dog and consider two cases. In the first case (scenario B), the dog is your own. She is well trained mature and obedient. (For the purpose of both scenarios B and C we are going to assume the dog makes no noise). Before you close your eyes, instruct the dog to "Sit and Stay[16]" in a location that will not interfere with your route. Being a mature and obedient animal, you are confident that she will stay seated where you told her to. Now try the task again. Does the experience differ from scenario A? You know there is now an element in the room with the potential to move, which could interfere with your task. But you have a high level of confidence that the dog will remain seated where she was told: you trust her. There should really be little difference in your experience of the task in this case.

Now let's change the dog (scenario C). The dog is now young and excitable and in the process of being trained. Go through the same process: check the layout of the room, tell the dog to "Sit and Stay" out of the way, close your eyes and start to move. How do you feel? Do you trust the dog to remain where you put her? Or are you concerned that she may have moved and is about to trip you up?

The difference is one of trust. In scenarios A and B when you first closed your eyes, you had a high degree of confidence in your mental model of the room. Confidence that you could trust what your mental model was telling you about where the various objects in the room were relative to your own position. The objects were unable to move independently, and you had a high degree of trust that your dog would not move from where she was told to stay. The only real uncertainty was the extent to which you had confidence in your own location relative to the door and other objects. Your mental model allowed you to perform the task confidently for as long as you had confidence in what your mental model was telling you. Confidence that reduced as the task progressed, and uncertainty increased to the point where you felt impelled to stop and look around.

The difference in scenario C was that now your trust in using your mental model to guide your behaviour was much lower due to the uncertainty associated with the potential for the dog to move. With the consequence that you felt driven to open your eyes to sample the world around you much sooner and more frequently than in scenarios A and B.

This simple task, under these three conditions, illustrates what we mean by a "mental model" and how the information the model is based on needs to be constantly updated as events in the real-world change. Scenario C illustrates the critical role that trust in what out model is telling us plays, and how our sense of unease increases as our confidence in our mental model reduces.

Some more on paying attention

There is a long and rich history of high-quality psychological research into how Attention works.[17] It is undoubtedly a gross simplification to say that the human brain can only attend to and process (think about, whether consciously or sub-consciously) one thing at a time. The reality is more complex than that. Though what is certainly the case is that our capacity to pay attention to multiple sources of information and objects in the world is limited: we cannot pay attention to more than a few sources in the same moments (depending on the modality and cognitive resources needed) without experiencing some disruption to our ability to attend to one or other source. And we find it hard to maintain attention on a single source over an extended period, especially if there is little of interest happening in that source to engage our thought processes.

Understanding what is needed to provide effective support to the process by which we allocate attention to features of the system and the world is therefore of central concern in supporting effective supervisory control. Which is why it is such an important element in the HASO model.

Multiple resource theory

The Psychologist Christopher Wickens developed his Multiple Resource Theory (MRT) to explain how we seem, to some extent at least, to have the ability to not only pay attention, but to reason with information from different sources in something like the same time [9,10]. Provided, that is, that the competing information does not occur in the same perceptual modality (visual or spatial) and does not make demands on the same cognitive resources (spatial reasoning as opposed to verbal/linguistic reasoning).

Multiple Resource Theory has proven to be useful to Psychologists and Human Factors specialists involved both in conducting research and supporting the development of advanced complex systems. While its application has probably been most prominent in the development of military systems, its potential for application is much wider. So while it is a divergence from explaining the HASO model, it is worth taking a few moments to understand just a little bit about the MRT model. Doing so can help to appreciate some of the complexities we are dealing with when we try to design systems that support the allocation of attention needed for effective human supervisory control.

MRT was derived from the idea that human performance is based on a general pool of mental "effort" that can be shared across different cognitive processes. Figure 7.4 shows a schematic representation of the 4-dimensional version of the MRT model. Three of the dimensions are involved in processing information and preparing to act;

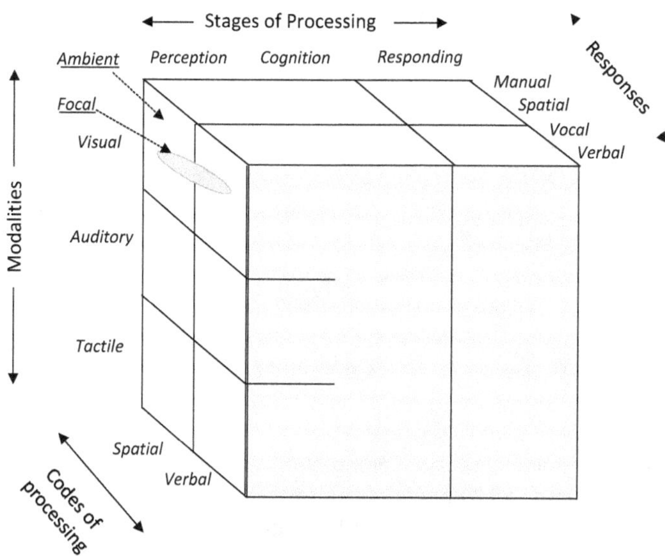

FIGURE 7.4 Wickens 4D multiple resource model [8]

- The ***stages of processing*** dimension is concerned with how sensory information flows through different stages as it is processed and used in the brain: Perception, Cognition and Responding. The figure illustrates that perception and cognition draw on different resources from those involved in responding to an event;
- The ***codes of processing*** dimension is concerned with the nature or format of the information that is being processed. The dimension indicates that verbal/linguistic information draws on different mental resources than does information that is spatially coded. This is consistent with a large body of research into what is known as "Working Memory" – which is where active thinking happens in the brain.[18]
- The ***modalities*** dimension is about how information enters the brain. As such, it exists only in the perceptual stage of processing, and not in cognition or responding. The figure indicates that the perception of visual, auditory and tactile information draw on different mental resources.

A 4th dimension was added some years after the initial model was developed. This dimension exists only in the visual modality and reflects the importance of whether visual information is being viewed primarily in the fovea or ambient areas (across the entire visual field). Foveal vision supports high levels of visual acuity necessary for object recognition, reading text, symbols etc. Ambient vision includes information from the peripheral visual field and supports spatial orientation and detection of movement, though not reading and recognition.

Note that on Figure 7.4, the ***Responses*** stage illustrates that responses or actions that are made manually or spatially draw on different resources from those made in a verbal or vocal format. Though responses are seen as being separate from the four resource dimensions involved in processing information.

The value of the MRT model is that it allows those concerned with trying to support high levels of human performance to predict where people are likely to have trouble performing two or more tasks in the same time frame due to the conflicting demands they make on the underlying pool of mental resources. As Wickens puts it: "…to the extent that two tasks use different levels along each of the three dimensions, time-sharing will be better". Time-sharing across the tasks will be "better", but not perfect: there may still be competition for common perceptual *resources*, even when processing the information uses different *modalities.*

Fortunately, only a small proportion of the information potentially available to us is relevant to our role as a supervisory controller at any moment. So the challenge becomes knowing;

a. *what* information is important at any time, and where to find it, and;
b. *when* to "sample" each source of information: that is, how often should we pay attention to each source of information we need to be effective as a supervisory controller.

That second challenge can be expressed from the opposite direction: for how long after sampling a source of information will the information acquired remain valid such that it can be used in thinking and decision-making? How quickly does its value to our situation awareness and decision-making decline to such an extent that we would be better off without it?

Those two questions, along with the challenge of maintaining attention over sustained periods when there is little to do, are at the heart of the challenge of being an effective human supervisory controller.

Some more on trust

The need to understand more about the characteristics of trust in automation was recognised by the NRC workshop that reviewed research needs for human supervisory control back in 1984 [4]. The discussion of human factors issue associated with the way trust in automation is developed, maintained and lost included in the 1984 report is as true today as it was at the time it was written.

The extent to which we build trust in an automated system is known to be influenced by characteristics of the system (such as our experience and beliefs about its reliability), ourselves (such as our confidence in our own abilities), and the situation we are in at the time (such as how busy we are and competing demands for our attention). A meta-analysis of studies of these issues in the context of human interaction with "robots" (though the definition of "robots" is not clear) by Peter Hancock and his colleagues [11] concluded that, of these three, system factors – the robustness and reliability of the system's abilities – have the biggest influence on whether are prepared to trust a system.[19]

Adults find it difficult to develop trust: that applies not only to trust in automation, but to trust in anything, people included (other than, perhaps, our own beliefs and judgement). Even more importantly, not only does it take a great deal of time to build trust, but, once it is lost, it is extremely difficult to get back.

Trust in Automation is a difficult topic. There have been many well documented situations (including, in aviation, incidents of 'controlled flight into terrain') where an underlying lack of trust in the reliability of sensor data has allowed cognitive bias to dominate decision-making. For example, where pilots or control room operators have decided that instruments must be faulty when they show readings that do not agree with what they believe, or want to believe, is the actual state of the system. A detailed discussion of trust is beyond the needs and purpose of this book. Our purpose here is restricted to presenting the generic HASO model as a useful way of thinking about and recognising the impact that system design features have on the ability of people to perform as supervisory controllers. So we will move on – though the issue of trust is one we will have to return to in the next chapter.

RECAP OF EXISTING MODELS OF HUMAN SUPERVISORY CONTROL

To briefly recap and summarise the models of supervisory control discussed so far;

- Figures 2.1 and 2.2 in Chapter 2 Illustrated how the introduction of automation to previously manual functions changes the role of the human from being a manual controller to being a supervisor, with all the implications that change brings. It illustrates that a supervisor needs information not only about the world that is being controlled, but about the state of the automation itself. This not only changes but can also significantly increase the cognitive demands on the human when compared with when the same functions are performed manually.
- Sheridan's hierarchical model, shown on Figure 7.1, emphasised the hierarchical nature of the sub-systems that make up an automated system of any non-trivial complexity. As well as making the distinct sub-systems explicit, the model identifies the different types of information and controls that need to be available in the human interface. It also highlighted the different ways in which the human supervisor interacts across the levels of the hierarchy;
- Moray's model (Figures 7.2a and b) provides a means of thinking about how knowledge (or, in my variant, the information it is based on) and control can be distributed between the different hierarchical layers in an automated system. The model explicitly distinguishes between whether information and controls are provided somewhere in the designed system or elsewhere, and whether the user is aware of them and understands how to use them.
- Endsley's Human-Autonomy Systems Oversight (HASO) model shown on Figure 7.3 summarises what has been learned from several decades of research into difficulties people face maintaining situation awareness when they are expected to support highly reliable automation, and the "out of the loop" (OOTL) problems that often arise. The HASO model identifies the relationships between characteristics of the design of a highly automated

system and the key cognitive functions and cognitive performance necessary to allow the human to provide the monitoring, decision-making, and intervention needed to be successful in the role of a supervisory controller. The model is intended to help those responsible for the design and implementation of systems with highly automated elements recognise and appreciate how the decisions they make in systems design and engineering can impact directly on the ability of people to provide oversight.

The HASO model is clearly the most detailed in terms of capturing the relationships between things that system designers can influence and key elements of the cognitive processes that drive human performance as a supervisory controller.

The models of Sheridan, Moray and Endsley all have value and can be used for different purposes and in different ways: Sheridan's model has been found to be a useful means of helping senior decision makers and engineers recognise the change in complexity of the role of the human that moving into the world of highly automated systems can bring. Moray's model directly lends itself to challenging a system design in terms of how it supports the user in acquiring and understanding the information controls they will need to perform their roles. It also lends itself to use in incident investigations, as a means of thinking about what was demanded of the users involved in an incident in making use of information and controls, and to what extent the system supported those demands. Endsley's model brings a deeper awareness of the mechanisms by which the cognitive processes that underpin supervisory control need to be supported in a system design and how decisions about the design of a system and the use of automation can impact on those processes.

Despite all of this, and recognising the extent and quality of research available, from my perspective as someone trained in cognitive psychology and who has spent a lifetime in applied Human Factors/ Cognitive Engineering, I'm left feeling that even the HASO model is some way short of providing a satisfactory account of what is involved when a human is expected to fill the role of a supervisory controller of a highly automated or autonomous system. What is needed is a more comprehensive model of the psychology underlying supervisory control performance. The following chapter will attempt to present exactly such a comprehensive model.

NOTES

1 Section 7.3. in [1] contains short (1–3 pages) concise summaries each written by a leading international authority in the field, of eighteen individual topics concerned with the ability of people to fill the role of human supervisory controllers.
2 I am grateful to Heriot-Watt University for the use of their Information Services to support this search.
3 Searches were also conducted using two (at the time) recently released "AI-powered" search engines: ChatGPT, and DeepSeek. Neither of the searches using these large language model engines identified any of the three models discussed in this chapter, though DeepSeek did identify [2].
4 David Mindell's excellent book 'Digital Apollo' [5] goes into great detail outlining the challenges NASA went through in balancing the roles and capabilities of automation and human astronauts in putting together the systems that eventually took the

Apollo spacecraft to it's successful moon landings. Mindell identifies the key role Tom Sheridan played in developing the human interface to the apollo guidance system: "... an MIT assistant professor with expertise in mechanical engineering and psychology who would come to define ideas in supervisory control and telerobotics". I would commend the book as essential reading for anyone interested in the organisational dynamics of how user interfaces are created in advanced complex systems. As well as anyone interested in inspired and startlingly brilliant user interface design solutions.

5 As well as the 1983 NRC report [3], there is a concise summary of the Sheridan et al model in entry 7.309 in [1].

6 Proactive operator monitoring as a task in human supervisory control is discussed in some detail in Chapters 8,10, 11 and 12.

7 See [6 or 7] for a detailed explanation of the nature and causes of OOTL problems.

8 https://journals-sagepub-com.hwu-ezproxy.idm.oclc.org/doi/full/10.1177/0018720819900402.

9 'The Big 4' is my term, not Endsley's.

10 For an introduction to the theory of Situation Awareness, and its application in system design, see [7].

11 Though that simplification is one that, during my professional career, I frequently found is not recognised, and its implications not understood, by engineers and others who have sought to make use of models such as HASO without the support of specialists who understand the psychology involved. That has especially been so when engineers try to use simplified models of human performance as a means of trying to provide quantitative estimates of the likelihood of serious human errors occurring in critical operations.

12 While there is no way of knowing, I have long believed that the extraordinary performance of Neil Armstrong during the final stages of landing Apollo 11 on the moon on 20th July 1989 probably reflects his having developed an unusually accurate mental model of the abilities and performance of the space craft. And that despite never having flown the vehicle in the real conditions he encountered on the day. For a vivid – not to mention sweat-inducing – description of the details of the final landing sequence, see David A. Mindell's book 'Digital Apollo' [5]. I would speculate that the top performers in a few professions – such as test pilots, some surgeons – are able to develop unusually technically accurate mental models of the systems they work with (and on). Though it is rare.

13 It will be most effective if you can do this for real. If you are unable to move about, you can do it as a thought-exercise, though without quite the same impact.

14 If you do fall, don't blame me. You chose to walk about with your eyes closed.

15 Assuming you've set yourself a not too trivial space to manoeuvre.

16 Or whatever other command will have the same effect – you don't want the dog to move while you are walking around with your eyes closed.

17 For a comprehensive introduction to the psychology of attention, see [12,13]. As of January 2025, Wikipedia also had a comprehensive entry on Attention: https://en.wikipedia.org/wiki/Attention.

18 For a gentle introduction to the Psychology of Working Memory, see [14]. For a more thorough technical treatment, see for example [15].

19 [16] provides an approach to designing systems that support high levels of trust.

REFERENCES

1. Boff, K.R. & Lincoln, J.E. Engineering Data Compendium: Handbook of Human Perception and Performance, Volume II. John Wiley & Sons. 1986.

2. Boussemart, Y. & Cummings, M.L. Predictive Models of Human Supervisory Control Behavioral Patterns Using Semi-Markov Models. *Engineering Applications of Artificial Intelligence.* 2011;24;1252–1262.
3. National Research Council. Research Needs for Human Factors. The National Academies Press. 1983. https://apps.dtic.mil/sti/tr/pdf/ADA129899.pdf
4. National Research Council. Research and Modelling of Supervisory Control Behaviour: Report of a Workshop. T.B. Sheridan & R.T. Hennessy (Eds.). The National Academies Press. 1984. DOI:10.17226/19376
5. Mindell, D. Digital Apollo. MIT Press. 2011.
6. Endsley, M. From Here to Autonomy: Lessons Learned from Human-Automation Research. *Human Factors.* 2017;59(1):5–27.
7. Endsley, M., Jones, D. Designing for Situation Awareness: An approach to User-Centred Design. 3rd Edition. Taylor & Francis. 2025.
8. Kahneman, D. Thinking, Fast and Slow. Allen Lane. 2011.
9. Wickens, C.D. The Structure of Attentional Resources. In R. Nickerson (Ed.), Attention and Performance VIII (pp. 239–257). Erlbaum. 1980.
10. Wickens, C.D. Multiple Resources and Mental Workload. *Human Factors.* 2008;50(3): 449–455. DOI:10.1518/001872008X288394
11. Hancock, P.A., Billings, D.R., Schaefer, K.E., Chen, J.Y.C., De Visser, E.J. & Parasuraman, R. A Meta-Analysis of Factors Affecting Trust in Human-Robot Interaction. *Human Factors.* 2011;53:517–527.
12. Styles, E. The Psychology of Attention. 2nd Edition. Psychology Press. 2006.
13. Nobre, A.C. & Kastner, S. The Oxford Handbook of Attention (Oxford Library of Psychology). OUP Oxford. 2018,
14. Psychology Today. Working Memory. 2025. Available from: https://www.psychologytoday.com/gb/basics/subpage/working-memory#:~:text=Reviewed%20by%20Psychology%20Today%20Staff,solving%2C%20and%20other%20mental%20processes
15. Cowan, N. What are the Differences between Long-Term, Short-Term, and Working Memory? Essence of Memory. *Progress in Brain Research.* 2008;169(Elsevier): 323–338. ISBN 978-0-444-53164-3. PMC 2657600. PMID 18394484. DOI:10.1016/S0079-6123(07)00020-9
16. Lee, J.D and See, K.A., Trust in Automation: Designing for Appropriate Reliance. *Human Factors.* 2004;46(1):50–80. DOI:10.1518/hfes.46.1.50_30392

8 What's missing from models of human supervisory control?

Chapter 2 introduced the basic ideas necessary to understand what it means to describe the human as a supervisory controller of a real-time system. Chapter 7 summarised three models produced by leading thinkers and researchers, providing different perspectives on the role. Each of those models emphasised characteristics that can help those responsible for designing and developing highly automated systems think about and understand the challenges involved in supporting the human in their supervisory role.

I make no claim to be a leading thinker in the field of Human Factors. I am not an academic and have not devoted my life to research, applied or otherwise. I have, however, spent a career applying the principles of Human Factors in the research, design, development, and analysis of complex systems across industries ranging from defence and aerospace to oil and gas, chemicals, rail, nuclear power and healthcare. I have been employed as the discipline lead for Human Factors in one of the world's largest companies, with responsibility for supporting capital projects valued in the tens of billions of dollars comply with the principles of Human Factors Engineering. In the course of that career, I have read and studied far more investigations into situations where things have gone wrong that have been attributed in some way to "human failure" than I can possibly recall. I have learned a great deal from studying incidents, from engaging with those involved in front-line operations, as well as from published research. And I have been active in feeding back what I have learned via training, in my first book and numerous other publications,[1] in contributions to numerous engineering standards and sources of design guidance, as well as in a series of "best practice" publications produced by professional bodies.[2] All with the aim of helping those responsible for designing, developing, and operating advanced systems understand what they need to do to avoid producing solutions that can lead people into injuring themselves or others or making design-induced mistakes that can have serious adverse consequences.

Drawing on all this learning and experience, I concluded the previous chapter by expressing the view that even Mica Endsley's HASO model is some way short of providing a satisfactory account of the psychology involved when a human is expected to fill the role of a supervisory controller of an automated system. A more comprehensive model is needed. The aim of this chapter is to present the technical and theoretical background for such a comprehensive model. The model itself will be presented in Chapter 9.

DOI: 10.1201/9781003679479-10

WHAT'S MISSING?

There is a large and long-standing body of scientific research into characteristics of how humans think, behave and perform complex tasks, or that influence how they make decisions and perform, that are not explicitly accounted for in the existing human supervisory control models. Characteristics that, over the course of my professional career, I came to recognise as being essential to any attempt to apply the principles of Human Factors to complex systems in a meaningful way: both in their design and development, as well as when investigating and understanding how and why things went wrong when people did not act or behave as expected and serious adverse consequences followed.

Before I can present my version of a comprehensive model, I need to explain what I think is missing from the existing models described in Chapter 7. To do that, we need to start by taking a closer look at precisely what is involved when someone performs the role of a supervisory controller of a continuous real-time activity. But before we can even do that, we need to be clear about a task that is central to supervisory control: proactive operator monitoring.

PROACTIVE OPERATOR MONITORING

A human supervisory controller must be *proactive*: that is, they must take the initiative and be active in putting in the effort and allocating time and attention to check on the status of the automation, looking for information that could suggest that there is, or could be, a need to intervene to support the automation either immediately or in the near future. Proactive operator monitoring (I will use the acronym POM from here on) means making the effort to pay attention to the automated system as often, and for as long, as is necessary to make that decision. As a reminder from Chapter 2;

- the term "human supervisory controller" refers to the ***role*** an individual adopts when they are put in the position of being expected to oversee a highly automated or autonomous process;
- Proactive Operator Monitoring (POM) is the key ***task*** that needs to be performed to be effective in the role of a human supervisory controller.

The alternative to monitoring proactively is to be *reactive*. Reactive monitoring means relying on prompts or alarms, usually from the system though potentially from other sources as well (colleagues, other systems, horns sounded by other motorists or whatever) that are effective in forcing the operator to attend to the automated system. Behaving in this way is not monitoring, it is simply responding. It may be there will come a time when most autonomous operations have become sufficiently reliable and resilient that they can be given the authority and responsibility to control their activities without the need for a human overseer. At that time the design of such systems may have developed to such a state that they can be relied on to draw the attention of an overseer tasked with supporting the system if needed, but whose primary responsibilities lie elsewhere. But for the great majority of systems of any complexity or where the consequences of failure are especially critical, that is still some time away. The great majority of systems seem likely to continue to rely on a human supervisory controller who holds overall responsibility for the performance of the system for some time to come. And as long as that is the case, the human must perform their role *proactively*, rather than *reactively*.

The SAEs taxonomy of driving automation, J3106 [16] includes the following note under the definition of the term "Monitor";

> The driver state or condition of being receptive to alerts or other indicators of a … performance-relevant system failure … is not a form of monitoring. The difference between receptivity and monitoring is best illustrated by example: A person who becomes aware of a fire alarm or a telephone ringing may not necessarily have been monitoring the fire alarm or the telephone. Likewise, a user who becomes aware of a trailer hitch falling off may not necessarily have been monitoring the trailer hitch. By contrast, a driver in a vehicle with an active Level 1 adaptive cruise control (ACC) system is expected to monitor both the driving environment and the …[automation's]… performance and otherwise not to wait for an alert to draw his/her attention to a situation requiring a response. [16]

This is essentially the difference between a driver who is reactive – and is therefore not monitoring - and a driver who is behaving as a proactive supervisory controller.

In an ideal world, a human supervisor who only has a single responsibility, which is to oversee, and if necessary, support, an automated operation, would do so continuously. Their attention would be permanently directed towards monitoring the system, even if they have no other active engagement, mentally or physically, in the performance of the system. That is precisely what is expected of drivers-in-control of self-driving vehicles, who still retain legal responsibility for safety even when the vehicle is in an autonomous driving mode. Unfortunately for those in a hurry to introduce autonomous vehicles, the human brain has not evolved to support such continuous monitoring when there is no other mental engagement.

The "POM scan"

The human's capacity to maintain attention on a single task, when there is little or nothing to do, and especially where there is little likelihood of having to do anything, is extremely limited. In general, humans, under normal conditions, are not capable of performing a demanding task that requires sustained attention and concentration, over more than a relatively short period of time - tending to seconds rather than minutes. No absolute limit can be put on how long someone might be able to maintain continuous attention on a monitoring task: it will depend on things like the training and experience and the individual's perception of the importance of the task, the organisational context, possibly time of day, as well as the nature, design characteristics, location and clarity of the information they are looking for. And it will be affected by personal factors, not least age, fatigue and possibly sex: there will be large individual differences for any specific monitoring task.[3] In the absence of significant research data, there is no way of generalising.

What we can be sure of, is that POM will rarely if ever be performed continuously over any significant period. Rather, it will involve an ongoing, though intermittent, series of actively induced checks of the state of the automation made by the supervisory controller. I am going to refer to these throughout the remainder of this book as "POM scans". A study for NASA led by Randall Mumaw and colleagues

[17] reviewed how pilots proactively monitor their flight path. The study developed a model of pilot monitoring based around maintaining their dynamic "Situation Model" which integrates current observations with information from the pilot's mental model held in long term memory. Based on both a review of relevant literature as well as interviews with pilots, Mumaw et al's model comprises three mental activities carried out sequentially as what they term a "Model Update Cycle"[4];

i. Formulating *Questions* about the current or future state of the flight path;
ii. Gathering and assessing *Evidence* to answer the monitoring question, and;
iii. Assessing the implications for *Action*.

In the case of pilots, the model update cycle needs to be performed regularly, and scheduled alongside other activities that need to be accomplished simultaneously. Mumaw et al's concept of the Model Update Cycle in pilots is similar to the more generic concept of "POM scans" described here.

POM scans can only be conducted intermittently. How frequently these intermittent scans are made, how long they last, and what information they consider will of course depend on many factors; from the dynamics of the process being controlled (principally how quickly it changes), the implications of failing to spot a threat and how much trust the individual has in the system, to how busy and distracted they are with other responsibilities and activities. Again, in the absence of research data, there is no way of knowing.

The heart of the POM scan is at least one decision. That is, in the moments the scan is being conducted, whether there is any indication of a threat[5] that could create a need to intervene to support the automation or to take over control. Secondary decisions might include whether there is any need to adjust the automated current settings; its' goals, constraints or perhaps what information is displayed to the user and how it is presented. The priority must be to decide whether there are any indications of a threat requiring manual intervention.

Of course, the decision can go either way. The supervisory controller can decide not to intervene in the presence of threats the automation is unable to manage. That may be because they didn't notice the signs indicating there was a threat; or because they make the decision not to intervene even though they saw the warning signs. On the other hand, a POM scan may result in the decision to intervene when there was actually no need to. Perhaps the supervisor misinterpreted the signs and came to believe there was a threat when none existed; or they noticed the threat and concluded the automation would not be able to cope when in fact it could.[6] Both failing to intervene and intervening when they didn't need to can have unwanted consequences.

The purpose of each POM scan, then, is to make a judgement about whether there are any immediate threats to the automated system continuing to operate autonomously, or if the operator needs to prepare to intervene.

HUMAN SUPERVISORY CONTROL BEHAVIOUR

Figure 8.1 illustrates, at a conceptual level, the possible behaviour of a generic human supervisory controller over time.[7] The timescale on Figure 8.1 is arbitrary: it could be

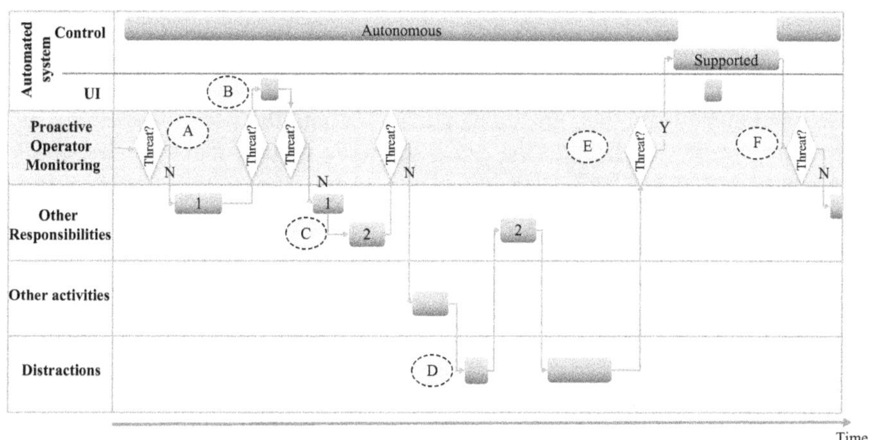

FIGURE 8.1 Conceptual illustration of supervisory control for a generic work activity

seconds, minutes or hours (even perhaps days in the case of space exploration). The actual timescale will be determined by the dynamics of the process being controlled, among other things.

The illustration on Figure 8.1 represents an individual who has the primary task of supervising the performance of a more-or-less autonomous system controlling some activity in real-time. This is the **"autonomous system"**, comprising both the elements that exert control over the process, as well as the user interface that allows the user both to exert control directly if needed, as well as to make changes to the control system's goals and constraints. The interface also allows them to change the layout and content of the information presented in the interface itself. Within its defined environment, the system is highly (though not perfectly) reliable. The key task of the supervisory controller is shown as ***"Proactive Operator Monitoring"***, which means performing the POM scans described above.

In addition, the individual has two simultaneous, though secondary, tasks they are responsible for (***"Other responsibilities"***). There are many situations where this is the case; control room operators in a process control environment or manufacturing facility, or medical staff overseeing equipment that automatically delivers medication to patients. In each case, while the individual is responsible for ensuring the automation performs as expected, they also have other tasks they are responsible for over the same time frame.[8]

As well as these other responsibilities, there will often be activities the individual might choose to perform, though they are not actual responsibilities (***"Other activities"*** on Figure 8.1): interacting with colleagues, taking phone calls, dealing with emails, and so on. And, of course, there are always potential distractions: a sudden distracting noise, overhearing an interesting conversation, paying attention to something being said on a background radio, or whatever (***"Distractions"*** on the figure). The challenge for the human supervisory controller is to juggle their mental and physical abilities across these competing demands for their attention, without failing to perform effectively the core task of monitoring the automated system.

At point A (to the left of Figure 8.1), the human has just completed a POM scan and made a judgement that there are no threats that require them to intervene. So, they can allocate their attention to performing another of their assigned responsibilities, and to maintain their attention on that task until their confidence in the last POM scan decision declines to the extent that they need to make another check. A scan at point B, again suggests no threat, though the user makes some adjustment on the user interface – changing constraints or the way information is displayed. They then, at point C, turn to perform another task. After initially working on task 1, the individual switches attention to a second task. This task switching captures the individual's attention, so it is a while before they again make the effort to actively check on the automation.

At point D, following another negative POM scan, the individual is distracted by some non-task related event. Over the course of the next period up to point E, their attention is captured by a combination of the distraction and the need to do something in task 2, such that they do not perform another POM scan over this period. When they do perform the POM scan, they make the judgement that there is a threat sufficiently important that they need to intervene in the automated system. Note that, up to this point, the automation has been performing autonomously. At this point, with the intervention of the human, the system becomes either supported automation, or fully manual, depending what action is needed. As part of the support, the user makes another adjustment in the user interface.

At point F, having completed their intervention and dealt with the perceived threat, the supervisor is happy to pass control back to the automation. After actively observing its performance, attention is the re-directed to task 1. And so it continues.

Human supervisory control in an autonomous vehicle

Given the current rapid pace of development and deployment of self-driving cars, which only seems likely to accelerate into the coming decades, as well as the emphasis given to the driver's role in them in Part 1 of this book, it is worth considering how the generic illustration of supervisory control behaviour illustrated on Figure 8.1 changes when we think about a driver-in-control of an autonomous vehicle.

At the time of writing, vehicle manufacturers and governments around the world rely on the assumption that there will always be a human driver in place capable of overseeing the performance of a self-driving vehicle and willing and able to take manual control if necessary: that is Level 2 in the Society of Automotive Engineers taxonomy of levels of automated driving [16].[9] Most importantly, they rely on the premise that the driver-in-control retains legal responsibility for the safe performance of the vehicle. Figure 8.2 attempts to illustrate some of what these assumptions mean for a driver-in-control. The discussion here is concerned with vehicles that have self-driving capability but that remain under the overall control of the driver: so the driver decides whether, and when, to pass control of the driving task to the vehicle systems, as well as when to re-take control.

Figure 8.2 illustrates two core mental tasks for the driver of a vehicle with autonomous capability;

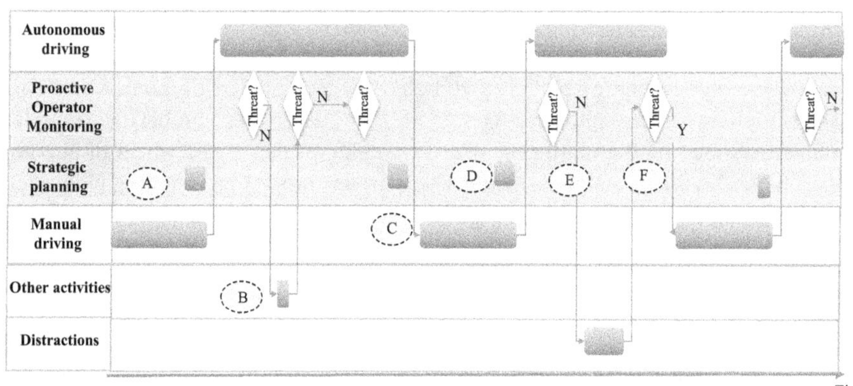

FIGURE 8.2 Conceptual illustration of supervisory control for a driver in control of an autonomous vehicle

A. Strategic planning of the driving task, including decisions on things like route and lane changes and over-taking, as well as whether and when to engage the self-driving function;
B. Proactive monitoring of the vehicles performance when autonomous driving mode is engaged.

The story starts at the left-hand side of Figure 8.2, with the driver in manual control of the vehicle. At point A, based on current driving conditions and with knowledge of the route ahead, the driver makes the decision to engage the autonomous driving function. As soon as it is engaged, the driver's role becomes that of a supervisory controller, proactively monitoring the vehicle and the road ahead, and repeatedly re-visiting the decision whether there is a need to re-take manual control.

At point B, having successfully handed control to the vehicle, and checked that it is performing as expected, the driver briefly performs some other activity: perhaps trying to find out the distance to the next fuel station, or making a hands-free phone call. After which the driver returns to proactively monitoring the vehicle and the road ahead.

At point C, strategic planning calls for an over-taking manoeuvre, which the vehicle is not capable of performing autonomously. The driver reverts to manual control until point D, when the overtaking is completed, and the autonomous mode is re-engaged. From point E to point F, the driver is distracted by some event not related to the driving task. The monitoring review at point F identifies a threat causing the driver to take manual control of the vehicle.

Note that the discussion around Figure 8.2 has, other than for very short periods, intentionally avoided allowing our driver to engage in other non-driving related activities while the automation is engaged. The significant issue the discussion highlights is the time and effort which the driver-in-charge of an autonomous vehicle is expected to apply to proactively monitoring and repeatedly reviewing the decision whether there are any threats that require them to take manual control. That is a task humans are singularly unsuited for and is at the heart of the Human Factors concerns around autonomous driving. Importantly, a great deal of experience, including a growing number of fatal accidents, make it clear that that is not how at least some human beings are likely to behave when faced with the task of being a driver-in-charge of an autonomous vehicle. The increasing number of examples available on social media platforms of drivers being asleep at the wheel of autonomous vehicles, as well as fatal accidents where drivers have been watching videos tell us the reality of what can and does happen.

Supervisory control is decision-making in uncertainty

To re-cap, as long as automation is allowed to control an activity, and up until the point that the human intervenes or is handed manual control, the supervisory controller is expected to be *proactive* in monitoring the performance of the system. The purpose of that monitoring is to answer the question "do I need to intervene?" Put another way, the question is; "Is there currently a threat that requires me to intervene either right now or in the immediate future?" This decision is at the heart of successful human performance as a supervisory controller. In Endsley's HASO model it is the element "Human Oversight and Interaction Performance" (see Figure 7.3 in Chapter 7). And as the HASO model also shows, the decision is influenced by two things, both of which vary in real-time: the individual's situation awareness, and their workload at the time.

If we probe a little deeper, we will realise there are two cases that can lead to the controller making the decision to intervene;

A. when there are strong, clear and unambiguous signs that the human needs to intervene. Hopefully - at least in a well-designed system - that will be because the system is presenting some form of information – perhaps an alarm, system message or other highly salient and unambiguous signal, presented in sufficient time for the user to be able to understand it and act - that the system needs support. Or it may be that signs are clear and unambiguous that the automation has already stopped working and has passed control back to the human – hopefully, again, before anything serious happens;
B. when there is no explicit information telling the user they need to intervene, but the user detects signs or indications either from the system itself or from the world it is operating in, that leads them to make the decision that they should intervene. Either;
   a. by detecting and interpreting signals the user believes (that is, their mental model tells them) are indicators of the systems performance, they make the judgement that the system is already not performing acceptably well or is unlikely to continue performing acceptably in the immediate future without their help, or;
   b. the user detects signals from the world the system is operating in indicating threat(s) to the continued safe performance of the system. Based on this, the user makes the judgement either that the system has not detected the threat, is not planning any activity to manage it, or that the system is not capable of dealing with it without their support.

Cognitively, there is a very big difference between cases A and B in the way in which the brain leads the human to the point of acting. Let's consider these cases from the point of view of the supervisory controller's situation awareness. In case A, there is no real decision to be made: the situation is clear, and the user needs to respond to what they have detected. The intervention is based on Level 2 situation awareness: the user has perceived the signals (Level 1 SA) and interpreted them (Level 2) to mean they need to intervene. Not really a decision at all, more a recognition triggering the need to act.

Case B, by contrast, is a decision made in the presence of uncertainty. Nothing untoward has happened yet, and there is nothing to tell the user unambiguously that, unless they act, it will. But, through observation and interpretation of what is currently happening (Level 2 SA), combined with what they believe the state of the system or the world will be in the immediate future (Level 3 SA), and what (from their mental model) they believe the automation is capable of, the user makes the decision that, without their intervention, the likelihood of something happening is unacceptably high. They make the decision that there is a threat they need to respond to.

If we think about this a little further still, we realise that there are three separate judgements the user needs to make before they come to the decision to intervene;

1. Are there clear indications of a current or imminent threat that will require action?
2. If so, can the automation be relied on to deal with the threat in an acceptable way?
3. If not, do I have the authority to intervene, and, if I do, am I likely to improve the chances of a successful outcome?

These options are illustrated on the decision diagram shown on Figure 8.3.[10] What matters here is to recognise that all three of these Case B judgements are made in the presence of uncertainty. There can be uncertainty in what does and does not constitute a threat: a fast-moving vehicle entering a congested steam of traffic on a motorway from a slip road might be a threat if it's driver does not slow down to create the time and space to join the traffic flow safely. Similarly, with other than the simplest or the most highly trained and expert users, there will often be uncertainty about what the system can do; a prompt on a user interface that the system is approaching its performance limits might be a threat depending on whether the system is capable of self-adjusting to avoid the issue. Or if I'm driving in icy or particularly wet road conditions and I see a sharp bend in the road a short distance ahead, I may be unsure whether my self-driving car will be able to slow down safely in these difficult driving conditions.

Issues around authority and ability will be discussed later in this chapter. But uncertainty can arise in industrial and commercial situations where there may be strict policies about the conditions under which people are allowed to override automation. If someone is unclear about what they are allowed to do, or if they are working in a culture where anything that slows down production (such as turning off a usually highly reliable and efficient automated process) can create problems with the individuals line management or bonus payments, there may be a reluctance to step in

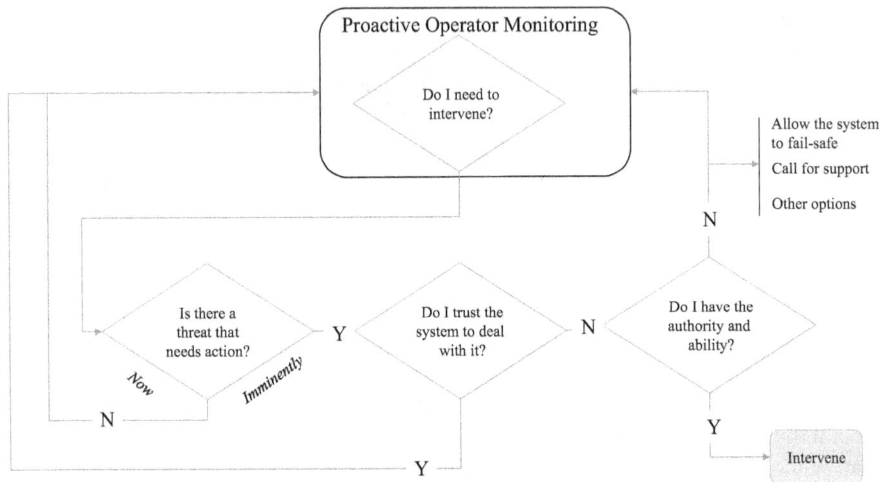

FIGURE 8.3 Decision diagram illustrating the logical judgements involved in POM decision-making

even if the user believes they should. Finally, much research has shown that lack of opportunity to practice and hone skills can. over time, cause supervisory controllers of highly reliable systems to come to lose confidence in their ability to take control.[11] They may come to the judgement that it is better to trust the system – even if they have doubts about whether it actually has the ability - than to try to intervene and find they are unable to do so effectively.

The first of the three judgements listed above is whether the supervisor has decided that there is a threat *that will require action*. That means the perceived risk associated with the threat is higher than a threshold the individual is prepared to accept. Expressing the judgement in this way recognises that threats carry different levels of implied risk, and that different people can have very different tolerances for risk. The point at which one individual decides that the risk associated with an identified threat is sufficiently high that some intervention is needed can be very different from someone else's judgement about the same threat.

To summarise, at the heart of the task of performing as a supervisory controller is the decision "do I need to intervene?" Put another way, "Is there currently a threat that requires me to intervene either right now or in the immediate future?" That decision requires the user to make several decisions or judgements, in real-time, in the presence of uncertainty. A comprehensive model of human supervisory control needs to recognise the psychological implications of having to make these real-time decisions in the presence of uncertainty.

SO WHAT IS MISSING?

So far, this chapter has reflected on the nature of the human supervisory control task, and especially the central role of the POM scan and the decisions and judgements involved in it. Based on that reflection, we are now ready to consider what needs to be added to the models discussed in Chapter 7. There are three significant topics that need to be accommodated in any model that seeks to give a satisfactory account of the psychology involved when someone fills the role of a supervisory controller;

1. The large research literature into the different styles of thinking the human brain uses when it is faced with making a decision or judgement. Specifically, that means the distinction between what are referred to as "fast" (or "System 1") and "slow" (or "System 2") thinking. And crucially it means recognising the impact the many cognitive biases associated with "fast" thinking might have on the effectiveness of human supervisory control performance, and;
2. What the psychological literature tells us about how people make decisions in the face of uncertainty.
3. The psychological implications of the distinction between where ability, accountability, control and responsibility lie at any moment. The importance of these concepts, and the use of the A2CR framework described by Frank Flemisch and his colleagues [21], was explained in Chapter 2. The issue is how an individual's understanding of the authority they have to exert control over a process and where they believe accountability and responsibility for the overall system performance lies might influence their decision making and behaviour.

Styles of thinking

In 2011, Daniel Kahneman published his book "Thinking, fast and slow" [22]. I am going to use Kahneman's overview of the more than 40 years of research that forms the basis of his description of System 1 and System 2 thinking as the basis for this section. Some psychologists take issue with Kahneman's simplification of the two systems, though for the present purpose, his descriptions are more than adequate. Academics working in very different fields including Professor James Reason – rightly famed for his groundbreaking contribution to understanding the psychology of human error [23] - and the Behavioural Economist Professor Richard Thaler [24] have described what are essentially the same ideas, though using different language and from very different theoretical and applied perspectives.[12]

Thinking and decision making (at least by non-expert decision makers) can be thought of in a simplified sense as comprising two distinct systems, or styles, of mental activity: referred to as 'System 1' (fast) and 'System 2' (slow) thinking. System 1 is fast, intuitive and efficient and works through a process known as "associative coherence". Faced with a problem, judgement or decision, System 1 draws on a near instantaneous mental network of associations in which ideas or feelings trigger other ideas and feelings. If it can find a solution that feels comfortable, or coherent, it will offer it to consciousness as a solution. System 1 is 'always on': we cannot turn it off. And it works automatically, requiring no effort or conscious control.

> System 1 provides the impressions that often turn into your beliefs and it is the source of the impulses that often become your choices and actions. It offers a tacit interpretation of what happens to you and around you, linking the present with the recent past and with expectations about the near future. It contains the model of the world that

> instantly evaluates events as normal or surprising. It is the source of your rapid and often precise intuitive judgements. And it does most of this without your conscious awareness of its activities. [22]

System 2 ("slow thinking"), by contrast, is slow, lazy and inefficient. But it is careful and rational. It takes conscious effort to turn on and it demands continuous attention: it is disrupted if attention is withdrawn. System 2 looks for evidence, reasons with it, takes the time to check, and questions assumptions. It is aware of doubt and sees ambiguity where there is more than one possible answer or interpretation of events. In a sense, System 2 is what you conclude and tell yourself consciously about what is going on, whereas System 1 is sub-conscious: you are not aware of it. So, if something arises in System 1 you may never have been conscious of it. Switching between System 1 and System 2 takes effort, especially if we are under time pressure.

Most of the time, System 1 works perfectly well. It is only because of its speed and efficiency that we can think, act and perform in ways that would simply not be possible if the brain had slowly and carefully to seek, process and think through all the information and options available to it. However, System 1 is associated with many cognitive biases, or systematic errors of judgement; it achieves its speed and efficiency by cutting corners, relying on simplifications, falling back on experience and expectations, and ignoring details that demand mental effort rather than thinking through the hard facts of what it is facing. It uses what Kahneman describes as '... *simplifying shortcuts of intuitive thinking'* [22]. System 1 does not recognize ambiguity, does not see doubt, and does not question or check. If the mental network can quickly produce an interpretation of the world or an answer to a problem that feels comfortable, it will take it;

> The measure of success for System 1 is the coherence of the story it manages to create. The amount and quality of the data on which the story is based are largely irrelevant. When information is scarce, which is a common occurrence, System 1 operates as a system for jumping to conclusions. [22]

Conclusions that, sometimes, can have disastrous consequences. As far as the performance of a human supervisory controller is concerned, it is the extent to which System 1 is prone to bias and irrationality, the tendency to jump to conclusions, and not to have doubt that has the potential to lead to unreliability.

The subject of cognitive bias first came to scientific prominence through research published by Kahneman and his colleague Amos Tversky in 1974 [25]. Since then, researchers around the world have identified and documented many systematic biases of irrational thought.[13] In several publications, I have discussed some of the potential operational implications of System 1 thinking on safety-critical front-line operations [2–4].

The scientific evidence behind the two styles of thinking is extensive and compelling. Although much of it may not be consistent with 'the-way-we-think-that-we-think' the evidence is overwhelming: everyone, irrespective of age, sex, culture, or the organizational controls surrounding how they are expected to work, must be assumed to be susceptible to System 1 thinking and of the systematic reasoning errors associated with it at some time. They don't of course apply all the time. That is inherent in the two systems. And different individuals, personality types, and even professions appear to be influenced by them in different ways. But when we are using a System 1 style of thinking, we should anticipate that our decisions will be influenced by cognitive bias and other sources of irrationality.

The law of least effort

Our knowledge of the two styles of thinking needs to be considered in association with another widely understood principle of Psychology: that people will find the easy way to get things done. The very powerful motivation we all as human beings have to make life easy for ourselves. This motivation is so strong that psychologists have even defined a law that governs it:

> The Law of Least Effort … asserts that if there are several ways of achieving the same goal, people will eventually gravitate to the least demanding course of action … laziness is built deep into our nature. [22][14]

Applying and using a system 2 style of thinking requires effort, while System 1 is always "on" and requires no effort. Unless an individual is willing to expend the energy necessary to think in a System 2 style – which the Law of Least Effort tells us is against our natural inclinations – we will tend to rely on System 1 most of the time.

That, then, is all we need to say here to summarise the two styles of thinking: a System 1 that is effortless, intuitive and always on; a system that most of the time supports efficient and reliable performance and enables expert judgement; but one that is prone to bias, irrationality and emotion, that doesn't see doubt or ambiguity and that jumps to conclusions. And a system 2 that is slow and takes conscious effort: but one that is rational, looks for evidence, doubts, questions and checks assumptions. All in the context of a human brain that actively seeks opportunities to avoid going to effort and that has a natural tendency towards finding the easy way to get things done.

That is a simplified summary of a great deal of richness and detail and of a large body of high-quality science. It is, however, sufficient for the purpose of this book. What is important is to recognize the power and speed of System 1 thinking and the effort it takes to overcome its weaknesses.

Implications of styles of thinking for a comprehensive model of supervisory control

The question then is whether what we know about the two styles of thinking, and in particular what we know about the biases, irrationality and emotion associated with 'fast' thinking, needs to be considered when we try to understand what is involved when we ask a human to perform as a supervisory controller over a highly automated system. Really, though, the question should be framed the other way around: why would it not be relevant?

As we have seen, at the heart of supervisory control is the ongoing and intermittent proactive operator monitoring (POM scanning) supporting a decision. A decision that requires a real-time assessment of the current and immediate future state of the world and of the activity being controlled, as well as the performance of the automation and how it might change in the immediate future. A decision the supervisory controller is expected to make in real-time and to revisit and review as often as is necessary. And a decision whose usefulness will start to decline from the moment it is made. That decision is whether manual intervention is needed now or is likely to be needed soon.

The actual frequency with which this decision is re-visited will of course be highly variable, depending on the nature and pace of the activities being controlled, the implications if things go wrong, competing tasks and distractions as well as differences between different people. Though what we can be reasonably sure of, certainly with a human supervisory controller who may have other things on their mind at the same time, and if they are involved with a system that is highly reliable and only rarely needs human support, is that the decision will generally be made quickly, with the expenditure of as little mental effort as possible.

Recall the context of the POM scans illustrated in Figures 8.1 and 8.2. The context was one of intermittent checks on the automated functions against a background of other responsibilities, other activities and distractions. Checks that, because they are required to be proactive, require the user to actively initiate: that is, they need to go to the effort of disrupting other activities and directing their attention towards them. And all this for a system that - at least if it is functioning within its intended limits - the user has come to expect to be highly reliable. Which means it rarely requires support from the human.

There are important characteristics of this description that unavoidably lead to the suspicion that the intermittent POM scans and decisions are frequently likely to be carried out using a system 1 style of thinking. These include;

- the need to actively make the effort to perform each POM scan and make a decision;
- the mental effort it can take to actively search for and evaluate potential threats;
- it may not be easy to make the judgement whether there any current or imminent threats: indeed, it may be difficult;
- the user's expectation that the automation can be trusted so the likelihood that there will be a need to intervene at any moment will be low.

As far as the performance of a human supervisory controller is concerned, POM scans and associated judgements and decision are likely to be made using our fast thinking, System 1 brains. Which means that the extent to which System 1 is prone to bias and irrationality, tends to jump to conclusions and not to see doubt needs to be recognised in any comprehensive model of human supervisory control performance.

This inevitably leads to the question of which of the many documented biases and irrationalities associated with System 1 thinking might influence the performance of our supervisory controller? Or, at least, which have the potential to exert a significant influence leading the human supervisory controller to the wrong decision? Trying to answer this can only be done by speculating. There is currently no solid research that explores decisions of the type, or in the context of interest here.

Wikipedia contains a well referenced summary of many of the cognitive biases that have been studied and reported in the psychological literature.[15] At least a few seem to have the potential for influencing the thought processes of a human supervisory controller faced with the task of deciding whether there is a need to intervene to support or take over control from an automated system. For example;

- *Confirmation* bias, a tendency to search for and take more account of information that confirms what we expect and already believe to be the case, rather than to consider information not consistent with what we expect or believe;
- *Commitment* a tendency to carry on with an existing course of action rather than making a change;
- *Availability*, the tendency to overestimate the likelihood of events with which are fresher or easier to remember, perhaps due to a dramatic personal experience or their emotional charge;
- *Automation bias*, a tendency to depend excessively on automated systems which can lead to erroneous automated information overriding correct decisions;
- *Anchoring*, or the *framing effect*, a tendency to draw different conclusions from the same information, depending on how that information is presented the first time it is encountered;
- The *sunk cost fallacy* where people will justify continuing to invest time and effort in an existing decision because of the cumulative prior investment made, despite evidence suggesting that the decision was probably wrong.

If confirmation bias is at work, someone who is not expecting to need to intervene based on the previous reliability of the system would seem likely to search for and favour information that suggests the automation is able to cope without support rather than looking for or believing information that suggests the opposite. And someone faced with several responsibilities, whose attention and cognitive resources have been engaged in a secondary task, might, through commitment bias, seem likely to be inclined more towards carrying on with the secondary task than interrupting it to carry out a POM scan.

Interestingly, cognitive bias has been suggested as a potential source of human error in Artificial Intelligence systems intended to support auto-contouring in radiotherapy. This is the where imaged and other data from cancer patients are fed into an AI-based system which automatically identifies organs-at-risk and delineates areas for radiotherapy planning. These systems do not work in real-time, so are fundamentally different from the task facing a human supervisory controller. Nevertheless, guidance recently prepared by the UK's Royal College of Radiologists [26] noted that two types of cognitive bias are of concern when radiologists must decide whether to accept the recommendations of the AI system: automation bias and anchoring.

Automation bias is described in the RCS report as "...a phenomenon whereby reviewers eventually favour output from the automated system, despite having evidence or knowledge that would suggest the automated system is wrong". More broadly, some people have an inherently high confidence in automation. Sometimes, people seem to have, or to develop, a tendency towards believing what technology is telling them even in the face of contradictory evidence.

There have been well documented examples, both from laboratory-based experimental studies, as well as from investigations into industrial incidents, where operators, including professional pilots, have continued to rely on automated systems despite having contradictory evidence available to them. In an important and broad review of research into Human Factors issues associated with automation,[16] published in 1997, Raja Parasuraman and Victor Riley [27] refer to concerns of the US Air Transport Association and the Federal Aviation Administration about the reluctance of pilots to take-over control from automated systems. They cite a laboratory-based experimental study using both students and professional pilots as subjects, which found that;

> ...although almost all students turned the automation off when it failed, almost half the pilots did not, even though performance on the task was substantially degraded by the failed automation and the participants were competing for awards based on performance

Another experimental study [28] found that;

> ...even after the simulated catastrophic failure of an automated engine monitoring system, participants continued to rely on the automation for some time...

Anchoring (or framing) is the tendency people have for their thinking to be unduly influenced by the first suggested solution to a problem they see. The RCS suggests that if the radiologist sees the suggestion from the AI auto-contouring system first, it may unduly influence their decision: "...the user would observe auto-contoured structures on the image data of a new patient and be unduly influenced as to their accuracy" [26]. Alternately, if they first viewed or made their own manual assessment, and only then reviewed what they AI system suggested, they may come to a different decision.

As well as the potential for POM decision making to be subject to irrational and biased thinking, there are at least two related and strong characteristics of System 1 thinking that have the potential to influence POM decision making: avoiding effort and answering simpler questions. The "Law of Least Effort", which leads us to find the easy way of doing things to avoid exerting mental effort if we can avoid it, is likely to discourage a user from going to the effort of carrying out POM scans. And when those scans are carried out, is also likely to discourage them from exerting a lot of effort in making the decision – especially if confirmation bias leads them to expect a particular answer.

Substitution

Another well-known characteristic of System 1 thinking is that, when we are faced with a difficult question or judgement, one which demands time and mental effort, System 1 will – subconsciously - tend to frame an easier question and to answer that instead. In the case of our human supervisory controller, the question we want them to answer, diligently and accurately, is; "Is there currently a threat that requires me to intervene either right now or in the immediate future?" In many situations, that might not be easy to answer: it will be difficult. So, unless System 2 is engaged, we should anticipate that the individual is likely, subconsciously, to answer a different question.

Perhaps something like: "Are there any warnings displayed". Or, if a system involves a variety of different indicators, all of which have the potential to indicate developing signs of trouble, though some are easier to see and understand (more salient and perhaps more cognitively compatible than others) they may rely on checking those that are easy to see and understand, rather than going to the effort of investigating the more demanding sources.

In summary, the conclusion that a comprehensive model of supervisory control needs to make allowance for the potential for irrationality and bias arising from the characteristics of System 1 thinking as well as a tendency to look for easy answers that avoid having to go to mental effort, seems unavoidable.

MACROCOGNITION

One final observation needs to be made about how people perform cognitive tasks in the real-world. Figure 8.3, (as well as Figures 9.2–9.5 in Chapter 9), used a decision tree-type diagram capturing the logical considerations assumed to be involved in performing the intermittent and ongoing POM scans. These diagrams are intended to represent the type of thought processes and issues likely to influence a human supervisory controller's thinking and judgement. They are not intended to imply that the actual thought processes going on in the head of a controller, whether consciously or sub-consciously, will proceed in anything like such a structured and logical way.

The simple information processing model of human cognition, illustrated on Figure 8.4, has been pervasive and useful not only in psychology but in engineering and related areas, since at least the 1970s. The model treats cognition as a kind of "black-box", using a 'brain-as-computer' metaphor where, in response to information about the world detected through sensory and perceptual processes, the brain reasons, thinks, makes associations and comes to decisions leading to action – action which may or may not be observable.

However, results and conclusions from many areas of scientific and applied research, not least the area of Naturalistic Decision Making made prominent through the work of Gary Klein and others [29], as well as emerging results from scientific research using neural imaging and other technologies, have made the limitations of the simple information processing model clear. As useful as the information processing model shown on Figure 8.4 has been - and continues to be - both in experimental psychology as well as in applied Human Factors, thinking and decision making in the real-world does not proceed in anything like the simple linear manner the model implies. When there is a need to more fully understand how the human performs complex cognitive tasks, the simple model is severely limited. Getting to grips with the psychology of human supervisory control is one of those situations when we need a deeper level of appreciation.

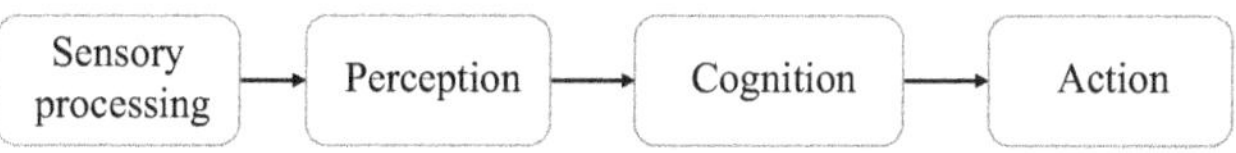

FIGURE 8.4 Simple linear information processing model

Instead, Klein and others have argued that when there is a need to understand how people perform complex cognitive tasks in the real-world, and the influence that the situation they are in can have, it is more useful and accurate to adopt what is termed a "macrocognitive" perspective. Here's an example.

Over the past five decades – stimulated by what was learned about the incident at the Three Mile Island nuclear plant in 1979[17] – the nuclear power industry has taken the potential for human error (also referred to as "human failure events") more seriously than probably any other industry. The nuclear industry has invested very significant time and resource in researching and developing methods and tools to help identify and understand the potential for human error in nuclear operations. A significant aim has been to try to quantify the probability of human error to support Probabilistic Risk Assessment (PRA) models of the risk associated with nuclear operations.

In response to a concern over the quality and consistency of Human Reliability Assessments (HRAs) conducted using different methods and by different analysts, the US Nuclear Regulatory Commission in 2006 initiated a programme of research to develop an improved HRA method that built on the strengths of existing methods, while addressing their limitations [30]. The new method was termed the Integrated Decision-Tree Human Error Analysis System (IDHEAS). IDHEAS was focused on overcoming the limitations of existing methods in understanding and assessing the complex psychological processes involved, as well as the influence of situational factors, when operators are faced with understanding what is going on in complex situations and deciding how to respond. Nowadays, nuclear operations are very highly automated and the human operators, certainly those working in control rooms, spend large amounts of their working time as supervisory controllers of the automated plant.

Based on a review of a large body of scientific literature from many disciplines, the study concluded that;

> ...an input-processing-output information processing metaphor was woefully inadequate to describe the true complexity and dynamics of human cognition. Human thought is not entirely serial or linear—a great deal of simultaneous, parallel, and circular processing occurs. Also, information-processing approaches cannot adequately account for the creativity, insight, illogical thinking, instinct, and moments of brilliance that people are prone to have … it has become clear that a macrocognitive perspective is eminently more useful. [30]

Rather than treating cognition as being a simple linear sequence, macrocognitive functions recognise that thinking and decision making, at least in complex real-world situations, if not in tightly constrained laboratory tasks, involves a great deal of parallel processing, and often occurs in a continuous loop, with feedback and overlap.

The report prepared for the US Nuclear Regulator [30] reviewed three existing models of macrocognition, as well as two other similar models. While there are minor differences, the models show a lot of commonalities. The NUREG study team adopted a model of macrocognition comprising the five functions shown on Figure 8.5.

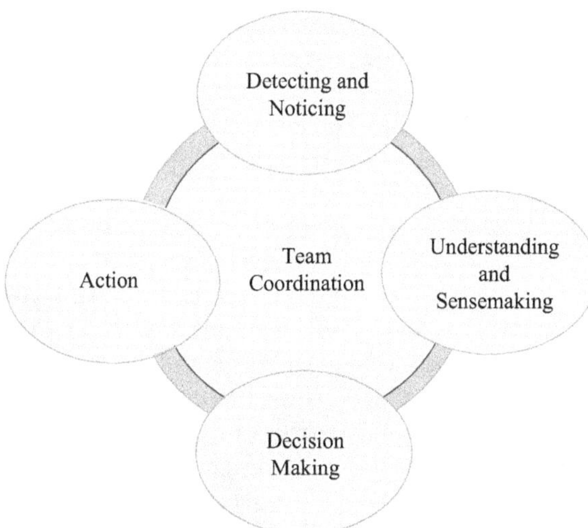

FIGURE 8.5 Macrocognitive functions adopted by NUREG (from [30])

Of these, it is the first two functions that are of most relevance to the comprehensive model of human supervisory control; (i) Detecting and Noticing, and (ii) Understanding and Sensemaking.

Detecting and Noticing

The macrocognitive function "Detecting and Noticing" integrates the functions of sensing things in the world, drawing on memory and experience to give those sensations meaning - which means perceiving them - and controlling where, when and to what the brain's limited capacity to focus attention is directed. Rather than treating sensing, perceiving and the control of attention as discrete processes carried out in series and independently from each other, treating them as a single macrocognitive function recognises the interdependencies and the fact that there is likely to be a great deal of parallel processing and interaction between them.

The outputs from Detecting and Noticing are used in the Understanding and Sensemaking function – though, again, this should not be seen as anything like a completed parcel of information being handed over for use. The processes involved in Detecting and Noticing will carry on updating and revising its products as they are used in later stages. Though a failure to detect or notice significant features in the world will of course adversely impact the ability to understand and make sense of the world.

Understanding and Sensemaking

Understanding and Sensemaking is the macrocognitive process of working out what the information that has been perceived means in terms that are relevant to ongoing or planned activities, beliefs and expectations. This might be thought of as equivalent to building situational awareness. The cognition involved ranges from the kind of effortless subconscious processes involved in System 1 thinking, to deliberate, effortful reasoning and questioning that is characteristic of System 2 thinking. Sensemaking is the basis for the doubt, questioning, challenging of assumptions and hypothesising that is required for deliberate rational thought.

The concept of sensemaking draws on many possible psychological processes and activities. It is not necessary – or perhaps even possible – to try to elaborate them for our purposes here.[18] It is sufficient to recognise that understanding what has been perceived and making sense of it in terms of ongoing real-time activities is again likely to be a repetitive and cyclical process involving a lot of parallel and distributed brain activity with feedback between processes. A process that involves trying to fit new information, perceptions and beliefs about the world into a mental awareness that makes sense and is coherent with other beliefs and perceptions.

To summarise, although the diagrams used in this chapter and Chapter 9 might superficially suggest that the mental activities involved in supervisory control proceed in a linear sequential manner, that is certainly not the reality. The diagrams should be viewed through a macrocognitive lens: representing mental activity and processes that are highly inter-dependent and likely to proceed in parallel with much iteration, interaction and feedback.

Decision making in uncertainty

In their book, "Designing for Situation Awareness" [31], Mica Endsley and Debra Jones discuss the role that uncertainty plays in influencing our ability to develop situation They discuss how uncertainty in the information available to our senses (information that is missing, unreliable, lacks credibility, is incongruent or in conflict, as well as information that may be out of date, ambiguous or 'noisy' can all affect situation awareness at Level 1 (Perception). Uncertainty associated with Level 1 SA inevitably affects our ability to develop situation awareness at Levels 2 (Comprehension) and 3 (Projection).

The second issue identified earlier as needing to be included in a comprehensive model of supervisory control was a recognition that the ongoing, though intermittent, proactive operator monitoring that is at the heart of the role depends on judgements that are likely to need to be made in the presence of uncertainty. Three separate decisions or judgements were identified;

i. whether there are clear indications of a current or imminent threat that will require action;
ii. if there are, whether the automation could be relied on to deal with the threat in an acceptable way (which might include automatically bringing the activity to a safe stop);
iii. If not, whether the supervisor feels they have the authority to intervene, and if so, whether they believe they are likely to improve the chances of a successful outcome.

The first of these is a judgement based on information the supervisor can detect through their senses - usually principally visual, though other channels can contribute – together with a judgement of whether the risk associated with the threat is sufficiently great that the individual believes it is going to need action (whether by the human or the automation) to reduce the risk to a tolerable level. The second and third of these judgements are reliant on the individual's knowledge, beliefs and experience, their trust in the automation, as well as their psychological and emotional state, including their self-confidence. There is a long history of psychological research that helps us understand how people make at least the first of these kinds of decisions in the presence of uncertainty. At the heart of this is a model of human decision making known as the Theory of Signal Detection (TSD). The central concepts from TSD have an important role to play in a comprehensive model of human supervisory control. Before looking at the implications for a comprehensive model, it is necessary to understand a little of the theory.

The Theory of Signal Detection (TSD)[19]

During the 2nd world war psychologists began investigating how to improve the performance of radar operators faced with detecting signs of approaching enemy aircraft on radar screens (the "signal", indicated by a bright, probably relatively large blob on the screen showing behaviour consistent with a moving aircraft) containing high levels of visual 'noise'.[20] The probability of there actually being an enemy aircraft at any time was typically low, though there was always a high degree of uncertainty. And of course, the implications of missing early signs of an incoming attack could be very high. This initial work stimulated much both theoretical and applied research into the same generic problem: how do operators remain vigilant when they are required to monitor radar displays over long periods for signs of infrequent, though high value signals in the presence of uncertainty and a high level of background noise? Among other things, this work led to the development of the Theory of Signal Detection (TSD) [32.33]. The theory has since been extensively research and widely applied to many situations.

TSD is based around two parameters that reflect the psychological processes involved in monitoring and responding to unlikely, though important, events in the presence of uncertainty. These are summarized graphically on Figure 8.6. The two parameters are:

- d' ("D prime") – a measure of how perceptually clear the 'signal' is; and,
- β ("Beta") – reflecting an individual's subjective bias towards or against treating perceived information as being a 'signal'.

Prime (d′) – how clear is the signal?

The concept of a 'signal' can be generalised to any significant information someone needs to search for using vision, hearing or another sense that could indicate a potential threat, with associated risk. In the case of a driver-in-charge of a vehicle in autonomous driving mode, for example, it might include information indicating risky behaviour or even potentially threatening intentions of surrounding vehicles, information that the road ahead is icy or risky, or information from the driver's user interface showing that the system has lost the ability to detect lane markers. These are all signals of potential threats to the continuing safety of the driver's own vehicle. Information that conveys signals (i.e. threats) nearly always exist in the presence

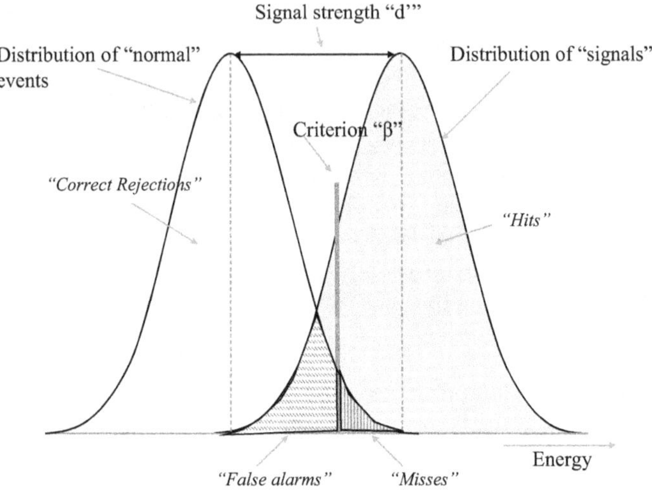

FIGURE 8.6 The elements of the Theory of Signal Detection

of information that are not "signals"; information that shows the normal (non-threatening) state of affairs: the range of normal movements and behaviours of other vehicles when they do not represent any threat, or the look of the road ahead when it is not icy or wet. In the context of TSD, these signs of normality, that are not a threat to the activity, are considered 'noise'. So, the challenge for the observer, our human supervisory controller, is to distinguish information that conveys signals (threats that may need action), from the normal situation that the automation can deal with without support.

The 'Energy' axis on Figure 8.6 indicates how perceptually clear, obvious, or easy to detect, a 'signal' is. The vertical axis is the probability of a signal of a given strength occurring. For any type of signal detection problem there is a distribution of normal – i.e. non-signal - events in the world: this is the routine statistical variation of events occurring when the world is 'normal'. The figure also illustrates a distribution of the sensory strength of 'signals' – the signs that the world is 'not normal', but that something is wrong, or is going wrong: that there is a threat.

The parameter d' is the distance between the two distributions and is an indication of how perceptually strong the 'signal' is compared with the normal background variation in the world – the 'noise'. The further the two distributions are apart, the perceptually stronger the signal is and the easier it will be to detect.

In the context of a radar operator staring at a screen, this is relatively straightforward, both conceptually and in practice: information and noise are both displayed as bright spots on the screen. Signals are indicated by combinations of their size, brightness, and behaviour over time. In the context of the human supervisory controller of a highly automated system, both the nature and the characteristics of "signals" as well as how to distinguish them from the normal state of the world, are likely to be significantly more complex, possibly vague and, at least in the early stages, difficult to define accurately. Conceptually however, the concept of d' – how easy it is to distinguish signs of potential trouble from the background of a normal world – is just as relevant to modern human supervisory control as it was to radar operators in the 2nd World War. The difficulty is in knowing exactly what information and signs the operator is expected to use to detect potential signs of trouble.

β ('Beta') - subjective bias

Just because signs of a threat are easy to detect does not in itself mean that the individual who detects them will act. That decision depends on the second TSD parameter, β, that reflects what is referred to as the individual's 'response bias'. Unlike d', which is a measure of how perceptually strong or detectable the signal is and is independent of the observer, β is subjective. It can be influenced by many factors;

- how likely the individual believes it is that a threat condition requiring action could exist at the moment of observation;
- how easily the individual can think of other explanations to explain what they have detected that don't imply there is a genuine threat;
- whether the observer believes they have the capability and the authority to act;
- whether they believe they have the skills and experience to act;
- what the individual believes would be the cost (wasted time and effort, disruption to other activities, possible financial cost, peer opinion, self-image, etc.) of raising a false alarm (i.e. acting as if a threat was detected when actually there was no threat present);
- what they perceive the benefit would be in correctly detecting and acting on threats. Benefits can be tangible as well as intangible, ranging from protecting the individual's personal safety, protecting others, or contributing to an organisation's goals and targets, to feelings of self-esteem created by demonstrating professional skill and judgement;
- irrationality rising from System 1 reasoning errors (biases).

These two parameters then, d' and β, have been understood, researched and used in applied settings by psychologists for decades. They offer a powerful means of understanding the psychology of how people make decisions in the presence of uncertainty: especially when, as in human supervisory control, they are expected to detect and respond to potential threats that are infrequent, probably unexpected, and that may be difficult to detect from the background of normal activity.

'Receiver Operating Characteristics' (ROC)

Take another look at Figure 8.6. It implies four possibilities:

i. if the perceptual signs fall to the left of where the observer sets their β they will not take any action. Most of the time these signs will belong to the distribution of normal events in the world, and the observer will have 'correctly rejected' them;
ii. in a small proportion of cases, the perceptual strength of true signals that something is wrong will overlap with the distribution of normal events.

However, because they fall below where the observer sets their β, the observer will not act on them but will treat them as part of the distribution of the normal course of events. These are 'missed' signals;

iii. if the perceptual signs fall to the right of where the observer sets their β, they would be expected to treat the event as a genuine signal and to act. In the distributions shown on Figure 8.6 in most cases these will be indications of a genuine signal, so they are considered as 'hits': the observer will have correctly responded to a genuine signal;

iv. but again, in a small proportion of cases, signs that belong to the world of normal events fall above where the observer has set their response threshold. In these cases, the observer would be expected to treat the event as a signal that something is wrong, and to intervene accordingly. These are 'false alarms': the observer acted when, in reality, there was nothing wrong.

Understanding the relationship between these four possibilities can reveal a great deal about how an individual approaches and performs a signal detection-type task. Figure 8.7 illustrates what is known as a 'Receiver Operating Characteristics' (ROC) curve. For a given strength of signal, the figure shows how the relationship between the probability of correctly detecting signals of a threat – a 'hit' - and the probability of a false alarm varies depending on the strategy or criteria the observer adopts. The strategy depends on three things: (i) how likely the observer thinks a signal is, (ii) the costs of being wrong, and (iii) the pay-off if they get it right.

Point 'A' on Figure 8.7 represents a strategy to maximise the likely of detecting a threat, while recognizing that there will be a high false alarm rate. Point B shows the opposite – when the aim is to minimise the number of false alarms, though recognising that a lot of signals of genuine threats will not be responded to. Consideration of the ROC curve shows that, if the signal strength does not change, expecting observers to detect more threats necessarily implies a higher rate of false alarms. On the other hand, for a given state of beliefs about value and probability, only a change in the

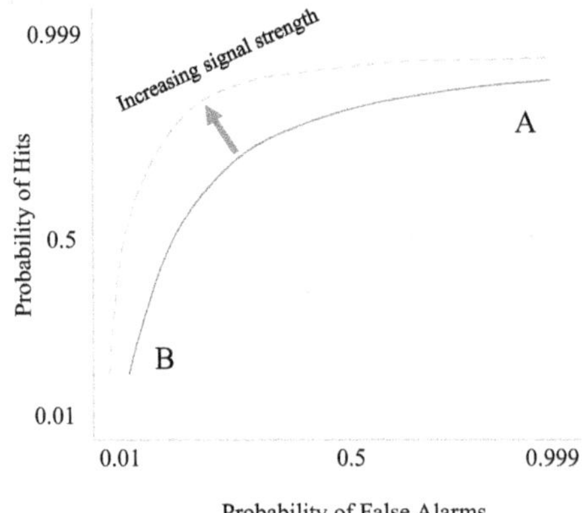

FIGURE 8.7 A Receiver Operator Characteristics (ROC) curve

strength of the signals will change the probability of detecting actual threats. The two curves on the figure illustrate that with strategy B, increasing the strength of the signal, improves the proportion of threat signals detected, without increasing the false alarm rate, while increasing signal strength with strategy A makes little difference the number of hits detected.

In summary, an ROC curve illustrates that, to increase the number of hits for a signal of a given perceptual strength, the operator MUST increase the rate of false alarms. Without going into the mathematics of the model, TSD implies that for a signal of a given strength, an observer is more likely to act as if it is a real signal if it has important consequences and if it is judged, in the experience of that observer, to be inherently probable. If the observer believes it is either unlikely or of minor importance, they will be less likely to perceive it as a signal. These subjective probabilities are estimates of value formed during training and experience.

Effective application of the principles of Human Factors in the design of highly automated systems (in fact, of any type of systems) and their user interfaces improves the ability of operators to detect and react to weak signs of trouble without increasing the rate of false alarms, by making signals perceptually strong – making d' big.

The preceding is a very much simplified account of the Theory of Signal Detection adapted to reflect the context of human supervisory control. The theory provides a great deal of sophistication and richness that can be applied to many complex issues. However, this explanation is sufficient for the purpose of this chapter.

There are clearly many overlaps between the concept of β and many of the characteristics and sources of irrationality associated with a system 1 style of thinking. For our purposes here, there is no benefit in seeking to explore the relationship between how β is established and System 1 thinking. The important point is to recognise that just because a human supervisory controller detects signs of a potential threat, does not necessarily mean they will proceed to act. While d' is objective, and a characteristic of the properties of the system (in its broadest sense), β is subjective and can be influenced by many factors. Factors both personal to the observer (such as self-confidence and experience, as well as things like fatigue, stress and anxiety), those arising from social and cultural norms, as well as working practices, incentives - and indeed disincentives- to act created by employers.

Not least, β can be influenced both by whether the individual believes they have the authority to act if they suspect they have detected a threat and by where overall responsibility for the performance of the automated system lies. Making a manual intervention in the performance of a normally highly automated system that is usually reliable and efficient can sometimes have serious adverse effects. In many industrial systems, manual intervention can increase risk or loss of control over safety, as well as disrupting production and reducing the efficiency of a production operation. So, even when a human supervisory controller does need to be in place, there can be a reluctance to give the individual the authority to manually intervene unless there is a strong case. And a manual intervention that turns out to have been unnecessary will rarely be appreciated by shareholder-oriented enterprises. Similarly, issues of where responsibility for the performance of the activity being controlled by the automation lies – or, more importantly, where it is believed to lie - can be an important driver of the behaviour of a supervisory controller.

A2CR

These latter two issues of accountability and responsibility are the subject of the final issue that needs to be recognised before we will be ready to build a comprehensive model of supervisory control. Chapter 2 included definitions of the four A2CR 'cornerstone' concepts; authority, ability, control and responsibility by referring to the work of Frank Flemisch and his colleagues [21]. The A2CR framework is based on recognition that, in other than the simplest of systems, there can be a wide range of different relationships between whether the human or the automation has the ability and authority to control a process or activity, as well as where responsibility for performance lies. That balance can change from moment to moment, depending on the system's objectives, capabilities, and the current state of the world. Inconsistency in the balance between these four concepts – for example, if the human or automation is given authority to control a function but lacks the ability - can lead to problems.

It is also important that the way the four concepts are implemented is consistent with the beliefs and expectations of the individuals filling the role of human supervisory controllers. Lack of confidence in their abilities, uncertainty about whether they have the authority to act to support or take over from the automation, or concern over their responsibilities and liabilities if they do – or do not - decide to intervene all have the potential to influence their effectiveness and reliability as a supervisory controller. For all of these reasons, the cornerstone concepts of ability, authority, control and responsibility need to be recognised in a comprehensive model of human supervisory control.

SUMMARY

The stage is now set to present a comprehensive model of supervisory control behaviour. This chapter has discussed what is missing from the existing theoretical models (which were reviewed in Chapter 7). The chapter has explored in some detail what is involved in performing POM scans, and the decisions or judgements that are at the heart of it. It has explained why a comprehensive model needs to reflect the potential impact of system 1 (fast) thinking and of both how easy is to detect signs of potential threats (d'), and the role of subjective bias (beta) in influencing whether an individual will choose to intervene if they do detect signs of trouble. And it has argued that a comprehensive model needs to be cognisant of the impact that the controller's beliefs about their abilities, the authority they have been given, and the responsibilities they hold can all have on the POM decision whether to intervene to support automation. It is now time to present the model.

NOTES

1 See for example, references [1–7].
2 Including references [8–15].
3 There is a long history of research into what is referred to as "vigilance" – continuous monitoring for rare and unexpected signals such as is required by POM. For an introduction to this literature, see [18].

4 The titles given to these three activities were published in a subsequent conference paper and use slightly different terminology from the original NASA report, which termed them Gaps, Information and Actions. See [19].

5 The meaning of the term "threat" was defined in Chapter 2. Generically, it can be thought of as anything that could lead the human supervisor to decide to intervene with the automation to maintain the integrity of the activity. Threats can be internal or external to the system being controlled.

6 As I found in my experience transitioning to owning a car that could drive itself described in Chapters 3 to 5, there can be many situations when the automation is able to cope in the sense that something I perceive as a threat would not actually disturb its performance. But it deals with it in a way that the I am not prepared to accept. In the case of my car, I have noticed many times that while the car is fully able to slow down and stop in the presence of what I see as threats such as stationery traffic ahead, or drivers cutting close in front of me in heavy traffic, it does so in a way that involves coming much closer to the other vehicles, combined with much heavier braking, than I am comfortable with.

7 This example is of course highly simplified. In the real-world, certainly in systems of any complexity, such as industrial process control rooms, the control room operator will have a much richer set of interactions with the control system than are suggested in this example. The point of the example however is to emphasise the behaviour of the human in monitoring and exerting control, as opposed to interactions that modify how control is exerted by the automation, but do not themselves exert control.

8 Of course, in the case of the driver of a car which is in autonomous mode, the driver should have no other simultaneous responsibilities beyond ensuring the safe movement of the vehicle. Though in the case of many professional drivers, there can clearly be a range of responsibilities with the potential for real-time interference with the driving task, from dealing with incoming work-related phone calls to keeping up with delivery schedules.

9 In a presentation to a meeting organised by the US National Academy of Engineering in 2017 [20], Professor Raj Rajkumar of Carnegie-Mellon University presented the intriguing question whether autonomous vehicles might be overseen by a remote operator, so that the occupant of the vehicle does not need to pay attention to its operation. This would be a high-pressure job like that of an air traffic controller, and the labour costs would be high. Such jobs may eventually migrate to countries with lower labour costs. But then the laws of physics come into play, since a delay of several tens of milliseconds introduced by communicating with a distant operator could mean the difference between a vehicle stopping or driving off the road.

10 I need to be clear that, in making this kind of decision, certainly in a real-time system such as we have been considering, the human brain is extremely unlikely to go through anything like the kind of logical decision-making process illustrated on Figure 8.3. This is discussed later in this chapter.

11 In her famous paper on the Ironies of Automation, Lisanne Bainbridge wrote in 1983: "…a formerly experienced operator who has been monitoring an automated process may now be an inexperienced one" [34].

12 In Chapter 11 of my first book [1], I included a discussion reconciling apparent differences between the way Kahneman and Reason describe these systematic reasoning errors. This was further developed in an invited paper given to the Applied Human Factors and Ergonomics conference held in Los Angeles in 2017 [4].

13 Reading through the list of published biases can give the impression that there seems to be a lack of rigour or at least a low threshold in what is considered in the scientific community to be a cognitive bias – a "systematic error of intuitive thinking" having general relevance to the human condition. It is also striking how many reported biases seem to be restricted to the domain of those involved in psychological research.

14 In [1], I included this as one of four 'Hard Truths' of Human Performance' and included two examples that illustrate how the principle can lead people to unsafe – in one case fatal – behaviours.
15 https://en.wikipedia.org/wiki/Cognitive_bias.
16 A paper they dramatically titled "Humans and Automation: Use, Misuse, Disuse, Abuse".
17 This incident was summarised in Chapter 6.
18 Though see [30] for an explanation of how NUREG chose to represent the psychological processes involved for their needs.
19 This section is a modified extract from material that appeared in Chapter 8 of [1].
20 The term 'noise' is used here in a general sense meaning sensory inputs with perceptual characteristics that can be in many ways similar to signals but are in fact un-related to the signals – the 'snow' that used to appear on a TV screen when the signal is lost.

REFERENCES

1. McLeod, R.W. Designing for Human Reliability: Human Factors Engineering for the Oil, Gas and Process Industries. Gulf Professional Publishing. 2015.
2. McLeod, R.W. Implications of Styles of Thinking for Risk Awareness and Decision Making in Safety-Critical Operations. *Cognitia*. 2016 Fall;22(3):4–8.
3. McLeod, R.W. Human Factors in Barrier Management: Hard Truths and Challenge. Process Safety and Environmental Protection. 2017. Available from: www.psep.icheme-journals.com/article/SO957-582(17)30012-5/pdf
4. McLeod, R.W. From Reason and Rasmussen to Kahneman and Thaler: Styles of Thinking and Human Reliability in High Hazard Industries. In R.L. Boring (Ed.), Advances in Human Error, Reliability, Resilience and Performance (pp. 107–119). Springer Books. 2018.
5. McLeod, R.W. & Bowie, P. Bowtie Analysis as a Prospective Risk Assessment Technique in Primary Healthcare. *Policy and Practice in Health and Safety*. 2018. DOI:10.1080/14773996.2018.1466460
6. McLeod, R.W. The Awareness of Risk, Complacency and the Normalisation of Deviance. In Process Safety Management and Human Factors: A Practitioner's Experimental Approach. 35–49. Butterworth-Heinemann. 2020.
7. McLeod, R.W. Approaches to Understanding Human Behaviour When Investigating Incidents in Academic Chemical Laboratories. *ACS Journal of Chemical Health and Safety*. 2022. DOI:10.1021/acs.chas.2c00020
8. CIEHF. Human Factors in Barrier Management. Chartered Institute for Ergonomics & Human Factors. 2016. Available from: https://ergonomics.org.uk/resource/human-factors-in-barrier-management.html
9. CIEHF. Learning from Adverse Events: A White Paper. Chartered Institute of Ergonomics & Human Factors. 2020. Available from: https://ergonomics.org.uk/resource/learning-from-adverse-events.html
10. CIEHF. Human Factors in Highly Automated Systems. Chartered Institute of Ergonomics & Human Factors. 2012. Available from: https://ergonomics.org.uk/resource/human-factors-in-highly-automated-systems-white-paper.html
11. IOGP. Human Factors Engineering in Projects. 2nd Edition. IOGP Publications Library International Oil and Gas Producers Association. Report 454. 2022.
12. IOGP. Cognitive Issues Associated with Process Safety and Environmental Incidents. IOGP Publications Library International Oil and Gas Producers Association. Report 460. 2012.
13. IOGP. Crew Resource Management for Well Operations Teams. IOGP Publications Library International Oil and Gas Producers Association. Report 501. 2014.

14. IOGP 626. Managing Fatigue in the Workplace. IOGP Publications Library International Oil and Gas Producers Association - IPEACA. Report 629. 2019.
15. Centre for Chemical Process Safety. Bowties in Risk Management: A Concept Book for Process Safety. Wiley. 2018.
16. SAE. Surface Vehicle Recommended Practice: Taxonomy and Definitions for Terms Related to Driving Automation Systems for On-Road Motor Vehicles. J3016. SAE International. 2021.
17. Mumaw, R, Billmanm, D. & Feary, M. Analysis of Pilot Monitoring Skills and a Review of Training Effectiveness. NASA/TM-220210000047. 2020 December.
18. Davis, D.R. & Parasuraman, R. The Psychology of Vigilance. Academic Press. 1982.
19. Billman, D., Mumaw, R, & Feary, M. A Model of Monitoring as Sensemaking: Application to Flight Path Management and Pilot Training. *Proceedings of the Human Factors and Ergonomics Society Annual Meeting.* 2020;64(1). DOI:10.1177/1071181320641058
20. National Academy of Engineering. Autonomy on Land and Sea and in the Air and Space. The National Academies Press. 2017.
21. Flemish, F., Heesen, M., Hesse, T., Kelsch, J, Schieben, A., & Beller, J. Towards a Dynamic Balance between Humans and Automation: Authority, Ability, Responsibility, and Control in Shared and Cooperative Control Situations. *Cognition, Technology & Work.* 2012;14:3–18.
22. Kahneman D. Thinking, Fast and Slow. Allen Lane. 2011.
23. Reason, J. Human Error. Cambridge University Press. 1991.
24. Thaler, R. Misbehaving: The Making of Behavioural Economics. Allen Lane. 2015.
25. Tversky, A. & Kahneman, D. Judgment under Uncertainty: Heuristics and Biases. *Science.* 1974 September;185(4157):1124–1131.
26. RCS. Clinical Oncology: Guidance on Auto-Contouring in Radiotherapy. Royal College of Radiologists. 2024.
27. Parasuraman, R. & Riley, V. Humans and Automation: Use, Misuse, Disuse, Abuse. *Human Factors.* 1997;37(2):230–253.
28. Miller, C., Parasuraman, R. Designing for Flexible Interaction between Humans and Automation: Delegation Interfaces for Supervisory Control. *Human Factors.* 2007;49:57–75.
29. Klein, G. Sources of Power. How People Make Decisions. MIT Press. 2017.
30. Office of Nuclear Regulatory Research. Notes on Building a Psychological Foundation for Human Reliability Analysis. NUREG-2114 INL/EXT-11-23898. 2012.
31. Endsley, M. & Jones, D. Designing for Situation Awareness: An Approach to User-Centred Design. 3rd Edition. Taylor & Francis. 2025.
32. Wikipedia. Detection Theory. Available from: https://en.wikipedia.org/wiki/Detection_theory
33. Keiser University. Signal Detection Theory: What It Is, Why It Matters, and How to Apply It. Available from: https://www.keiseruniversity.edu/signal-detection-theory/
34. Bainbridge, L. The Ironies of Automation. *Automatica.* 1983;19(6):775–779.

9 A comprehensive model of human supervisory control

The comprehensive model of supervisory control set out in this chapter recognises and builds on the key elements of Mica Endsley's HASO model described in detail in Chapter 7. It acknowledges the key role features of the system design, both the explicit user interface as well as characteristics that emerge from the automation interface paradigm, play in influencing the performance of a human supervisory controller. As with Endsley's model, it is based around the 'Big 4' cognitive constructs – situation awareness, mental models, attention allocation and trust[1] - and recognises the impact competing tasks and other demands, as well as the individual's beliefs about the situation and the context they are in at the time can have on the controller's ability both to allocate attention effectively and on the extent to which they will be prepared to intervene.

The comprehensive model builds on the 'Big 4' constructs in several ways;

- drawing on ideas in Moray's information and control model, it recognises the importance of sources of information from outside the boundaries of the system design;
- it recognises three different states of awareness that can results from a POM scan[2] that may lead a controller to interact with the automation;
- it acknowledges the impact that incentives and motivations, whether explicitly set by an employer or other body, or implicit in the individual's beliefs and values, can have on how they allocate attention to objects and tasks in the world;
- it recognises that, even if the human supervisor detects a potential threat, a number of factors can influence whether that awareness actually leads to intervening: the individual's subjective tolerance for risk; their confidence in the abilities of the automation to cope; whether they believe they have the authority to intervene and the ability to do so effectively; and whether they believe they are likely to be held responsible for any adverse consequences if they do or do not intervene;
- it captures the impact of the individual's beliefs both about the cost of intervening, as well as the potential benefits if they do intervene, and;
- it suggests elements that could potentially be prone to the irrationality and biases associated with system 1 thinking.

DOI: 10.1201/9781003679479-11

THINKING IS NOT LIKE THE DIAGRAMS

The figures that illustrate the comprehensive model in this chapter, set out a logical flow of information using decision tree and other types of elements. Superficially, it might be concluded from those figures that the model suggests the cognitive processes behind human supervisory control involve the same kind of logical flow. Such a conclusion would not only be superficial: it would be wrong.

In industries such as medicine, defence, finance and many others, professionals increasingly use automated decision support systems, based on machine-learning, natural language models and other AI technologies, to interpret, make sense of and recommend action based on complex and multi-dimensional data sets. The role of the human in these kind of systems is to draw on the recommendations made by the automation to decide how to proceed. These kinds of decision support systems do not operate in real-time. When decisions are especially important, the process of making them is often required to follow formal and explicit processes, to use a structured logical approach, and to document the reasoning. A process that is consistent with a System 2 type of thinking.[3]

Human supervisory control, and the POM decisions that are at the heart of it, does not proceed in anything like such a formalised way. As Figures 8.1 and 8.2 in Chapter 8 illustrated, the continual need to monitor what is going on, to make decisions about whether there is a threat at the time a POM scan is carried out, and, if so, whether to intervene, are made in real-time, probably largely sub-consciously, and with no explicit or formalised process. Chapter 8 also emphasised that the cognition involved in performing real-world tasks is often better thought of in terms of *macrocognitive* processes: many different mental processes working in parallel, iteratively and with much interaction and feedback between them, to perform higher level (macro) cognitive tasks. Furthermore, it seems unavoidable that a human overseeing the control of real-time activities who has other simultaneous responsibilities is likely to conduct POM scans using a System 1-type thought process. System 1 thinking works through the association of ideas[4]: a near instantaneous mental network of association in which ideas or feelings trigger other ideas or feelings. If the network quickly produces an interpretation for what it is experiencing, or an answer that feels comfortable, it will take it. That is how the cognitive processes represented by the figures in this chapter should be understood. The figures are intended to illustrate the components of the comprehensive model and to identify the type of processes and issues that could influence a human supervisory controller's thinking and judgement. The figures are not intended to suggest that the actual thought processes going on in the head of a controller, whether consciously or sub-consciously, proceed in anything like such a structured and logical way.

The real purpose of the model, and the figures that illustrates it, is to draw attention to features that are potentially under the control of those with influence or responsibility for the design, development, procurement, manufacture, licencing, purchasing, use or support of highly automated systems that rely on people filling the role of a supervisory controller. Those features range from the design of user interfaces and other support systems, aspects of the culture, and working arrangements in organisations that deploy highly automated systems, to the personality, perceptual and cognitive capacities and appetite for risk of the people who may be authorised to use or support those systems.

A word on KISSing

The comprehensive model presented in this chapter is more complex than any of the models described in Chapter 7.[5] When I first concluded that even Endsley's HASO model lacked important features needed to satisfactorily understand the complexity of human supervisory control, I realised I was about to break a principle that had almost become sacrosanct in much of the applied Human Factors community. A principle that had guided so much of the work I and numerous others working across many industries had done or contributed to over many years; preparing standards and guidance as well as tools and methods to help non-specialists understand and apply some of the principles of Human Factors in their work without having to rely on calling in specialists. The principle to "Keep It Simple Stupid" (KISS).

Human Factors has often struggled to be taken seriously as a professional discipline with something unique and of value to offer. There can be many reasons for this resistance. Probably the most common being a desire among those responsible for budgets to avoid committing funding to a topic they are not convinced adds value. Sometimes they equate Human Factors with "common sense" and, therefore, take the view that you don't need special qualifications or experience to do it. Sometimes there is a lack of clarity about the differences between the subject matter Human Factors professionals engage with and expect to have responsibility for, and the competencies other more established professions or disciplines provide. Obvious examples being the application of ergonomics in the layout of working environments, where competent architects, designers and other disciplines frequently have overlapping skill sets. Or the design of user interfaces to computer systems, where computer engineers, graphic designers and others can have, or believe they have, equivalent skills.

Company boards and executive committees sometimes approve projects with budgets and delivery schedules that constrain their directors and managers even if they do recognise the importance of Human Factors. Schedules and budgets are set that make it impossible to comply fully with engineering standards and best practices that require Human Factors effort. So corners are cut. Human Factors effort is frequently on the wrong side of those cuts.

Faced with this resistance, the importance of keeping our explanations, guidance, tools, and methods simple if we want to have any hope that professionals not trained in Human Factors are going to be persuaded that the discipline had something to offer has become sacrosanct. To give just a few examples;

- The simple information processing models of how the human brain processes information and makes decisions (i.e. Detect, Interpret, Decide, Act). While this has provided a useful framework for thinking about the nature of human tasks (and indeed, is used in this book), it is not anything like an accurate description of how experienced people actually perform skilled work. The reality is much more complex. Reliance on such a simple model frequently leads engineers and others to under-estimate the complexity of human performance, leading to overly simplistic assumptions about human behaviour and the likelihood of people making mistakes.

- Simple analysis tools that attempt to quantify the likelihood of a human making an error performing a task. Many engineers have been led into performing calculations of the form: if a task comprises three sub-tasks that must be performed in sequence, and the probability of an error performing each sub-task independently is estimated as being 1 in 10 (i.e. 0.1), then the probability of all three tasks being performed incorrectly is 1 in 1000 (0.001). Such simplistic reasoning ignores the complexity of human performance. In this case, both the potential for the first errors to be identified and corrected, as well as the fact that task steps are frequently not independent: if an error is made on one task in a sequence, the chances an error on subsequent tasks can often be significantly higher.
- The Generic Error Modelling System (GEMS) developed by Professor James Reason [1]. GEMS provides a conceptual framework for understanding human failures based around three generic error types; skill-based slips and lapses, rule-based mistakes, and knowledge-based mistakes. Anyone who takes the trouble to understand Reason's explanation of the model will appreciate the depth of psychological complexity associated with it. Applied with skill and understanding, the model can provide great insight not only into why human performance may have failed in the way it did on a specific occasion, but what to do to prevent failures occurring in similar circumstances in future. Unfortunately, with the simplifications of the GEMS model in widespread use, it is common for non-specialists to apply the labels of the generic error types to classify the kind of errors they have observed. Few however, other than some specialists, properly appreciate the complexity of the model or understand what the error types they have identified mean or imply.

This is a problem that has plagued the Human Factors discipline for a long time. Human Factors professionals often feel under pressure to try to make the technical basis of what they do easy to understand by people from different disciplines. With the result that they over-simplify what are inherently complex matters. This is done in good faith, seeking to make the content understandable and useable by people who lack subject specific training and experience. The problem is that over-simplifying what are inherently (often psychologically) complex topics not only gives a misleading impression that an issue is understood but leads to solutions and actions that are not effective in resolving the issue. As the American journalist Henry Louis Mencken put it; "For every problem there is a solution which is simple, clean and wrong".

So, on the one hand I have a concern that suggesting a more comprehensive model of the psychology behind human supervisory control is a move against the need to keep things simple (stupid). On the other hand, the topic is sufficiently important and psychologically complex that it demands a more complex model to adequately describe it. I am encouraged and guided by another famous quote, this time from Albert Einstein; "Everything should be made as simple as possible, but not simpler."

I have tried to ensure my comprehensive model meets Einstein's challenge.

THE COMPREHENSIVE MODEL

The comprehensive model is based on eight types of elements, most of which are assumed to exist or take place inside the head of the human acting as a supervisory controller.[6] The elements included are;

i. states of mind, knowledge, or beliefs;
ii. mental states, such as fatigue, anxiety, stress, or strong emotions;
iii. decisions or judgements that demand real-time attention and cognitive effort;
iv. outputs from situation awareness that reach consciousness about the state of the automation or the wider world;
v. perceptual-motor activity that requires allocation of real-time attention and cognitive resource;
vi. a process controlling the allocation of attention across different sources of information in real-time;
vii. the human's mental model of how the automation works, its' dynamics, abilities, reliability, limits, and so on;
viii. the human's mental model of the world the automation operates in, particularly key objects and actors that that might interact or interfere with the performance of the automation.

The model also recognises the influence that elements that are not a part of real-time cognition can have on the decisions and judgements involved in human supervisory control as well as locations where irrationality and cognitive bias might operate. Figure 9.1 shows the key to these different types of elements included in the comprehensive model.

An individual's situation awareness at any moment will contain representations of many things in the world, at varying levels of detail and accuracy: the relative physical

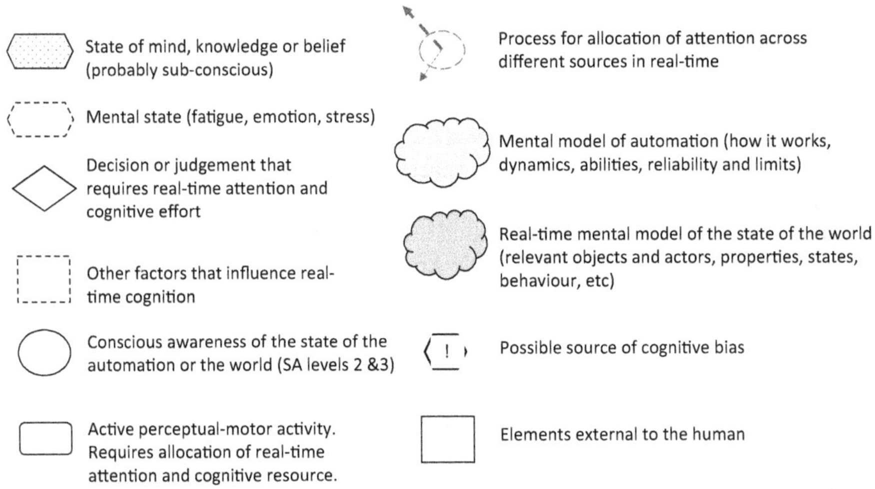

State of mind, knowledge or belief (probably sub-conscious)

Mental state (fatigue, emotion, stress)

Decision or judgement that requires real-time attention and cognitive effort

Other factors that influence real-time cognition

Conscious awareness of the state of the automation or the world (SA levels 2 &3)

Active perceptual-motor activity. Requires allocation of real-time attention and cognitive resource.

Process for allocation of attention across different sources in real-time

Mental model of automation (how it works, dynamics, abilities, reliability and limits)

Real-time mental model of the state of the world (relevant objects and actors, properties, states, behaviour, etc)

!

Possible source of cognitive bias

Elements external to the human

FIGURE 9.1 Key to shapes used on comprehensive model on Figures 9.2–9.6

locations of different objects, their properties, current and anticipated behaviour, and possibly the emotional impact they have on the observer, among many other things. Some of these (a relatively small number, probably bounded by George Miller's famous dictum that memory can only hold 7+/-2 "chunks" of information at any one time) will exist in conscious memory (they will be activated in one of the elements of Working Memory). Others may exist in a pre-cognitive state.

From the point of view of the behaviour of a supervisory controller, the comprehensive model identifies at least six judgements involved in a POM scan that contribute to overall situation awareness of the current and anticipated state of the automation and the process or activity under control;

i. how well is the automation currently performing (is the activity proceeding within acceptable limits?
ii. how close is the activity to the limits of the automation's abilities?
iii. how close it is to the limits of the automation's authority[7]?
iv. is the automation aware of current threats known to the human?
v. if so, is the automation showing an intention to deal with those threats?
vi. are there objects, actors or events in the external world that have the potential to become threats to the successful performance of the controlled activity?

Awareness includes beliefs and knowledge about the properties, limits and abilities of the automation. That includes beliefs that reflect the individual's confidence in the system and knowledge about the limits of their own authority. Knowledge can be declarative, procedural or implicit. In the context of real-time human supervisory control, these beliefs and knowledge may not enter conscious thought: if asked, the individual may not be able to express the belief or knowledge in the same way as it was taken into account during what may be the rapid, sub-conscious performance of a POM scan.

Figure 9.2 provides an overview of the four core processes included in the model;

1. Automation Situation Awareness;
2. Attention;
3. Threat Assessment and Monitoring, and;
4. The decision whether to Intervene.

Figures 9.3–9.6 provide expanded details of the content of each of these core processes.

Attention: the POM scan

Figure 9.3 shows the detail of the Attention process. Being proactive in conducting a POM scan requires the individual to *actively* direct their attention towards potentially relevant sources of information. The success and timing of that active attention switching should be stimulated by declining confidence over time (which may be experienced as a growing sense of unease) in the value of information gained from the preceding scan – represented in the user's mental model - about the current

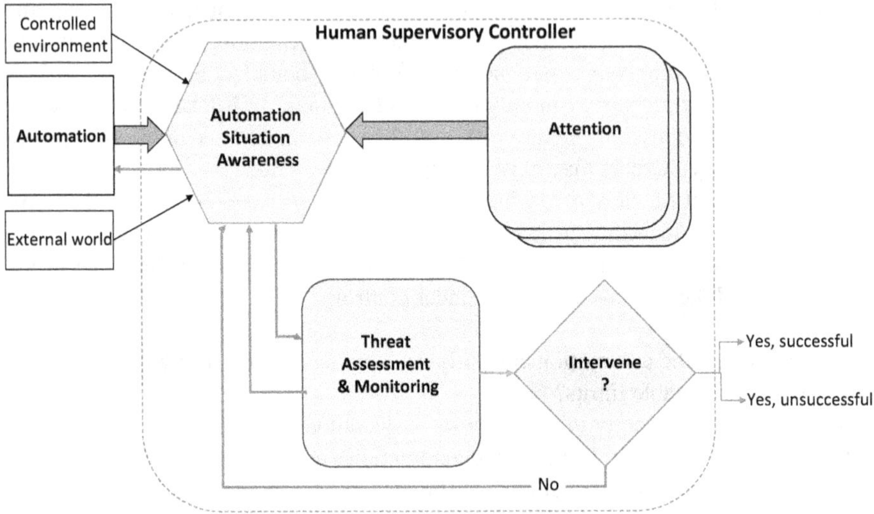

FIGURE 9.2 Top level overview of a comprehensive model of human supervisory control

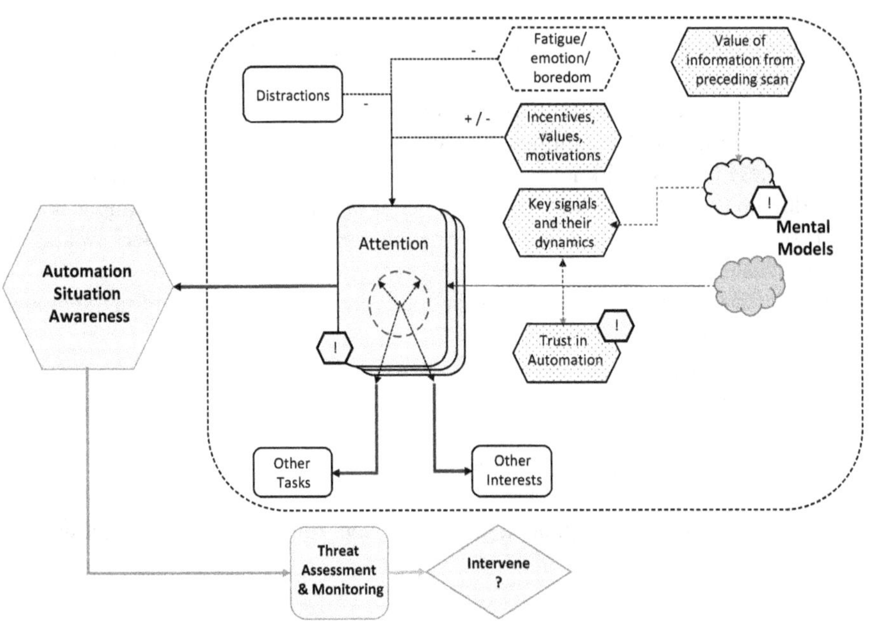

FIGURE 9.3 Details of attention process

situation. The rate of declining confidence in that information depends not only on how quickly things change in the real-world, and on how accurately the mental model represents the dynamics of the process being controlled, but also on how much trust the individual has in the automation. The initiative and willingness to actively perform a POM scan will also be affected by the extent to which the individual is mentally engaged in other things. And of course, distractions are likely to interfere with actively monitoring the automation.

Actively switching attention to check the state of the automation requires mental effort. The willingness to go to that effort can be influenced – both positively and negatively - by incentives and motivations that may either be internal to the individual (a desire not to cause an accident) or external, set by another body (such as a company bonus scheme that penalises shutting down automated processes). Boredom, fatigue and even an individual's emotional state may all act to reduce an individual's willingness and ability to direct attention to the automation actively. And as was discussed in Chapter 8, humans have a deep motivation to avoid going to mental effort if they can avoid it (the "Law of Least Effort").

AUTOMATION SITUATION AWARENESS

Figure 9.4 shows expanded detail of the process Automation Situation Awareness. When attention is allocated to performing a POM scan, there are likely to be up to three possible sources of relevant information the user needs to focus attention on. These are indicated on the left of the figure;

1. Information explicitly represented in the automation user interface, including alarms, prompts and other indicators intended to capture the human's attention.
2. Information about the state of the environment the controlled process is operating in. Some of this will be explicitly represented in the user interface, but others will not, so must be monitored directly, out with the user interface. In some situations, this kind of information may be available from third parties.

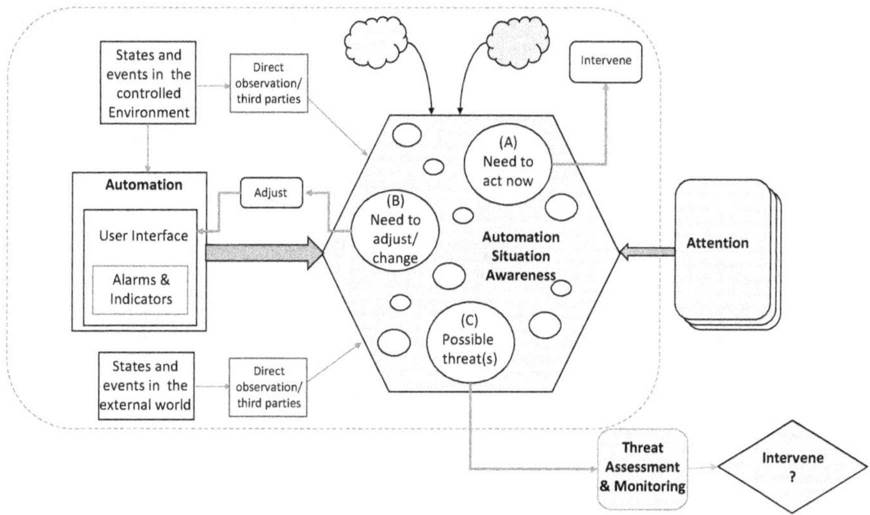

FIGURE 9.4 Details of Automation situation awareness process

3. Information about other objects and actors in the external world that have the potential to become threats. These are not explicitly represented in the user interface, so must be monitored directly or via third parties.

How effective the individual is at locating and understanding information from these sources will be influenced partly by the individual's mental models of the behaviour, properties and capabilities of both the automation and the world, as well as by their ongoing real-time awareness of the state of the activity. Figure 9.4 implies that in most (at least reasonably well designed) systems, information represented in the user interface is likely to dominate as the source of the automation situation awareness generated by POM scans.

Figure 9.4 illustrates the three outcomes[8] of POM scanning that could lead the controller to actively interact with the automation user interface;

A. detecting signs showing that the automation has failed or is asking for human input;
B. recognition of a need to amend the automation's goals or constraints, change the information available in the interface, or how it is presented;
C. a belief that there is one or more threat(s) that either is currently causing the automation not to perform effectively or has the potential to cause a problem but that the system either is not showing awareness of and/or is not demonstrating any intention to take action to manage.

The first two of these can be expected to lead to action directly. In the first case, there is no decision to be made: the user must only detect that the automation has failed or is explicitly asking for support. Having detected that, they need to step in and support or take over control from the automation; they need to take on the role of being a manual, instead of a supervisory, controller. Numerous studies and incident investigations across different industries, have demonstrated that this second case is a situation that can be full of risk. Even skilled and highly trained people can be overwhelmed and unable to respond successfully when they are suddenly and unexpectedly required to take over manual control of an activity when automation that is usually reliable suddenly stops working. This is a complex situation, usually involving loss of situation awareness both of the state of the activity being controlled as well as what the automation is doing and why, and perhaps a loss of skill and self-confidence. And there can be many compounding issues, not least user interfaces and training that have not considered the reality of what a human is going to need in those moments of sudden and unexpected transition from being a human supervisor to a manual controller. One of the most disastrous and dramatic examples of a sudden loss of supervisory control occurred on 1st June 2009, when Air France Airbus A330-230, flight AF 447 crashed into the Atlantic whilst en-route from Rio-de Janeiro to Paris with the loss of all 228 passengers and crew on-board.[9]

This book however is concerned with understanding the psychology of the role of the human supervisory controller. At the point where action is needed, whether because the automation has failed, or because the human has decided to intervene, the human is no longer a supervisory controller: they have reverted to being (hopefully) a manual controller. As such, the Human Factors and psychology of what is needed to ensure the human has a good chance of responding successfully in these situations is beyond the scope of this book.

In the second case leading the human supervisor to act, they only need to make some adjustment and do not need to take over control of the activity. They do not need to become a manual controller. But they do need to interact with the user interface under conscious attentional control.

The third outcome of interest from POM scanning is the awareness that there is a possible threat. There remains a lot of psychology between forming a belief that there is a potential threat and deciding to intervene (or not). As was discussed in Chapter 8, this third decision is often likely to need to be made in the presence of uncertainty.

How clear is the signal?

Signs that lead the human supervisory controller to suspect a threat are often going to exist in the presence of uncertainty. Uncertainty whether what seems to be a threat might simply be the normal variation of behaviour of elements in the controlled or external world: something that falls within the range of events and behaviours the controlled system has been designed to deal with. The issue of interest here, and the one that developers and users of highly automated systems should be interested in, is how to minimise the uncertainty: how to make the signs of potential threats as clear and obvious as they can be.

It can be helpful to think about this in terms of one of the parameters at the heart of the Theory of Signal Detection (TSD).[10] Conceptually, the TSD parameter d' ("D prime") reflects how clear the signs are that an observed feature actually indicates a state or event that requires action as opposed to reflecting the normal variability of the observed world.[11] Or in the case of our human supervisory controller, how clear are the signs that what has been detected is actually an indication of a possible threat that the automation is going to need human support to deal with? Put simply, the challenge for those who develop and deploy highly automated systems, is how to make d' for threats as big as possible?

Trying to answer this challenge needs to start by being clear about three things;

1. Exactly what events or conditions constitutes threats that may require human intervention?
2. What are the indicators that would allow a supervisory controller to detect those events or conditions?
3. What is needed in design and otherwise to be confident those indicators will be noticed and understood in sufficient time for action?

What are the threats?

Answering these questions can be surprisingly difficult. Let's start with the question of what constitutes a threat that may require human intervention. This itself comprises two more detailed questions:

a. what, in reality, are the things the automation is unable to deal with without human intervention, and;
b. what does the human supervisor believe those things to be?

We might hope that those responsible for designing and developing automated control systems will know what events and circumstances represent threats to the performance of the automation. And that they will provide effective solutions to keep the human informed in a timely manner when those threats are detected. Unfortunately, we cannot always be confident that will be the case.

A project team that designs and develops automated control systems should have a clear understanding of the limits of the system's capabilities, both in terms of its functionality, and therefore abilities, as well the states of the world it is capable of operating in. In the case of automated vehicles, this is what the Society of Automotive Engineers (SAE) in publication J3016, its taxonomy of levels of automated driving [2], defines as the "Operational Design Domain (ODD)": "Operating conditions under which a given driving automation system or feature thereof is specifically designed to function, including, but not limited to, environmental, geographical, and time-of-day restrictions, and/or the requisite presence or absence of certain traffic or roadway characteristics". Other industry bodies use different, though similar terms, though ODD is sufficiently widely understood that I will use it here.

As long as the ODD is specified sufficiently clearly, it seems that it should be a relatively straightforward matter for those who develop automated systems to identify conditions and their signifiers that would tell the human when the limits of the ODD are being approached. As well, hopefully as how quickly they are being reached, and what is expected of the human supervisor if the limits are reached. Though what we really want is for the human supervisor to know about approaching limits not only before they are reached, but *in sufficient time* for them to pay attention, orient themselves, and be ready to intervene if it turns out they are needed.

Many current systems tell their users when the limits of the ODD are reached, though whether the human is given that information sufficiently early seems more of a challenge. Modern industrial process control systems provide control room operators with the ability to display and monitor graphics showing the historical trend of process parameters (temperature, pressure, etc) over time.[12] Such graphics generally give clear indications of the normal operating state, as well as showing high and low operating limits. So the human can see how parameters are trending over time, and whether, and if so, how quickly, they are changing relative to acceptable operating limits. Similar types of displays are commonly used in complex IT infrastructures to allow system managers to track performance metrics such as response times, CPU usage, and application behaviour to help detect issues early. In the case of automated driver assist features such as Lane Keeping in road vehicles, on the other hand, the vehicle will generally have a display showing whether it is able to detect lane markings.[13] The human can see when lane markings are being detected and the system is working. Though if the automation loses track of the lane markings, the system will instantly stop working with a simultaneous change in the drivers display – though with no advance warning.[14]

The second part of the question of knowing what constitutes a threat is understanding what the human supervisory controller *believes* are threats. And therefore, what they are likely to pay attention to and search for during their POM scans. This can be a very different answer to the actual threats understood by the system's developers. And the answer will almost certainly change as the human progresses through the process of transitioning from being someone who manually controls an activity to being a supervisor of an automated system. A driver, such as me,[15] who, with no prior training or knowledge of the system design features, is suddenly put in the position of supervising a vehicle that can drive itself, initially sees threats everywhere: vehicles ahead, to the side and overtaking, curves in the road, wet weather, fog, speed limit signs, and so on. Indeed, it took concentration, attention, and trust in the vehicle manufacturers for me to turn the system on at all. Over time however, as knowledge and experience of the system's capabilities grew, the nature of what I believed to be threats became clearer. I gained confidence in things like the car's ability to hold its position in a lane while maintaining a safe separation from the vehicle in front, it's braking and cornering abilities and so on. So I felt able to relax, at least a little, the degree of attention I gave to monitoring for those issues.[16]

The point of this discussion is that the proactive operator monitoring (the POM scanning) that is the core of the human supervisory control task will be driven largely by what the individual *believes* to be threats to the successful performance of the automation. These are not necessarily the same as the threats those responsible for designing and developing the system identified. That means effort is needed both on the part of system developers as well as by those who deploy or sell highly automated systems to close the gap. To ensure the human has an informed and adequate understanding of what constitutes threats to the automation's performance that can drive their POM scanning behaviour.

Making threat indicators as clear as possible

The third question in seeking to make d' as big as possible, is how to ensure the signs that indicate the existence of potential threats are easy to detect and understand? Clearly there can be no generic answer to this. It will depend on the nature of the system, as well as the users and the context in which the information is used. Following best practice in designing user interfaces, in particular to support situation awareness will always be important.[17] But understanding some general principles can help.

First, this issue goes beyond simply ensuring the user interface to the automation is well designed, clear and "easy to use" according to usability style guides or other sources of guidance. It must recognise that indicators of potential threats may need to be attended to and understood in the context of intermittent and time-limited POM scans when the human is likely to be engaged in other activities not to do with the automation.

Second, it needs to be clear that, as the aim is to support proactive operator monitoring, it is not simply a case of relying on alarms or other system-generated means of drawing the user's attention to the information. That would be to rely on the human behaving reactively. Alarms and other forms of alerting have their place: but this is not it.

Third, in a system of any complexity, signs of threats are unlikely to be reflected exclusively in information contained in the user interface. The user is likely to have to direct their attention to a variety of other sources. Figure 9.4 shows two other sources; those that provide information about the state of the environment the automation is working in that is not represented in the user interface; and sources of information from the world external world about threats that have the potential to interfere with or interact with the automation or the controlled environment.

In his model of the range of the distribution of information and control features associated with supervisory control behaviour, Neville Moray identified that, sometimes, the information the human needs is simply not available to them.[18] And sometimes information might be available, but the user is not aware of it or does not understand what it means. Moray's information model included;

- information not reflected in the user interface, but directly available from other sources;
- Information available via the user interface, but that the human has only partial awareness or understanding of;
- information reflected in the user interface, but that the user has no awareness of or does not understand;
- information the user is not aware of, but that could be accessed from other sources;
- information the user does not know about, and that is not fully represented in the user interface;
- information that is not available to the user through any source;
- information that is only available to the user from prior knowledge, training, or experience.

All of these raise issues that those who develop highly automated systems, as well as those who deploy and employ people to support them can and should be aware of. Issues that could be used, at least, when assessing the quality of support they are providing for the human supervisory control task.

The core issues in seeking to make d' associated with threats as big as possible are knowing;

i. what information the user needs to be able reliably to detect and understand the implications of potential threats;
ii. how much of that information can be provided in the user interface, and how that information should be presented to ensure it is both effective in capturing the human's attention and compatible with how the brain is likely to use it during intermittent POM scans, and;
iii. where else the information is likely to exist other than in the system's user interface.

The first of these can only be satisfied if those responsible for the system design go to the effort of properly analysing the role of the human as a supervisory controller of the system they intend to develop and deploy. That analysis must include consideration of the context in which the system will be used. Conducting user trials and usability tests in a laboratory or system development environment will be of little value if the system is to be deployed in an industrial operation by people busy with other responsibilities while simultaneously performing the role of a supervisory controller. Similarly, basing design decisions on the feedback of people who understand the system and are given dedicated time to use it, will have little value if the system is going to be supported by people given little training or support.

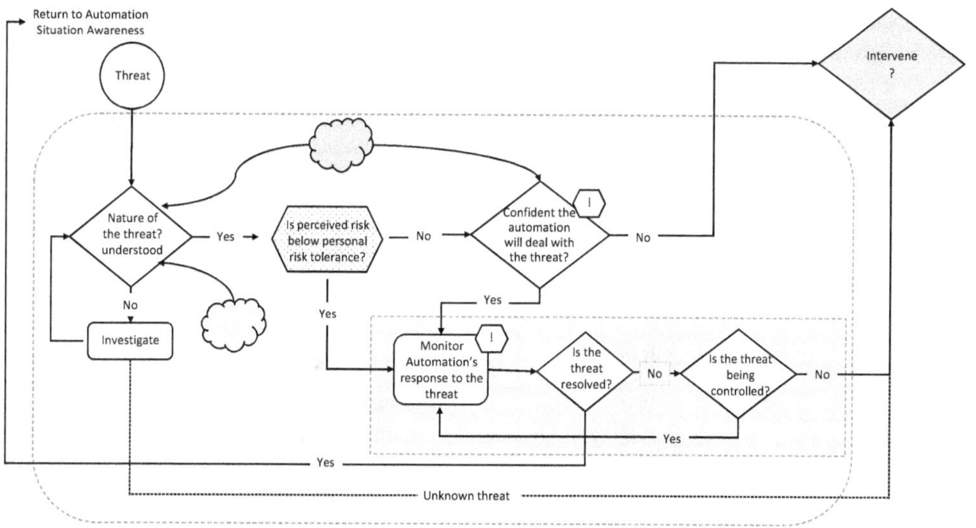

FIGURE 9.5 Detail of the threat assessment and monitoring process

Assessing the threat

What is likely to happen when a POM scan leads the human supervisor to suspect there is a possible threat currently causing the automation, while not actually failing, not to perform effectively? Or, when the human supervisor detects a threat that the system is not showing awareness of, or not showing any intention to take action to manage? This is the domain of the third process in the comprehensive model: Threat Assessment and Monitoring. Figure 9.5 illustrates the key cognitive elements and processes likely to be involved.

Bear in mind that elements are included in the model because they suggest the possibility of intervention, either in design, or at some stage in deployment or use of the automation. Intervention that could help optimise the chances of the human supervisory controller being as effective as they can be in the role.

The first step in assessing whether the suspected threat demands intervention is to decide, drawing on the individual's mental models of the automation and the world, whether the nature of the threat is adequately understood. If not, some attempt is likely to be made to try to understand it better. That could involve simply observing how it behaves or develops over time. Or it could mean looking for more information, perhaps from another source, to support what is known. Though even if the supervisor still isn't clear what the threat is, in a real-time system they will, nevertheless, need to move towards deciding whether they are going to need to intervene.

The supervisory controller is only likely to be stimulated towards action if they assess the risk associated with the threat as exceeding a level they are subjectively comfortable with; i.e. their subjective risk tolerance level. This is not of course in any sense an objective assessment; it may be no more than a sense of ease or unease or discomfort. If the perceived risk is lower than a level the user is comfortable with, they are still likely to continue monitoring and actively paying attention to the threat until either they are comfortable it has passed, or they decide they need to act: after all, it was a sufficient threat to capture conscious awareness, so is unlikely simply to be dismissed.

If, however, the sense of risk associated with the suspected threat exceeds the individuals subjective risk tolerance, the next decision is whether they are confident that the automation will manage the threat. This draws on the individual's mental model of the capabilities of the automation, including the extent of trust and confidence, they have developed in its abilities. If they are confident in the automation, they are again likely to keep an eye on the issue until either it has been resolved, or they are stimulated towards intervening.

Deciding whether to intervene

Figure 9.6 illustrates details of the Intervene? process. Faced with a threat they are not confident the automation can, or is, managing, a number of other considerations with the potential to influence the decision to intervene need to be recognised. The figure shows two main streams of thought. The first assumes the individual believes they know what action is needed. Though whether they intervene depends on whether they believe they have the authority and the ability to do so, and who they believe is responsible for any action taken.[19] The second stream of thought is to do with the individual's subjective bias for or against acting, which is influenced by their belief and perceptions of the costs involved in acting and the benefits likely to be achieved.[20]

Of course, it is possible that an individual may choose to intervene even if they don't believe they have the authority: they may decide, based on things like their wider morals, value system and what they believe their peers or colleagues would expect of them, that, in the circumstances of the moment, they should intervene rather than allow the alternative to happen. Similarly, they may decide, after all of this thought and deliberation - which could well have taken place over no more than

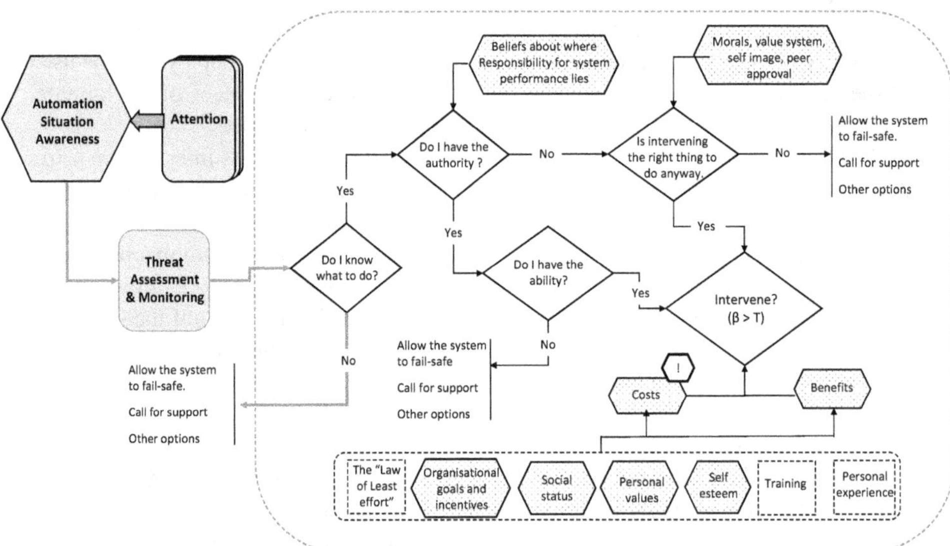

FIGURE 9.6 Details of the intervene? Decision process

a few moments – that, despite knowing that the automation is not coping, they still should not intervene. There may be other options available depending on the circumstances and potential consequences; from simply allowing the system to fail, allowing it to fail-safe, calling for assistance, or some other option.

COGNITIVE BIAS IN THE COMPREHENSIVE MODEL

Chapter 8 argued that where people are busy with simultaneous responsibilities other than monitoring an automated system, it must be assumed that the thinking, judgements and decision making behind POM scans are likely to be conducted using a "fast", or System 1, style of thinking. Which means that at least some of the many known types of irrationality and bias associated with fast thinking should be expected to influence the performance of the human supervisory controller.

TABLE 9.1
Possible elements of the comprehensive model of human supervisory control behaviour likely to be subject to cognitive bias

Element	Example
Mental models	Many sources of bias could influence the beliefs held in an individual's mental models. These could affect the way the mental model represents the content, characteristics, behaviour and properties of; (a) the controlled system including the automation, and; (b) the external world. Biases affecting the content of mental models would affect many of the thought processes, beliefs and judgements underlying supervisory control performance, including how information is interpreted. Biases might also impact the supervisor's ability to imagine adverse consequences if they did or did not intervene.
Attention	Confirmation bias would be likely to lead an individual to focus too much attention on sources where they expect to find information about anticipated threats, at the expense of other sources. It would also cause them to search for confirming information at the expense of information that was not consistent with what they believed was the situation.
Trust and confidence	Automation bias is known lead some people to have an unduly high level of trust in automation. Confirmation and Automation biases are likely to lead an individual to look for signs showing the automation is performing successfully, even when there is evidence it is not.
Threat assessment	Availability bias could lead an individual to have unduly high, or unduly low, confidence in the ability of automation to deal with a threat if they find it easy to think about a similar situation where automation did or did not cope with a similar threat.
Perceived costs and benefits	Prospect Theory[a] and Framing could lead an individual's decision to intervene to focus excessively on actions that would avoid what they perceive to be a loss, at the expense of what they perceive to be a gain even when there is no actual difference in value between the two actions.

[a] Prospect Theory states that people will invest twice as much effort trying to avoid what they perceive to be a loss, as they will to achieve what is perceived as a gain. Their work on Prospect Theory was cited as among the reasons Amos Tversky and Daniel Kahneman received the Nobel prize in Economics in 2002.

With the current state of knowledge. It is not possible to say with any confidence how irrationality and bias might influence human supervisory control. The potential for bias will be influenced by many factors, not least an individual's personal susceptibility to System 1 reasoning errors,[21] their state of fatigue, and the context in which judgements and decisions are made. Different people faced with the same situation are unlikely to be subject to the same kinds of biases or to the same extent. And there is no reason to assume that any individual will be suspect to biased thinking all the time. Nevertheless, if we want to understand why an individual may have made a poor or erroneous judgement, and certainly if we want to try to optimise the reliability of human supervisory control performance, the potential for irrationality and biased thinking needs to be taken seriously.

While the science base does not yet exist to be confident, a little informed speculation suggests some areas in the comprehensive model where irrationality and bias seem most likely. These are indicated on Figures 9.2–9.6 by the symbol ⟨ ! ⟩. Table 9.1 summarises some likely candidates.

SUMMARY

The comprehensive model of human supervisory control illustrated in Figures 9.2–9.6 focuses on issues developers and operators of highly automated systems need to pay attention to both during the design and development of automated control systems, as well as when those systems are being prepared for deployment. That means features ranging from the design of user interfaces, organisational and working arrangements to the wider role and characteristics of people expected to perform the role.

The model illustrates the kind of mental processes likely to be involved in performing the role of a supervisory controller up to the point where the individual decides action is needed to support the automation. After that point, the human is no longer a supervisory controller: they have reverted to being (hopefully) a manual controller. How people behave, and what is needed to ensure the human has a good chance of responding successfully after the decision to intervene is made is beyond the scope of this book. From the point that the individual steps in and acts, the supervisory control model no longer applies.

The logical decision-type representation of the model should not be taken to suggest that the processes and activities that go on inside the head of a human supervisory controller are assumed to follow anything like the explicit and logical structure shown. They do not.

NOTES

1 See Chapter 7 for an introduction to the "Big 4".
2 "POM scans" were defined in Chapter 8 as being "…an ongoing, though intermittent, series of actively induced checks of the state of the automation made by the supervisory controller".
3 Although personally I would question whether that is what happens. Even such formal decision-making processes can be influenced by the irrationality and bias associated with system 1 thinking. For an example, in [3] I explored the potential impact of system 1 reasoning errors on judgements and decisions about risk that are made in slow time using formal risk assessment matrices (RAMs).

4 Also termed 'Associative Activation', or 'Associative Coherence'.
5 See Chapter 7 for discussion of previous models.
6 Bearing in mind the caution discussed above that the thought processes involved do not proceed as logical flows as the figures and description below might suggest to the unaware.
7 For example, if an autonomous vehicle is driving on a motorway in the UK, where the manufacturer declared the automation as being intended for use, and the user can see that the road ahead changes to a non-motorway state, then they should know that the authority implied by the manufacturer's declaration will come to an end.
8 These were introduced in Chapter 8.
9 Chapter 6 contains a summary of this crash, including a description of a number of Human Factors issues as well as lessons that can be learned for human supervisory control.
10 The Theory of Signal Detection was introduced in Chapter 8.
11 We will come to considering the implications of the other TSD parameter, the users subjective bias towards or against intervening known as β ('Beta'), when we look at the Intervene? process later in the chapter.
12 These process trend graphics are sometimes referred to as "Proactive Monitoring" displays.
13 Different vehicle manufacturers support these graphical indicators in different ways.
14 Interestingly, as I discussed in Chapter 5, even if the lane keeping system is not able to function in my Lexus, the vehicle will not necessarily drop out of the autonomous driving mode.
15 Part I of the book describes in some detail my experiences, and the reflections and learnings arising from them, of taking ownership (unknowingly) of a new car that had the ability to drive itself.
16 Chapter 5 includes reflections on how my view of my car's autonomous driving capabilities, as well as my own role as the driver-in-charge, have changed with experience.
17 The single most useful reference to help understand how to design interfaces that support situation awareness, is the book by Mica Endsley and Debra Jones [4].
18 Moray's model is summarised in Chapter 7.
19 Chapter 8 included a discussion of the need for a comprehensive model of supervisory control to take account of the concepts of ability, authority, control and responsibility, and of the importance of understanding the extent of consistency between them.
20 Chapter 8 also discussed the use of the Theory of Signal Detection (TSD), including the role of the subjective bias (β) as a way to explore the psychology involved when decisions must be made in the presence of uncertainty, as in the case of this decision to be made by the supervisory controller.
21 Shane Frederick's 'Cognitive Reflection Test' [5] has been suggested as a way of identifying people with a strong tendency to be robust against System 1 reasoning errors. The test comprises three questions, including the famous 'bat and ball' question; "A bat and a ball cost $1.10 in total. The bat costs $1 more than the ball. How much does the ball cost?"

REFERENCES

1. Reason, J. Human Error. Cambridge University Press. 1991.
2. SAE. Surface Vehicle Recommended Practice: Taxonomy and Definitions for Terms Related to Driving Automation Systems for On-Road Motor Vehicles. J3016. SAE International. 2021.

3. McLeod, R.W. The Impact of Styles of Thinking and Cognitive Bias on How People Assess Risk and Make Real-World Decisions in Oil and Gas Operations. Oil and Gas Facilities. October 2016. Society of Petroleum Engineers.
4. Endsley, M. & Jones, D. Designing for Situation Awareness: An approach to User-Centred Design. 3rd Edition. Taylor & Francis. 2025.
5. Frederick, F. Cognitive Reflection and Decision Making. *Journal of Economic Perspectives.* 2005;19(4):25–42. DOI:10.1257/089533005775196732

Part 3

Assuring the reliability of human supervisory control

The ideas behind the summary produced in 1951 by the scientist Paul Fitts [1] of the things people could do better than the machines available at that time, and vice-versa, continue to have traction after more than seven decades. While the content and details have been revised and updated many times since the 1951 original, the key premises behind "Fitts List" remain as pervasive as ever:

1. It is in principle possible to identify generic functions and abilities that either the people or technological elements of systems are consistently and significantly better at than the other.
2. System functions should be allocated to people and "machines" based on those relative strengths and weaknesses.

The first of these has been an increasingly slippery slope as the power and capabilities of technology have increased over the decades. In recent years, AI-based technologies equipped with intelligent reasoning, machine vision and learning abilities have increased the abilities of technology well beyond what was conceivable in 1951.[1] Simultaneously, the idea of uniquely human abilities has become increasingly difficult to defend.

However, the reality is that other than in possibly a few highly specialist applications such as perhaps developing systems for deep space exploration, or other similarly boundary-stretching projects, the continuously improving capabilities of advanced - and increasingly AI-based - technologies mean that in the great majority of cases, it

DOI: 10.1201/9781003679479-12

is the capability of the technology that matters. Systems and products are going to continue to be developed, as they have for many decades, that seek to exploit the capabilities of whatever technology can do, if it is associated with some sort of competitive advantage, irrespective of the relative theoretical capabilities of people and technology. Faced with that reality, there is no need to search for the optimum balance in authority between people and technology. In market driven economies, technology is always going to win, and the human will continue to be relied on to fill the gaps. Which means that products and systems designed to provide automated control over real-time activities will continue – increasingly so - to rely on people filling the role of supervisory controllers to meet customer expectations of ensuring products and systems are safe, reliable, and effective at delivering whatever improvements are sought.

What matters then is that sufficient attention is given during the development and deployment of those products and systems to be confident that the people who are going to be relied on to fill the supervisory role;

a. can detect and recognise situations that threaten the capability of the system to exert effective control without human support;
b. understand when there is a need for them to intervene either to adjust, or to take over control of the activity;
c. can do what is needed of them to take over control; and,
d. are motivated and willing to act in the situation that exists at the time.

Rather than focusing on the relative abilities of people and technology, the point therefore becomes understanding *what* it is that the people are being relied on to do, and ensuring those people have a reasonable chance of possessing the skills, knowledge, and motivations to do it in the *situation* and *context* likely to exist when they are relied on.[2]

Part 1 of the book contained a narrative account of the experience of becoming the owner of a car having the capability, in limited circumstances, to drive itself autonomously. It used that narrative as the basis for exploring some of the implications and lessons arising from changing the long-established role of being the manual controller of a vehicle to being the supervisory controller of a motorised computer. Although Part 1 was based on experience with an autonomous car, the lessons are equally relevant to other situations where automation is being introduced to take over all or some real-time control tasks previously performed manually.

The second Part explored the psychological complexity associated with performing the role of being a human supervisory controller of an automated system. The Part presented a conceptual framework illustrating how the human's role changes. It then discussed three conceptual models of human supervisory control prepared by leaders in the field. The Part concluded by presenting what was considered a (more) comprehensive model of human supervisory control. One that built on the previous models and sought to incorporate important influences on human performance not accounted for in previous models.

The material in the first two parts of the book was intended to raise awareness among those who develop, authorise, and operate highly automated real-time control systems of exactly what they are expecting of the people they continue to rely on to assure the safety and effectiveness of those systems. The material suggested ways of thinking about the relationship between people and technology that can be beneficial in generating insight and deeper understanding of what needs to be provided to support the role of the human as a supervisory controller of a highly automated real-time activity. In particular, the first two parts sought to emphasise three key points;

i. allowing automated systems to control real world systems fundamentally changes the role of the human being from being a manual controller of a system to being a supervisory controller;
ii. filling the role of a human supervisory controller depends on people being effective performing the task of proactively monitoring the performance of the automation. That means, while simultaneously being involved in other competing real-time activities, being able to actively and intentionally direct limited attentional resources to sources of information in the world or the user interface that indicate the current and likely future situation. It means understanding the meaning of that information and recognizing when the automation is likely to need support. It also means being both willing and able to step in to take control, even in challenging circumstances, in the presence of uncertainty, and with little time to prepare, if, based on what they perceive as the current and imminent situation, they believe it is necessary;
iii. that the change from manual to supervisory control, and the need to monitor proactively, fundamentally changes the psychological demands on the human. Because of that, it demands a fundamental rethink of the information and other support that needs to be provided to the human when systems with autonomous capability are being designed, deployed, and approved for use. When introducing systems that rely on people acting as supervisory controllers, it is not sufficient simply to add new information or controls to an existing user interface designed for manual control: a fundamental reconceptualization of the user interface, and the underlying user support model, is likely to be needed.

This final part of the book focuses on the needs of those responsible for designing, developing, and introducing highly automated and autonomous products and systems. Specifically, the Part is focused on situations where the performance of the human functioning as a supervisory controller is going to be relied on as a defence or barrier against some form of significant adverse event. The Part explores the risks associated with human supervisory control and provides practical suggestions to help provide confidence that everything that could reasonably be expected has been done to ensure the individual(s) filling the role of the supervisory controller have every chance of being as effective in the role as is expected.

ADVERSE EVENTS

I'm using the term "adverse" events, rather than terms such as "accidents", "incidents" or others to reflect the fact that unwanted events can take many forms depending on the nature of the system or product and the context in which it is used. These can range from concerns over safety, exposure to health risks, environmental damage or disruption to an operation or service, failure to meet an enterprises objectives, financial loss or reputational damage. Here are a few examples;

- in automated vehicles, whether cars, trucks, trains or aircraft, the worst-case adverse event would be loss of control of the vehicles motion leading to a crash. Though there are other types of adverse event of concern involving loss of control of other aspects which automation has been given the authority to control, such as the vehicle's geographical location (which was the event leading to KAL flight 007 being so far off course leading to it being shot down in 1983[3]), fuel management or energy efficiency;
- in industrial manufacturing processes, the most serious adverse events generally involve loss of control over hazards (chemicals, fuel, pressure, etc.) involving major risks to health and/or safety. Events that lead to loss or disruption of an industrial or commercial process (such as inadvertently causing equipment to slow-down or shut-down) would also be seriously adverse in terms of lost production and therefore profit, though would not necessarily directly be safety related;
- in oil and gas exploration, adverse events in an automated drilling operation might involve loss of control over pressurised gas in an underground reservoir. As with manufacturing, other types of events might affect costs and production schedules, without being directly safety related;
- in online enterprises, or businesses such as banks that place heavy reliance on online services to support their customers, failure of automated processes to manage customer's requests or personal data can be seriously adverse both for the reputation of the business as well as for the ability of its customers to go about their affairs;
- in utilities such as energy or water companies, failure of automated systems has the potential for massive adverse consequences ranging from loss of power or water to cities or other communities, to polluting water supplies or rivers.

STRUCTURE OF PART 3

For this Part of the book, we are going to assume that the developers of a highly automated real-time system – whether it has autonomous capability - need to be able to demonstrate that they have done everything that could reasonably be expected of them to provide effective support to people filling the role of human supervisory controllers of their new system.

The Part is organised into three chapters;

- using the comprehensive model presented on Chapter 9 as a framework, Chapter 10 identifies six principal risks associated with human supervisory control. The chapter explains the nature of each of the risks and makes suggestions about the kind of questions that should be addressed to be confident the people involved will be able to fill the role effectively and reliably;

- Chapter 11 focuses on the key requirement that a human supervisory controller is proactive in monitoring for signs of potential trouble, rather than simply reacting to prompts from the system or some other source. The chapter summarises the concept of safety defences and considers what would be needed for someone filling the role of a supervisory controller to be considered capable of meeting those criteria and therefore of being relied on as a "Barrier". And the chapter argues that, provided sufficient effort is put into assuring both the design of the user interface that supports the role, as well as the context and situation the individuals involved are likely to be in at the time, the function of Proactive Operator Monitoring (POM) can come close to having characteristics needed to be relied on as part of a systems defences against adverse events;
- Chapter 12 presents a structured method, referred to as the Proactive Operator Monitoring Assessment Tool (POMAT), that can be used to informally evaluate how well the combination of the design of an automated system and the context and situation it is expected to be used in are likely to support reliable proactive operator monitoring.[4]

NOTES

1 See [2] for a recent discussion of the current relevance of the original list.

2 It is important to understand the meaning of the terms Situation and Context as they are used here. In its White Paper on Learning from Adverse Events [3], the Chartered Institute of Ergonomics & Human Factors (CIEHF) defined the terms as follows: The term Situation refers

> ...the set of circumstances particular to the specific time and place in which an adverse event occurred. Situational factors are essentially factual and, in principle, are discoverable as 'evidence' of the situation the individuals involved were in at the time the event occurred.

The term Context was defined as

> ...the meaning ascribed to a situation by the individuals involved and the long-held beliefs they hold about the situation they are in. It means factors likely to influence what people believe about the immediate situation and what is expected of them.

3 There have been many other similar events. In his book 'Taming HAL' [4], Asaf Degani gives a detailed account of the nature of the breakdown between the crew and the automated autopilot that led to the grounding of the cruise ship Royal Majesty close to Nantucket Island in the US in 1995.

4 The original conception of this Part of the book included the intention to include two other tools that can be useful to developers of automated real-time control systems; (i) an expanded description of a tool first suggested by myself and Dr Nora Balfe (see [5]) which briefly described an approach to performing a structured analysis of the allocation of authority and control between the human and automated elements of a system to highlight areas where the human is likely to need specific support, and; (ii) a more detailed development of the concepts of the "Gulf of Resilience" and the "Gulf of Authority" suggested in the CIEHF White Paper on 'Human Factors in Highly Automated Systems' [6] (with acknowledgement to the work of Professor Don Norman). Unfortunately, it was not possible to develop this material in time.

REFERENCES

1. Fitts, P.M. (Ed.). Human Engineering for an Effective Air-Navigation and Traffic-Control System. National Research Council. 1951.
2. de Winter, J.F.C. & Hancock, P.A. Reflections on the 1951 Fitts List: Do Humans Believe that Machines Now Surpass Them? *Procedia Manufacturing*. 2015(3):5334–5343 DOI:10.1016/j.promfg.2015.07.641
3. CIEHF. Learning from Adverse Events: A White Paper. Chartered Institute of Ergonomics & Human Factors. 2020. Available from: https://ergonomics.org.uk/resource/learning-from-adverse-events.html
4. Degani, A. Taming HAL: Designing Interfaces beyond 2001. Palgrave Macmillan. 2001.
5. McLeod, R.W. & Balfe, N. A Human Factors Approach for Analysing Highly Automated Systems. In N. Balfe & D. Golightly (Eds.), Contemporary Ergonomic. CIEHF. 2020. https://publications.ergonomics.org.uk/uploads/4_29.pdf
6. CIEHF. Human Factors in Highly Automated Systems. Chartered Institute of Ergonomics & Human Factors. 2012. Available from: https://ergonomics.org.uk/resource/human-factors-in-highly-automated-systems-white-paper.html

10 Key risks in human supervisory control

The concept of preparing a documented safety demonstration for new or significantly changed products, process or operations is common practice in many sectors of the economy. In the traditional high hazard industries – nuclear power, oil and gas, chemicals, manufacturing, aviation and air traffic management, rail, defence, and others - many countries have adopted regulations requiring some form of safety demonstration or "safety case" before a business will be given approval to operate a new venture or release a new product. Across Europe, EU safety directives in industries including rail, offshore oil and gas, and nuclear and others associated with major accident hazards require companies to provide evidence that risks associated with their activities are understood and are controlled effectively.[1] The United States takes a more prescriptive rather than regulatory approach. Although companies conducting hazardous activities in the US are also expected[2] to have prepared some form of evidence-based demonstration that they understand the risks in their operations and can carry out their activities safely.[3]

The situation is similar in the case of organisations developing products for commercial sale, but whose operations are not themselves associated with significant hazards. For example, the European Union's General product safety regulation [1] requires anyone seeking to sell products into the European market to "...place or make available on the market only safe products".[4] The regulations include requirements on manufacturers to ensure products are "...safe by design", and to conduct and document internal risk analyses. Importers and distributors of products produced outside the EU are required to ensure the manufacturers comply with these regulations. The EU regulation was adopted by the UK following BREXIT. UK regulations consider a "safe" product as "...one which under normal or reasonably foreseeable condition of use does not present any risk or only the minimum risk compatible with the product's use" [2].

What is meant by the terms "safety case" or "safety demonstration"? In 2012, the UK's Health Foundation published the results of a survey of approaches to safety cases in different industries [3]. Their purpose was to assess the potential value of adopting safety cases in healthcare settings. They included the following description of the purpose of a safety case;

> ...to provide a structured argument, supported by a body of evidence, that provides a compelling, comprehensible and valid case that a system is acceptably safe for a given application in a given context. The core of the safety case is typically a risk-based argument and corresponding evidence to demonstrate that all risks associated with a particular system have been identified, that appropriate risk controls have been put in place, and that there are appropriate processes in place to monitor the effectiveness of the risk controls and the safety performance of the system on an ongoing basis. [3]

DOI: 10.1201/9781003679479-13

That is a reasonable and thorough enough description for the purpose here. The degree of rigour and formality expected in safety demonstrations varies considerably between industries and countries. As does the extent to which government or other regulators take a role in reviewing and assessing the quality and comprehensiveness of the arguments made.

The purpose here is not to try to align with any approach, or any particular industry or country's requirements for the preparation of a safety case, or what it should contain. There is no need to get involved in a debate over whether the objective is to reduce risk to a level that can be shown to be As Low As Reasonably Practicable (ALARP), As Low As Reasonably Achievable (ALARA), or that risk should be reduced So Far As Is Reasonably Practicable (SFAIRP).[5] Rather, this chapter is concerned with the principle so far as it applies to the change in role of a human from being a manual controller of a real-time activity to being a supervisory controller of automation. That principle is that any organisation seeking to automate the control of a real-time activity previously controlled largely or entirely manually, should be able to demonstrate that the risks associated with the change are understood and have been managed effectively. Or to put that into the positive view of the role of the people in a system, they should be able to demonstrate that the reliability of people expected to fill the role of a human supervisory controller can be shown to be As High As is Reasonably Practicable (AHARP).[6]

The purpose of this chapter then, is to assist organisations involved in developing, implementing, and operating highly automated systems, including those with autonomous capability, that will move humans into the role of a supervisory controller. It does so by considering in some depth six principal risks to the performance and reliability of a human supervisory controller identified from the comprehensive model presented in Chapter 9. The chapter explains each of the risks and makes suggestions about the kind of issues that need to be addressed during the design, development, and implementation of the automation, as well as the kind of support those tasked with filling the role of a human supervisory controller are likely to need while they are transitioning from being a manual controller.

CONSTRAINTS

There are two important constraints to the scope of this chapter. First, our interest is limited to the role of the human in supporting the automation, rather than in the ability and performance of the automation itself. We are concerned only with ensuring that development and implementation of the product or system has taken sufficient and effective account of the needs and requirements of the people expected to fill the role and perform the tasks the system needs them to do.

The second constraint is that the book is focused on the effectiveness of the human only while they are filling the role of being a supervisory controller of the automation.[7] We are not here concerned with whether what happens after the human decides to take manual control is effective: i.e. we are not here concerned with the possibility that either;

- the human attempts to take over manual control, but does the wrong thing; or,
- the human attempts to take over manual control, does the right things, but is too late to be effective.

Clearly, both risks are extremely important to the effectiveness of any system, whether it has autonomous capability. It may seem unsatisfying that a book about the role of people in supporting highly automated systems does not address these two risks. There is however good reason for strictly enforcing the constraint. As has been demonstrated throughout Part II of the book, human supervisory control – irrespective of what happens after the controller is handed or decides to take over manual control – is sufficiently complex and demanding psychologically that it needs to be properly understood and addressed, independently from what happens during or after the reversion to manual control.

WHAT IS THE HAZARD?

A conventional safety demonstration in one of the traditional safety-critical industries would start by identifying the hazards associated with the use of a product or system or the conduct of an activity. The Human Factors elements of the demonstration would identify those tasks and activities that need to be carried out by people to ensure sufficient controls are in place to prevent uncontrolled release of those hazards. These are what are generally referred to as "safety-critical tasks". The Human Factors demonstration would then assess the risks associated with each safety-critical task and seek to demonstrate that actions have been taken to ensure the chances of those risks occurring are sufficiently low as to be considered acceptable.

Hazards are generally either the presence of some form of energy - pressure, temperature, electricity, chemicals, radiation, etc. - or activities that, if not adequately controlled, have the potential to cause harm - performing major surgery, landing an aircraft in the presence of extreme crosswinds, starting up a complicated chemical process, or controlling the pressure in a newly penetrated oil field and so on.

In terms of the automated products and systems of interest here, we are concerned with the ability of people to perform the role of a supervisory controller, and the potential consequences if they are not able to perform that role reliably and to a high standard. Conceptually, the change from being a manual to being a supervisory controller could be conceived as the potentially hazardous situation. A situation that occurs every time authority is passed from a human to an automated system to control all or part of an activity;

- in the case of vehicles with autonomous capability, the hazardous situation exists as soon as the driver engages the vehicle's autonomous features and hands control of the vehicle's motion to the automation. And it ends as soon as the driver takes over manual control of the vehicle;
- in an industrial process control system, the hazardous situation exists as soon as the automation is activated and continues to exist for as long as the automation controls the process;
- in aviation or shipping, the hazardous situation is initiated as soon as control is passed to the autopilot, and the aircraft's pilots or the ship's officer-of-the-watch are expected to monitor it to ensure it maintains the assigned route.

So we are really concerned with only a single hazardous situation: where authority is passed to automation to perform all or part of a real-time activity. More specifically, we are concerned with the period when the automated system is being introduced and the human needs to make the transition from performing a role that previously involved manually controlling an activity to a new role that is based on monitoring and overseeing the automation.

Our interest is in being able to demonstrate two things in connection with the introduction of a product or system that gives that authority to automation;

i. that the risks associated with making the transition to human supervisory control are adequately understood, and;
ii. that sufficient and effective attention has been given throughout the development and implementation of the product or system to reduce the chance of those risks leading to some form of serious adverse event to an acceptably low level.

Those involved in developing highly automated and autonomous systems must be able to demonstrate sufficient awareness and understanding of the range of threats, both internal and external, they are relying on the human supervisory controller to be able to detect.[8] More than that, they need to understand the characteristics of those threats. That means, not least, understanding;

- what kind of information would indicate the potential existence of a threat? I.e. what are they looking for? Are those indicators likely to be conveyed in a single piece of information, or will it involve the coincidence of different pieces?
- where that information might exist. I.e. will it be included in the user interface to the automation? Or will it exist via other instruments, in the external world, or might it be known to other people performing a different role?
- what format the information is likely to exist in. I.e. will it be shown as a graphical image, text, or table on an instrumentation display? Does it require the human to mentally integrate pieces of information from different sources, and possibly different formats to understand what is happening?
- whether the human controller will need to do anything to convert the information available to them into a form that they can reason with and think about that are relevant to whatever activity is being controlled?

SUPERVISORY CONTROL AND SAFETY-CRITICAL TASKS

Once hazards and hazardous situations are understood, a conventional human reliability analysis would start by identifying what are considered "critical" human tasks necessary to manage the risks associated with each hazard. In its guidance to conducting human factors assessments in major accident hazard sites in the UK, the Chartered Institute of Ergonomics & Human Factors [6] defines a "critical" task as:

> … a task in which human action or inaction could initiate, fail to control or fail to mitigate a major accident, including operations, maintenance and emergency response…

In conducting Human Factors assessments of human reliability in safety-critical operations, it is now common practice to view the monitoring conducted by operators in control rooms as a "safety-critical task". For example, the CIEHF's guidance identifies the task of "Monitoring plant from a control room" as being a critical task that requires detailed analysis to establish the likely reliability of the people tasked with performing it. The guidance recognises that, because of its continuous nature and reliance on decision making, monitoring is not suitable for analysis using a conventional Hierarchical Task Analysis (HTA) approach. Instead, it suggests that the reliability of a human in monitoring a process plant can best be assessed by evaluating the usability of the control system, the human machine interfaces, and the workload in the control room. While user interface design and workload clearly play a role in influencing the effectiveness of human supervisory control, they are far from being a sufficient basis for assessing the reliability of human performance in the role of a supervisory controller.

Identifying human supervisory control as a "safety-critical task" in the way that term is generally used in safety management is to push the boundaries of what is intended when reference is made to a critical "task". It reflects a lack of understanding of the characteristics and complexity of human supervisory control (see the models of human supervisory control in Part II of the book). Human supervisory control is a complex perceptual and cognitive activity, often conducted over significant time frames. It is too complex to be assessed using the kind of approaches to Safety-Critical Task Analysis[9] or Human Reliability Analysis[10] generally adopted in high hazard industries. An alternative approach is needed. One that reflects the nature and complexity of the psychology involved in filling the role of being a human supervisory controller. The comprehensive model of human supervisory control developed in Chapter 9 provides the basis for such an alternative approach.

WHAT ARE THE RISKS?

The comprehensive model of human supervisory control is built around four core cognitive processes;

- Automation situation awareness;
- Attention;
- Threat assessment and monitoring;
- Deciding whether to intervene.

The detail of these processes provides a framework to consider the risks associated with human supervisory control. The top level of the model – which is shown on Figure 10.1 - identifies six generic risks associated with these four core processes that could lead the human not to intervene in the presence of a threat when they would have been expected to;

1. They were not paying attention.
2. They were paying attention but did not detect or appreciate the meaning of information about the potential threat.

3. They detected the potential threat but did not understand it or its significance.
4. They placed too much trust in the ability of the automation to deal with the threat.
5. They realised the automation could not deal with the threat without support, but despite that decided not to intervene.
6. The controller makes some form of error in setting or adjusting the limits, data, or modes to be used by the automation.

Note that all six of these risks are based around things the human did or did not do. Phrasing them in this way might seem to conflict with a principle that is at the heart of much of the professional practice of Human Factors: the principle that, rather than simply blaming people, the things they do and the way they behave need to be understood in terms of the situation and context they were in – or are put in – by the overall socio-technical system of which they are usually a part. In phrasing the risks of things the human did or did not do in this way, there is no intention to imply that the individual would have been careless, negligent or otherwise at fault (though there is, of course, always that possibility). The content of the remainder of this chapter explores the many factors – including the design and layout of equipment, the environment, and jobs as well as working conditions and organisational incentives – that can motivate or otherwise influence the way people behave.

The remainder of this chapter will draw on a knowledge of the research base associated with human-automation interaction, what has been learned from studying many incidents, as well as a lifetime of experience applying Human Factors knowledge in industrial settings. That experience includes innumerable discussions and workshops with people who work at the front line of complex systems. Drawing on that knowledge and experience, the chapter will explore the kind of reasons that could lead to each of these generic risks becoming a problem.

There is a very large variety of different applications where the control of some real-time activity could be automated, changing the role of the human from being a

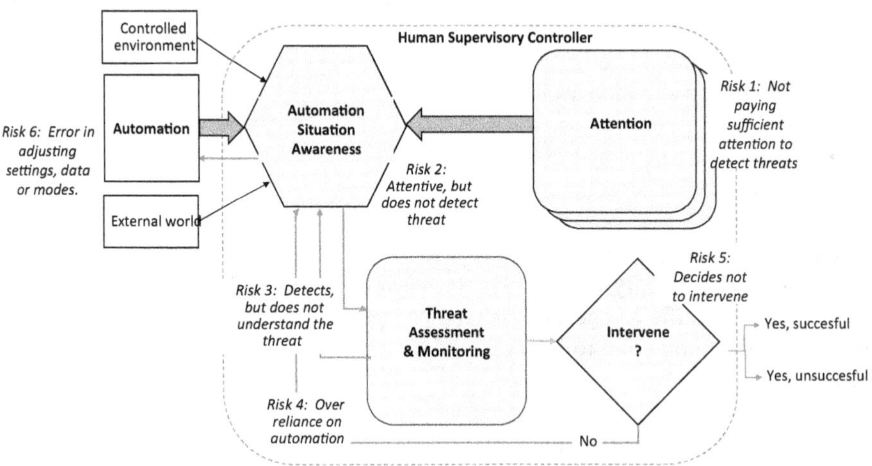

FIGURE 10.1 Summary of risks associated with human supervisory control

manual controller to a supervisory one. And there is probably even more scope for different solutions in the way automation is implemented, the characteristics, capabilities, training and experience of the people who will fill the supervisory controller role, the user interface designs, and other facilities intended to support them, the legal and regulatory context in which the system or product will be used as well as the situations and contexts in which those people will find themselves performing the role. For at least all of those reasons, it is not possible to be prescriptive about all the reasons that could lead to any of the generic risks summarised in Figure 10.1 occurring. The discussion in the remainder of this chapter will therefore focus on illustrating the kinds of psychological factors that can lead to a failure of the core tasks of monitoring, detecting, assessing, and deciding whether to intervene in response to potential threats. An organisation engaged in developing and implementing a system or product that relies on people performing as supervisory controllers will need to assess the nature of the product or system they are involved with, the people they expect to use it, and the kind of situations and contexts it will be used in to develop an understanding of the generic risks in a way that is customised to their application.

The discussion of these risks will be based around reliance on a single human supervisory controller. That was the perspective adopted throughout the review of models of supervisory control in Part II of the book, as well as the comprehensive model presented in Chapter 9. There are though, situations where the supervisory control function might be performed by more than one person. That might be formally, such as a team of individuals having well-defined roles and responsibilities who work together towards achieving common goals, or informally, where the supervisory responsibility is held by one individual, though others could informally play a role, such as drawing the controller's attention to something they may not have noticed. Much of the research behind the concept of macrocognition discussed in Chapter 9 was based on teams of people working collectively to perform complex cognition-based tasks.[11]

Before embarking on considering the six risks identified on Figure 10.1, we need a brief reality check. Although this chapter is structured as a sequential and logical walk-through of the key risks associated with the four processes identified in the comprehensive model of supervisory control, we need to remind ourselves of the macrocognitive perspective of how the human brain performs complex cognitive tasks in the real world. That is, that this structured and sequential consideration of the causes of the various kinds of risks, and the logical structure of the figures setting out the details of the comprehensive model (Figures 9.3–9.6 in Chapter 9), are not intended to suggest that the actual thought processes going on in the head of a controller, whether consciously or sub-consciously, will proceed in anything like such a structured and logical way. Rather, as the macrocognitive perspective emphasises, thinking and decision making, at least in complex real-world situations, if not in tightly constrained laboratory tasks, involves a great deal of parallel mental activity distributed across the brain's structures, and often occurs in a continuous loop with feedback and overlap.

With that warning front and centre, the following sections will explore the characteristics of the six risks identified on Figure 10.1.

Supervisory control risk 1: not being attentive

The first risk shown in Figure 10.1 is associated with the process **Attention**. Details of the Attention process are summarised on Figure 9.3 and the associated text in Chapter 9. The risk is that the human supervisory controller was not paying attention to sources of information that could have allowed them to detect the potential threat. Note that this includes two things; i) not paying attention at all, and ii) not paying attention often enough. There can be many reasons for someone in a supervisory control role not paying sufficient attention to the task, including;

i. the individual was not present;
ii. they were present, but inattentive (or even asleep) due to high levels of fatigue;
iii. they were allocating attention to monitoring the automation, but not often enough;
iv. they were distracted;
v. they were motivated or incentivised to focus their attention on something else;
vi. paying attention to the automation was insufficiently engaging to encourage them to give it the level of attention it needed: i.e. it was boring;
vii. they were performing reactively, waiting for the system to tell them if they needed to pay attention to it

Not being present

There have been many incidents where the individuals expected to be in place and able to detect problems have either not been there or have been inattentive. On 10 March 2025, the container ship MV *Solong*, travelling under autopilot at a speed over the ground of about 16 knots (18.4 mph) collided with the oil/chemical tanker MV *Stena Immaculate* which was at anchor just off the Humber Estuary in England with a cargo of over 220,000 barrels of aviation fuel. One crew member on the *Solong* died in the collision. At the time of writing, the UK's Marine Accident Investigation Branch (MAIB) is still conducting its investigation into the collision. However, in an interim report published in April 2025 [10], the MAIB reported that: "Neither Solong nor Stena Immaculate had a dedicated lookout on the bridge." That is, the Solong was under autopilot control at the time of the collision, but there was no human on the bridge acting as lookout. No one was present to monitor the automation.

Present, but inattentive

Not paying attention due to fatigue is a significant risk in many activities. Management of fatigue has been a major issue for some years in industries that operate continuous, 24-hour operations that rely on staff working shifts, including night shifts. The natural circadian rhythm that causes us to want to sleep in the early hours of the morning, combined with long working hours and insufficient opportunity for restorative sleep, can lead people to being inattentive, and even falling asleep, when they should be monitoring the automation. In their investigation of the explosion that led to fifteen fatalities and 180 injuries at the Texas City oil refinery in 2005, the US Chemical Safety Board (CSB) concluded that the ability of the control room operators on shift at the time was probably impaired by the fatigue arising from acute sleep loss and cumulative sleep debt [11]. Some of them had worked 12-hour shifts for up to 39 consecutive days.

There have also been many reports in the media of drivers of self-driving cars being asleep at the wheel while the car is in autonomous driving mode.[12]

Not monitoring often enough

The third reason for a human supervisory controller not paying attention to the automation is that, although the individual was proactively directing their attention to the monitoring task, they were not doing it sufficiently often to be able to detect developing threats in time to deal with them. That is, the threat emerged in between what were described in Chapter 9 as proactive operator monitoring scans ("POM scans") and was not detected until it was too late. To understand the reasons why this might happen, we need to explore the psychology behind how an individual, faced with competing tasks and responsibilities, is likely to decide, proactively, to direct their attention towards monitoring the automation.

Chapter 7 discussed the importance of the mental models all humans build of the environment around them and the properties of the systems they interact with. The comprehensive model of human supervisory control in Chapter 9 identified several activities that draw on the content of the individual's mental model in detecting and responding to potential threats. The aspect of the mental model that interests us here is how well it reflects the individual's beliefs about how quickly things or events happen or change in the world being controlled by the automation. More specifically, we are interested in how quickly the confidence an individual has about the state of the world based on the information acquired in a POM scan decays over time. It is the growing sense of unease as the confidence in that information decays that drives the individual to decide to proactively direct their attention to perform another scan.

The representation of the dynamics of the world held by an individual's mental model determines how often the individual believes they need to check on the state of the world to detect signs of a significant event or change. If things in the world are believed, and actually are, changing very slowly, the individual is likely to allow more time to pass between each time they check on the world, confident in the knowledge that nothing significant is likely to have occurred since the last check. On the other hand, if the individual believes, incorrectly, that the world is changing slowly, but in fact things are happening quickly, then they are unlikely to monitor the world often enough to detect emerging threats.

As far as the dynamics represented in the mental model are concerned, we need to recognise two distinctly different types of worlds;

i. a world that is stable and regular and where events unfold in a way that makes is possible to identify predictable patterns, and;
ii. a world that has a high level of instability and irregularity, and where it is more difficult to predict when events might occur.

An example of the first kind of world would be a driver on a straight stretch of quiet motorway where the road ahead is visible for a reasonably long distance, and where oncoming traffic is separated by physical barriers. The driver might feel confident to take their eyes and attention off the road for a few seconds perhaps to look at the scenery or change the music. Or the Officer of the Watch (OOW) on the bridge of a ship under autopilot control on passage in the open sea. This kind of world is often found under normal operating conditions in industrial processes, where the entire process is contained and constrained and where the dynamics of the underlying process are highly linear, or predictably non-linear. In this first type of world, things change, and threats develop, relatively slowly. So the information gained in a POM scan will continue to have value for some time. Consequently, experienced people can schedule POM scans at reasonably regular intervals.

An example of the second kind of world would be a driver on a small city road, perhaps in a busy shopping area, where there is a lot of traffic in both directions, some trying to join or cross the road, with pedestrians also trying to cross the road, and perhaps children playing on the pavement. Or the OOW on the same ship's bridge, but this time under passage in a constrained busy shipping area such as the English Channel with not only commercial ships following each other in controlled shipping lanes, but frequent and unpredictable small vessels trying to cross at 90 degrees to the shipping lanes. In industrial process control environments, these kinds of situations can arise when processes are being started up or shut down, or when some unusual or abnormal situation has occurred. This kind of world is much more unstable and irregular. It is much more difficult for the supervisory controller to predict when threats might materialise. The value of information acquired in a POM scan decays quickly, so the supervisor needs to attend to the automation and the world much more often, with little opportunity to attend to other tasks.

Of course, many real worlds will include some combination of both types, varying between the two as an activity progresses.

We can get a flavour of how an incorrect mental model can lead to a supervisory controller in the first of these two worlds not paying attention often enough by considering Figure 10.2. The centre of the figure represents a hypothetical world. In this world, when events that are a threat occur, they take a minimum of five seconds to develop to the point where some form of adverse event might occur. For example, in an automated vehicle, let's assume it takes at least five seconds between the car's sensors being unable to detect a lane marker (the threat event) and the car hitting a vehicle in the adjacent lane (the adverse event). The top of Figure 10.2 represents mental model X that correctly captures this five second delay between the threat and the adverse event. The bottom of the figure represents mental model Y that, incorrectly, represents the minimum delay between the threat and the adverse event as being eight seconds.

Consider two scenarios. Towards the left of Figure 10.2, threat A occurs (and can be detected by the supervisor) five seconds into the scenario. If it is not responded to, it will lead to an incident five seconds later, at 10 seconds. Under mental model X, the supervisory controller conducts monitoring scans (POM scans) every five seconds, as dictated by their mental model of the properties of the system. The first POM scan is conducted at around one second, and the second at around six seconds. So the supervisor detects the event on the 2nd POM scan. Under mental model Y, the first POM scan is also performed at one second, and the second at nine seconds. So under

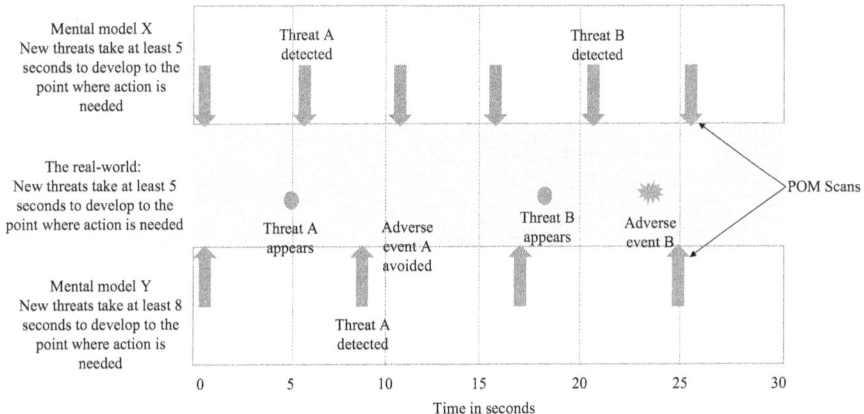

FIGURE 10.2 Illustration of mental models driving the timing of proactive operator monitoring scans (POM scans) in a stable, predictable world

mental model Y, the threat event is also detected before it can lead to the adverse event.

A second threat event occurs 18 seconds into the scenario, with the potential to create an incident at 23 seconds. Under mental model X, this is detected by a POM scan carried out at 21 seconds, allowing the supervisor to take action to avoid the incident. Under mental model Y, however, the next POM scan will not be carried out until around 25 seconds, two seconds after the incident. So, unless something else happens to draw the supervisor's attention to the threat the car will hit the vehicle in the adjacent lane.

This is, of course a highly simplified example: not least the fact that POM scans will never be carried out in such a regular clockwork manner. But it serves to make the point. Which is to illustrate how the dynamics of the world represented by a supervisory controller's mental model, in a world which is stable and predictable, will influence how frequently they are likely to proactively pay attention to what is going on in the automated activity.

In summary, when an individual is monitoring proactively, which is what we want supervisory controllers to do, how often they choose to attend to the automation rather than to something else will be determined largely by how often their mental model stimulates them to attend. Specifically then, one reason for the supervisory controller not paying sufficient attention to detect an emerging threat, is that their mental model of the dynamic properties of the world being controlled assumes things in that world change or events happen more slowly or infrequently than they actually do. Because of that, the individual does not pay attention to information sources as often as they need to.

Another reason why a supervisor may not monitor the automation often enough is to do with trust in the automation. There is a sizeable body of research dealing with what is termed "automation-induced complacency" [12,13] in people filling the role of a supervisory controller of highly automated systems. Automation-induced complacency arises when the human has become over-reliant on the automation: they trust it too much. Consequently, the supervisor does not monitor sources showing how the automation is behaving as often as they should. Automation-induced complacency is essentially a sub-optimal allocation of the operators' attention, arising from unrealistic – indeed, over-optimistic - beliefs about how reliable the automation is.

Distracted

We all know that it is very easy for anyone involved in carrying out an activity to be distracted by people or events not associated with the task. That is especially the case; i) when an activity is performed over an extended period, and ii) when the task is not highly mentally engaging: i.e. it does not demand a high level of focused attention and concentration. In the modern world, probably the most common source of distraction is interference by phone calls, emails, notifications, or text messages that can occur at any time.

In industrial and commercial settings, not being attentive to monitoring an automated activity can be common where the individual is expected to divide their time and attention between supervising the automated activity and other tasks they are responsible for. Though it is the opposite of what is expected in autonomous vehicles, where, when the automated driving function is engaged, as far as the law is currently concerned the driver has no other simultaneous responsibility other than ensuring the movement of the vehicle does not put the occupants of the car or anyone else at risk.

An individual with responsibilities that compete for their attention at the same time might not pay attention as often as they should because of distraction arising from other tasks that demand too much of their time. So they simply do not have the time available to proactively monitor the automation for potential threats.

Motivation and incentives

Sometimes, when people are responsible for several tasks at the same time, those tasks can have, or can be perceived to have, different priorities: i.e. the individual believes some tasks to be more important than others. This prioritisation can influence how often the individual gives their time and attention to lower priority tasks, as well as how long they spend and how deeply they mentally engage in those lower priority tasks.

The relative priority between tasks is sometimes a result of explicit instruction: perhaps the individual is told by those they work for that, on that day or for the next hour, it is important to complete or focus on one task at the expense of others. Sometimes the prioritisation is explicit in bonus payments or other incentive or penalty schemes. Or it may be implicit, in an understanding of what an employee thinks matters to their managers or the company they work for.

In the events leading up to the explosion and fire at the Buncefield fuel storage depot in England in 2005, for example,[13] two fuel transfers were being carried out simultaneously using different pipelines. The contract for delivering the fuel on one of the lines however included penalty payments in the case of a failure to complete the transfer on time that did not apply on the other line. This had the potential to incentivise control room operators to focus their attention on the transfer over the pipeline that contained the penalty clause.

Insufficiently engaging

One of the main reasons for those in human supervisory control roles not being proactive in paying attention occurs because of the reliability and consistency of modern automated systems. Systems that perform reliably to a high standard over long periods of time with little or no human input. Unless ways can be found to keep the supervisory controller mentally and meaningfully engaged in what is going on, such that they are willing and able to go to the effort of deliberately and actively paying attention, they will soon become bored and find other things to do to fill their time.

Finding ways to keep supervisory controllers sufficiently engaged in monitoring an automated activity that they remain alert and attentive, while being able to maintain awareness of what is going on, is a major challenge. As discussed above, the risk of simply giving them other things to do is that those other tasks either take priority or take up so much of the supervisor's time that they compete with attending to the automation. Some modern process industrial control systems explicitly include screens, facilities and tasks intended to keep the control room operator engaged as a proactive monitor. Building in some form of activity for the controller to do to help them maintain attention on and awareness of the automation will only be effective however, if those activities are recognised as being meaningful and are mentally engaging.

Performing reactively

The final reason for people not being sufficiently attentive to monitoring the automation, is that, rather than being proactive, deliberately and actively choosing to take the time to direct their attention to look for potential signs of threats, they rely on a strategy of performing reactively. They only monitor the automated activity when something happens to draw their attention to it. Usually that will be some form of alert or prompt generated by the automation, though it can also be awareness of something happening in the controlled environment or the external world: such as, for example, another person drawing their attention to a threat.

Supervisory control risk 2: not detecting the threat

The second major generic risk that could lead the human supervisory controller not to intervene in the presence of a threat is associated with the process **Automation Situation Awareness**. Details of the process are summarised on Figure 9.4 and the associated text in Chapter 9. Specifically, the risk is that the human was paying attention but did not detect or appreciate the meaning of the information about the potential threat.

Again, there can be many reasons why this might happen. To list just a few;

- **the controller might not know where to look to detect threats, or what information they are looking for;**
- the information they need might not be available;
- **the information might be perceptually unclear, not meaningful, or ambiguous;**

- even if the information is (theoretically) available and clear, it might be difficult to access, or represented in a format that requires effort to make it meaningful and able to be reasoned with;
- **the controller might not make the mental connection between things they can detect and the possibility of something going wrong;**
- they may not be aware of the existence of the automated function, understand how it works or what it does. So they might not realise that information they have detected is associated with a threat to the automation;
- there is too much uncertainty associated with the available information, so they decide not to act on it until they have more or better information;
- **their thinking might be affected by the systematic reasoning errors and cognitive biases associated with System 1 thinking.**

To illustrate the kind of psychological issues that can defeat what might seem a straightforward expectation that someone who is alert and monitoring proactively would detect threats, the following sections will consider the four reasons in the list above emphasised in bold text.

Not knowing where to look or what to look for

Information about the existence of potential threats can come from many sources. Often it will be available from a systems user interface. However, as the loss of the two BOEING 737 MAX aircraft in 2018 and 2019 demonstrated,[14] at least with modern autonomous components that are deeply embedded and perform only a part of a complex control system, that is not necessarily the case. The risk is that the individual acting as human supervisory controller, who is expected to proactively check for signs of trouble, does not know either where to look, or what they are looking for.

Neville Moray's model of information and control in a supervisory control system[15] identified a variety of ways in which the information the supervisor needs might be distributed around a system. For example;

- information available directly from the system's user interface;
- information available from other sources, but not from the user interface;
- information that is not available from any source;
- even if it is available from the user interface or other sources, the user may have only limited awareness or understanding of what it means;
- some information might only be available from prior knowledge, training, or experience.

The message from Moray's model is that those involved in designing and developing highly automated systems need to take a broad view of all the sources of information that might help the human supervisory controller detect and interpret potential threats. They need to ensure the individuals filling the role are aware of where to find information and what it means. And they need to be clear about what they intend the user to understand by the information they do decide to represent on a user interface.

Information that is perceptually unclear, not meaningful, or ambiguous

To say that an item of information is not perceptually clear means it is not easy for a user to perceive or extract meaning from it: text on a screen that is too small to read comfortably, or drawn in a colour that makes it difficult to see easily against the background; graphical icons that are visually complicated and rely on discriminating fine detail; sounds or audio messages whose volume, or frequency content makes them difficult to discriminate and understand against background noise.

On 5th October 1999, at Ladbroke Grove in West London, a Class 165 train leaving London Paddington station passed a STOP signal and crashed into a High-Speed train approaching Paddington on the same line. Thirty-one people died, and 417 were injured. Analysis of the tragedy identified that, among many other contributing factors, the driver of the Class 165, had been faced with a visually complex scene comprising steel gantries holding up many signals each relevant to different lines [14]. The curvature of the track against a background of bridges, and the need to support bi-directional running through the junction led to a complex signalling infrastructure meaning it was not always obvious which signal applied to which line. In addition, the 4-lamp signal head for the signal the driver needed to read had an unconventional 'L' design, as opposed to the normal design where the four signals are arranged vertically. Faced with this complex visual field, and looking into bright sunlight at a low angle, the driver needed to identify and understand the state of the signal relevant to the line he was on. The information the driver needed was not perceptually clear. It seems he identified and responded to the wrong signal – one that was showing green, meaning it was safe to carry on, rather than red, meaning stop immediately.[16]

To say information is not meaningful often means that, while it may be easy to locate and perceive, it is presented in a way that is not easy to understand in terms that are relevant to the goals and tasks the user is trying to perform. This often happens, for example, in systems that display raw data about a process or activity (perhaps pressure, speed, temperature, data rates and so on). The data are only meaningful when they are transformed into a form that can be thought about in relation to task objectives: for a pilot (if we assume the absence of stall alarms) knowing the speed and angle of attack of their aircraft are only relevant if they also know at what speed the aircraft is at danger of stalling.

Figure 10.3 illustrates this issue. Imagine an automated control system is being used to fill a tank with some sort of liquid. The maximum level the tank is allowed to be filled to is 750 litres. The figure illustrates two ways the level of fuel in the tank, and the rate at which it is changing, might be displayed to an operator in the role of the system's human supervisory controller. Figure 10.3a simply displays the current level of liquid in the tank (700 litres). The operator can get an idea of how quickly the level is changing by watching the display over time to see if the number increases or decreases. There is nothing to tell the operator how close the level is to the tank's capacity. This is indirect information: it is only meaningful to the operator monitoring the automation if they; a) know the capacity of the tank, b) mentally compare the displayed level with the tank's limit and work out the difference, and c) watch the display over time to see if the level is increasing or decreasing, and at

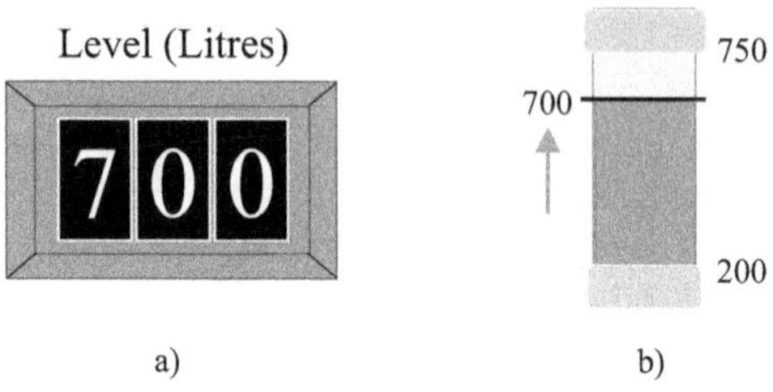

FIGURE 10.3 (a) information needs to be mentally transformed to be useful; (b) information that is directly meaningful by visual comparison

approximately what rate. While this might not seem a difficult task, for a supervisory controller who has other responsibilities and needs quickly to check that the automation is not in danger of over-filling the tank, while possibly simultaneously monitoring other parameters, it is not only unnecessarily demanding, but it is prone to making mistakes.

Figure 10.3b, by contrast, displays the information this supervisory controller needs in a way that is much more direct and easier to monitor. In this case, the display;

- shows the maximum and minimum levels the tank should be filled to (the figures on the right-hand side) – there is no reliance on memory;
- visually shows the difference between the current fill level (700 litres) and the maximum allowable level;
- visually shows if the level is increasing or decreasing (by the direction of the arrow);
- indicates the rate at which the level is increasing or decreasing (the long line means a high rate; a low rate might be shown by a short line).

Displays that are conceptually similar to Figure 10.3b are often referred to as "at-a-glance" displays: they allow the operator to monitor the underlying process "at a glance". These kinds of displays are becoming increasingly common in many industrial processes. The situation illustrated on Figure 10.3a however can still be found on many user interfaces in industrial and commercial systems as well as consumer products. Users are presented with raw data, and it is left to them to perform the mental transformations needed to turn it into information the brain can think and reason with.

Presenting users with information in a way that is not meaningful also occurs in interfaces where graphical icons, colour, abbreviations, and acronyms are over-used or badly designed. In Part II, I described my experience as the owner of a new car which can drive autonomously. The driver's user interface in my car contains at least 85 different graphical icons, as well as many abbreviations and acronyms. These are all apparently expected to be meaningful to a driver who has been given no training on the car whatsoever. Many of them were not perceptually clear.

Ambiguous information means information that, in the mind of the user, could mean more than one thing, depending on the context or even the mode the system is in at the time. To take an obvious example there is a long history of operators involved in real-time control tasks being flooded with "nuisance" alarms. Alarms or messages generated by the system that are effective in drawing the user's attention, but that are generated too frequently and often when there isn't any problem the user needs to know about. They are referred to as "nuisances" because they occur so often, are distracting, but are usually not important (until, of course, they are). To a supervisory controller – faced with alarms or other system messages that they believe are "nuisances" - such alarms are ambiguous: they might be an indication of an actual problem...but they might not. The controller needs to decide whether to treat them as being real and act, or assume they are not real, and ignore them – or wait for further confirmation before deciding whether to act.

The ambiguity exists in the head of the user. Information that means different things depending on the system mode is not in itself ambiguous, though it can certainly lead to misunderstanding. If the user believes the system is in one mode, when in fact it is another, the information will not in itself be ambiguous; the user will be in no doubt what they think it means, even if they are wrong because the system is not in the mode they think it is.

These issues of information that is perceptually not clear, not meaningful and/or ambiguous, have at least two implications;

1. The controller will not be able easily to detect and understand the significance of signals about potential threats.
2. They may be reluctant to accept that something is actually an indication of a threat until they are confident they are not making a mistake.

The issue is that even if the supervisory controller is attentive, and does detect signs of potential trouble, that information may not be sufficiently clear and meaningful to generate an awareness of a potential threat.

A supervisory controller may need to draw on information from many sources, not only the user interface. Though even user interfaces are not always well designed. There is a large literature, including many sources of guidance about how to design user interfaces that present information clearly and is easy for the user to detect, understand and reason with. If the developers of a system expect the human users to understand the state of the system and identify potential threats through the user interface or other features they have designed into the system, they need to ensure the information they include in the user interface is clearly presented and relevant to the task: i.e. that it is perceptually clear, meaningful, and unambiguous.

Essentially, the issue is that a human supervisory controller, faced with information that is unclear, lacks meaning or may be ambiguous has to decide whether they believe they have detected signs of a potential threat. Or whether what they are noticing is just variability in the normal world. A conceptually useful way to think about this is in terms of the two parameters defined by the Theory of Signal Detection (TSD) which was (discussed in Chapter 8);

- how perceptually clear, or strong, is the information indicating a potential threat is (what is termed d' in TSD);

- where is the user likely to place their subjective response bias towards or against treating information they are unsure about as indicating a potential threat (this is the TSD parameter β).[17]

In terms of the risk we are discussing in this section – that of a supervisory controller not detecting or appreciating the meaning of information indicating the existence of a threat – the two TSD parameters provide a framework to help assess and make the case that the risk has been understood and mitigated[18];

1. Those involved in designing and developing highly automated systems should ask whether they have done everything they reasonably can to ensure information that will help users detect signs of potential threats is presented in a way that makes it is as easy to detect and understand as it reasonably can be. That is, have they designed their systems in a way that makes d' for information associated with threats as large as possible: that signs of potential trouble will be perceptually strong and clear.
2. Those with responsibility for implementing and supporting those systems should consider how things like working practices, incentive schemes, sales and marketing strategies, training material etc. can influence the bias users show towards or against treating information they are unsure about as potential signs of a developing threat. A sales and marketing strategy that emphasises – and perhaps even over-states – the reliability of an automated system might encourage users to adopt a bias against intervening even if they think they have detected a threat. In industrial and commercial settings, users of automated systems need to be aware of the extent to which management and operational policies as well as design decisions can influence the response bias adopted by their human supervisory controllers. Organisational arrangements, policies and incentive schemes that emphasise production and efficiency will tend to encourage users to adopt a high threshold: so, if they are unsure, they will be inclined to wait until they are very confident before considering something as a genuine threat. Which means they will tend not to react to genuine threats unless the signs are extremely clear. On the other hand, incentives and policies that have little tolerance for risk will tend to encourage controllers to adopt a low response bias. They will be inclined to see a lot of potential threats, many of which will be false alarms. Though they will rarely miss something serious.

Not seeing the possibility of something going wrong

Being able to interpret information from a user interface or other sources as indicating a potential threat to an activity is a characteristic of level 3 in Mica Endsley's well-known model of Situation Awareness.[19] Awareness at Level 3 means being able to interpret the implications of knowledge of the current state of the world (which is awareness at Level 2) into the future and to predict what could happen.

We all rely on Level 3 awareness all the time. If we could not mentally project into the future, it would be impossible to perform even the simplest tasks, never mind complex ones like supervisory control. Imagine trying to make a cup of instant coffee without Level 3 situation awareness: putting coffee, and perhaps milk and sugar into a cup requires little ability to see into the future (though they are, of course, directed towards achieving the goal of drinking coffee in the future). But what about filling the kettle, turning it on and … then what? Without the ability to predict the future, we would be unable to see the relationship between turning the kettle on and being able to pour hot water into our cup a few minutes later.

While the principle is the same, a human supervisory controller being able to predict the implications of something they detect in the world in terms of potential threats to an automated control activity is a cognitively much more difficult task. The ability of a supervisory controller, who is alert and attentive, to make good predictions of possible future events depends on at least three things;

i. a sufficient understanding of the automated function, what it does, its capabilities, and the limits of its abilities;
ii. a sufficient understanding of the properties and dynamic behaviour of the process or activity being controlled, and;
iii. the mental capacity to see the relationship between something they have detected in the world and the possibility of things turning out badly.

Concerns over lack of transparency about not only what automation does, how it works, what mode it is in but sometimes even its very existence have been well documented over many years.[20] Transparency refers to the extent to which the user can detect (whether visually or through other means) what the automation is doing, and to predict what it is likely to do in the future.[21] A lack of transparency makes it difficult for the supervisory controller to understand how the automation is likely to react in the face of a possible threat. While training can help the supervisor understand the basics of how automation works, in a real-time system of any complexity, that knowledge can be of little help to the controller faced with predicting how a usually reliable real-time control system might respond in the face of an unusual event or situation.

Having a reasonable understanding of the properties and dynamic behaviour of the activity being controlled, as well as relevant aspects of the world it exists in, is also necessary for the supervisory controller to be able to see the implications of something they have detected. This is partly down to the experience and competence of the individuals performing the role of the supervisory controller. Though if the activity or world is especially complex, or something unusual has happened, it can be very difficult even for a well-trained person to predict the implications.

The third requirement is that the supervisory controller has the capacity – indeed, even the willingness - to form a mental picture of events unfolding in an unwanted way. That is, being able and willing to visualise the possibility of something going wrong. Some years ago, along with Dr Laura Fruhen and Professor Rhona Flin, I contributed to a review of the psychological literature to try to identify the psychological attributes of the mental state referred to as "Chronic Unease" [17]. We were trying to establish a psychological basis for a programme aimed at helping those responsible for front-line safety in oil and gas operations develop a sense of wariness towards risk. Professor James Reason contrasted the mental state of chronic unease with that of complacency, where the absence of negative events can lead people to "forgetting to be afraid".[22] Our literature review identified five important psychological attributes, one of which was termed "Requisite Imagination". Requisite imagination refers to the ability to imagine and visualise future scenarios of things going wrong. People seem to have this ability to varying extents. This might be partly related to personality types, but it can also arise due to complacency based on not having experienced things going badly before or putting too much trust in technology.

Whatever the cause, a controller being willing and able to mentally see the possible relationship between something they have detected and things going wrong is an important and necessary attribute of a good human supervisory controller.

Cognitive bias

Chapter 8 considered the need for any proper understanding of the psychology of human supervisory control to recognise the impact that System 1 thinking, including the many cognitive biases and systematic errors of judgement associated with it, as well as a deep human resistance to exerting mental effort could have on the performance of someone filling the role. The comprehensive model of supervisory control (Chapter 9) speculated on some of the ways System 1 reasoning errors and biases could impact on the four core processes included in the model.

In the absence of directly relevant research, we lack knowledge of precisely how these systematic reasoning errors and biases might affect the ability of a human supervisory controller to detect or appreciate the meaning of the information available to them about potential threats. Though from what is known, it seems extremely unlikely that the judgement and reasoning involved would not be influenced by them in some way. Two likely candidates are;

- expectation, and confirmation; where, perhaps from training or prior experience, or from events in the immediate past, the supervisor has some reason to expect a particular event. So their attention and reasoning are focused on looking for evidence that meets or confirms the expectation. That focus comes at the expense of being less able to consider other information that – while it may indicate a genuine threat - is inconsistent with what is expected;
- commitment; where, because the individual has made a mental decision in the immediate past to commit to a course of action or belief, their attention and reasoning become focused on supporting that commitment, at the expense of being open to consider other evidence that might suggest they should be focusing on another threat.

There are many other possible types of bias that could cause a supervisory controller not to become aware of a potential threat or to treat what they do detect seriously. The bad news is that it is difficult to overcome the biases and irrational thinking associated with System 1 thinking. Simply making people aware of the tendency for bias or trying to train them to be resilient against their influence, has little or no effect. Some people seem to be mentally better equipped to resist being captured by System 1 reasoning errors than others. This is the basis of Shane's Frederick's famous "Cognitive Reflection Test"[23] [21] which seems to be able to identify people who are less prone to these kinds of reasoning errors. Though trying to select people as supervisory controllers based on their results on the Cognitive Reflection Test or some other psychometric test, seems unlikely to be practical other than in perhaps a few very special situations.

One strategy that has promise is based on the fact that forcing people, whether through the design of a user interface or by some procedural task, to exert a little bit of mental effort in performing a task can be effective in making them engage System 2, overcoming the biases associated with System 1 thinking. This is a strategy that has always seemed to me to have a lot of potential. While I have come across a few examples that seem to be successful, it is not yet something that has been widely adopted in the design of user interfaces to highly automated systems – or indeed, any other type of product or system.

Supervisory control risk 3: not realising the significance of the threat

The first two risks were associated with the **Attention** and **Automation Situation Awareness** processes identified in the comprehensive model of supervisory control developed in Chapter 9. The next two are both associated with the process **Threat Assessment and Monitoring** (see Figure 10.1). Details of this process are summarised on Figure 9.5 and the associated text in Chapter 9. The risk is that having become aware of an event or situation in the world that has the potential to be a threat, the supervisor still does not consider intervening or responding. Consideration of the detail of the process (Figure 9.5), suggests at least two possible reasons why this might occur;

i. they do not understand the nature of the threat or its potential impact on the automated activity;
ii. the perception of the risk associated with the perceived threat is below a threshold necessary to stimulate the individual into action: i.e. either the likelihood that what has been perceived as a potential threat will turn out to be real, or, if it does happen, the impact it is likely to have on the controlled activity is assessed as being not significant.

Supervisory control risk 4: placing too much trust in the automation

The fourth risk shown on Figure 10.1, and the second associated with the process **Threat Assessment and Monitoring,** is that, even if the threat is understood, and recognised as being a risk to the automated activity, the human supervisory controller believes, incorrectly, that the automation will deal with the threat.

As has been discussed earlier in the book,[24] a lot of both research evidence as well as applied experience has found that people – whether in the role of supervisory controllers or who interact with automation in other ways - frequently come to develop too much trust in the technology [12,13]. Sometimes they even continue to believe in its abilities and reliability after they have seen it fail. This has been referred to as an "automation bias": when an individual develops a high level of confidence and trust in the reliability and abilities of automation, even when they may have personal experience or knowledge that the automation is not as capable or reliable as the bias leads them to believe.

Our interest here is in the specific situation where the supervisor has identified a threat, understands its significance and has decided that it is a serious risk, but then places their trust in the automation to deal with it. When, in reality, the automation does not have the capability of dealing with the threat. This is different from the situation where the supervisor doesn't pay sufficient attention to monitoring the automation because of a belief in its reliability. Research has demonstrated that people who place too much trust in automation tend not to monitor the automation as often as they should. In the situation we are interested in here, the user has paid attention, has detected the threat, understands it, and has judged that it is a genuine risk to the controlled activity. But they believe that the automation will manage it. A belief that is not justified, and that has the potential to lead to an adverse event.

To give a simple example, Part I described the cognitive challenges and feelings of unease I experienced transitioning to the ownership of my self-driving car. I described many situations where, while I was getting used to the car's autonomous capabilities, I had to make uncomfortable judgements about the car's abilities. For example, on one occasion with the car's autonomous capability engaged, I could see stationery vehicles about 400 metres ahead. But I didn't know if the car would become aware of them in time to slow down and stop safely. On another occasion, again with the car in autonomous mode I could see an unusually sharp corner on the motorway ahead, together with speed restriction signs. But would the car detect it in time to take the corner at a safe speed? Initially, my reaction was to step in and take over so I knew the car would be under control. But having gained experience and confidence in the car's abilities, nowadays I would leave the car in control. But what happens when there is ice, or perhaps a spill of oil on the road surface? Is the car still capable of manoeuvring safely? I don't know (though I'm inclined to assume it's not). The risk is that I decide the car is capable when in fact, because of the unusual circumstances, it is not.

Of course, as the details of the Threat Assessment and Monitoring process of the comprehensive model shown in (Figure 9.5 in Chapter 9) illustrates, an individual who believes the automation will deal with a threat that they know is real and serious, is unlikely simply to ignore that threat and blindly rely on the automation to handle it. They will continue to monitor the automation's response until they are confident it has been managed. And if they lose confidence it is being managed quickly or effectively enough, they are likely to move towards intervening.

Supervisory control risk 5: deciding not to intervene

The fifth risk identified on Figure 10.1 is associated with the process **Intervene?** Details of this process are summarised on Figure 9.6 and the associated text in Chapter 9.

The specific risk is that the supervisory controller realises the automation cannot deal with the threat without support, but, despite that, they still decide not to intervene. The immediate question must be, why would someone in this situation not intervene? Not to do so sounds, at the least, careless, if not reckless or even negligent.

To begin to understand why it is a risk that those who develop, implement, or regulate highly autonomous systems need to take seriously, and that cannot simply be dismissed by blaming users, we need to do two things.

First, we need to remind ourselves of the warning and reality check made at the start of this chapter (as well as in Chapter 9). We need always to remember that thinking and decision making in any remotely complex real-world situation involves a great deal of simultaneous cognitive activity: it does not proceed in anything like a structured and logical way. So while, perhaps with hindsight, we could describe the situation as involving someone who is aware there is a threat, knows it is serious (i.e. likely to create a problem with significant impact if it is not controlled) suspects the automation is not capable of dealing with it unaided, but, even so, does not intervene, the user's mental state in the reality of the real world is unlikely to be anything like as clear as this description makes it sound.

In terms of the risk we are considering, it is not a question of someone having consciously thought through all the stages in sequence, as this chapter has done; deciding where and when to pay attention; becoming aware of a threat; considering whether they understand it; and recognising that the automation is not able to deal with it. While there may be a serial time-base of events dictated by the laws of the natural world, most of the cognitive activity involved in generating situation awareness, making judgements, reasoning, and making decisions and inferences will be happening in parallel, with many interactions and feedback, and probably involving both conscious and sub-conscious thought.

Local rationality, situation, and context

The second thing we need to do to be able to understand this risk, is to recognise that for most people, the way we think, behave and act in real-time, in the real world is always influenced by both the *situation* and the *context* we are in at the time. The concepts of situation and context are extremely important in understanding human behaviour and decision making, especially when people don't do what it is expected they would, and unwanted events follow. We need to explore these two concepts a little further before returning to consider the fifth supervisory control risk.

The situation and the context people are in at the time they think, make decisions and act are the elements of what Human Factors professionals refer to as "local rationality". Local rationality is about trying to "get inside the head" of the people involved and to understand how they are – or were - likely to think about their tasks and the world they are in. It's about standing in the shoes of the individuals involved and trying to see the world as it seemed to them at the time. As the former pilot and widely respected thinker and writer on safety management Sydney Dekker puts it;

> …people in safety-critical jobs are generally motivated to stay alive, to keep their passengers, their patients, their customers alive. They do not go out of their way to deliver overdoses, to fly into mountainsides or windshear; to amputate wrong limbs…In the end, what they are doing makes sense to them at the time. It has to make sense, otherwise they would not be doing it. [19]

A few years ago, I led a team of highly experienced specialists on behalf of the Chartered Institute of Human Factors and Ergonomics (CIEHF) in preparing a

White Paper on the subject of 'Learning from Adverse Events' [20]. Our objective was to help organisations understand the Human Factors perspective to investigating and learning from adverse events, and to identify a few key principles organisations could apply to capture the human contribution to those events. In the white paper, we provided definitions of the concepts of situation and context as they applied to understanding human behaviour[25];

> Situation is the set of circumstances particular to the specific time and place in which an adverse event occurred. Situational factors are essentially factual and, in principle, are discoverable as 'evidence' of the situation the individuals involved were in **at the time the event occurred….**
>
> Context refers to the meaning ascribed to a situation by the individuals involved and the **long-held beliefs** they hold about the situation they are in. It means factors likely to influence what people believe about the immediate situation, and what is expected of them. Context is derived from prior knowledge, from what we have been led to believe, as well as our experience of similar situations in the past. It includes the influence of the values, priorities and goals of the organisation and society. By its nature, defining context involves some speculation: trying to ascribe meaning to people's intentions and behaviours that are not observable, and not available as factual evidence of what may have motivated decisions or behaviours.

Situational factors may have some impact on whether our human supervisory controller decides to intervene in the situation we are interested in here. Though contextual factors are likely to be much more important: what beliefs does the individual hold that could influence the decision whether to intervene? We have already established some of those contextual factors: the controller *believes* there is a threat; they *believe* it is significant; and they have made a *judgement* that the automation is not going to be able to deal with the threat. These only exist in the mind of the supervisory controller. And none of them need necessarily be true: what they have decided is a threat may just be a misunderstanding or misjudgement of something they have noticed; because of other controls in the system unknown to the user, even if it does happen, it may have little effect; and the threat may actually be well within the automation's abilities, though the user doesn't know that. What is important is that they establish a context in the mind of the controller. A context in which they need to decide whether (and how) to intervene.

While many other contextual factors may also come into play (perhaps the individual has noticed the automation behaving in unexpected ways recently, distraction through personal or other worries, their values or sense of self-worth, or their perception of their role and status relative to peers), there are at least three other important contextual factors that can bear on whether the individual decides to intervene;

- whether they believe they have the *authority* to intervene;
- whether they believe they have the *ability* to intervene;
- where they believe *responsibility* for the consequence of their decision is going to lie.

There are operations and activities where people are so highly trained that they can respond to events in a way that overcomes the influence of the situation and - especially - the context they are in at the time. The purpose of the regular and thorough training military personnel, pilots, operators of nuclear plants and others undergo is to try to ensure they will follow a predefined set of behaviours in response to defined threats or other prescribed events. For our purposes here we will assume that we are dealing with human supervisory controllers who have not had such rigorous training.

Why would they not intervene?

With those two reality checks in mind, and an awareness of the way that situational and contextual factors can influence the way people think and make decisions in real-world tasks, we can return to the question posed at the start of this section: why would someone not intervene when they know there is a significant threat that they believe the automation cannot deal with? Figure 9.6 suggests several reasons, including;

- not knowing what to do;
- believing they do not have the authority;
- believing they do not have the ability;
- making a judgement that the benefits of intervening do not justify the perceived costs.

Not knowing what to do

Perhaps the most obvious reason for someone not acting in the risk situation we are considering is that they simply don't know what to do. Even if they have been trained, if a system lacks transparency, or after long periods of the automaton performing reliably with little for the human to do, the supervisor may have lost awareness of the state of the controlled activity or the system or what the system is currently doing. Unless the threat is one that the user recognises is a trigger they have been trained to respond to with a predefined procedure, they may not know what they need to do to respond to the threat.

Though experience has shown that, even skilled professionals such as pilots in the role of supervisory controllers who have been trained to respond to specific threats do not always follow that training and intervene effectively. A powerful example is the way the pilots of the Air France Airbus flight AF447 failed to follow the procedure they had been trained on only a few months previously in the event flight computers lost air speed data resulting in their aircraft crashing into the North Atlantic on 1 June 2009.[26]

Believing they do not have the authority

A second reason the supervisory controller may decide not to intervene in the risk situation we are considering is that they believe they do not have the authority to do so. It might be expected that anyone put into the role of being a human supervisory controller of an automated real-time control activity would have the authority to step in take over control from the automation should they decide they need to. However, in a world where AI-based automation is increasingly reliable, and being used to perform ever more complex, sophisticated, and critical functions, that is not always the case.

Automation is increasingly being used to control functions that can be critical not only to safety, but to the security of enormous sums of money or personal or commercially sensitive data. Increasingly, organisations are taking the view that their operations are safer, more secure, and more reliable when they do not allow people to intervene in automated processes. It has been the case for many years in nuclear power, as well as other types of operations where real-time processes are run by automated process control systems, that in the event of failure or abnormal situations, humans are expected to keep their hands off the controls until the automation has had the opportunity to bring the system to a safe state. A similar issue was behind the widely reported incident at the Wimbledon tennis championships in 2025 when a chair umpire did not intervene after the Electronic Line Calling system failed to call a ball that had clearly landed out of play "out": under the policy put in place by the organisers, chair umpires had no authority to overrule the system.[27]

Unless the boundaries are extremely clear, the confidence and reliance that is placed on automated systems in commercial and industrial operations has the potential to create a lack of clarity and uncertainty in the mind of human supervisory controllers about when they have the authority to intervene. Lack of clarity and uncertainty that, if it also exists in an organisational environment where controllers are held responsible for any disruption to real-time operations, can act as a disincentive to intervening even if the supervisor has doubts about the abilities of the automation to deal with a threat. In some real-time process control operations, control room operators are actively discouraged from intervening to slow-down or shut down a process. The potential for blame and being held responsible for a loss of production can act as a disincentive to someone intervening when they should.

Of course, if the human decides to overstep their authority, intervenes, and successfully prevents an adverse event, the individual is unlikely to be found at fault. They might even be praised. Though they may still be required to explain why they intervened to disrupt a usually reliable automated process. And if the threats the individual detected and responded to are not also understood by those who want that explanation, the individual may still find themselves in a difficult position. Much experience, as well as the knee jerk reaction of the international media and many politicians to want to find someone to blame, shows that more credibility can sometimes be given to judgements made with hindsight than to those made by the individuals who were actively involved and had to make decisions in real-time, often under pressure.[28]

The point is, that to mitigate the chances of the risk we are considering here actually leading to an adverse event, those expected to fill the role of the system's human supervisory controller need to be very clear about when they do and do not have authority to take over control from the automation. And they need to know that when they choose to exert that authority, they will be supported by those they are accountable to, whether that is their employers or the law.

Believing they do not have the ability

The risk we are interested here is not whether the human supervisory controller actually possesses the ability to take control to manage the threat they have identified.

It is whether they *believe* they have that ability. That means not only actually possessing the competence in terms of the skills and knowledge but also having sufficient confidence in their abilities to be willing to act.

People who have been moved to a supervisory controller role can find it difficult to maintain the skill they need to be able to take control over real-time activities they previously controlled manually. It has been recognised for decades that automation removes or reduces the opportunity for operators to practice and retain the skills needed to be able to exert manual control should they need to. It is difficult to put the issue more succinctly than Lisanne Bainbridge did in her famous, and very widely cited paper describing the "Ironies of Automation" in 1983:

> ...a formerly experienced operator who has been monitoring an automated process may now be an inexperienced one. [22]

This is of course compounded by the issue of where the individual believes *responsibility* for their actions lies. Again, what matters is what they *believe* in the moments they need to decide, not what the facts may be. A human supervisory controller faced with the situation of interest here is facing something of a double-edged sword. On one hand, an individual who, it turns out, lacks the necessary skill is likely to fail to control the situation, or may even make things worse, and be held accountable for their failure. On the other hand, the individual who is inclined not to intervene due to a lack confidence in their abilities, may believe they will be held equally liable for their failure to act.

Many organisations, in aviation, manufacturing industry, defence and other areas, recognise this issue of loss of ability and invest significant time and resources trying to maintain skill levels in their human supervisory controllers. Though experience has shown that even investment in sophisticated simulator-based training systems can be a poor substitute for the realities of the situation and context people face when unexpected events happen in the real-world. At the least, what matters for those developing highly automated systems, though perhaps even more for those who implement, operate, and regulate them, is that the risk of the people who are relied on to intervene if needed not having confidence in their abilities, or being unclear about what they may be held responsible for, are understood, and treated seriously. It is not easy to ensure skills are retained when there is little opportunity to practice them in the real-world. But effort should be made both to ensure the risk is mitigated as far as it can be, and to identify any residual risk that cannot be managed.

It should in principle be possible to ensure responsibilities are clearly apportioned and understood. Though, as we are currently seeing in the world of self-driving vehicles, where manufacturers remain keen for the human driver-in-control to hold ultimate legal responsibility for the safe movement of their vehicle and are keen to deny that their vehicles have self-driving capabilities while including increasingly more autonomous functionality into their products, the issue of where responsibility lies can be far from clear. While it may gradually become clearer in law, it is not necessarily clear in the minds of the human supervisory controllers faced with deciding whether to intervene to take over manual control from a highly automated system.

Making a judgement that the perceived benefits of intervening do not justify the costs

The final possible reason why a supervisory controller may decide not to intervene even when they believe they have identified a serious threat that they think the automation is not able to deal with, is because they assess the costs of intervening outweigh the benefits. This is not, of course, a conscious decision made following a process of rationally thinking through the various costs and benefits. It will be an intuitive judgement, probably made using a System 1 style of thinking, leading to a reluctance to act.

As Figure 9.6 illustrates, many factors could influence how an individual views the various costs and benefits of intervening. Among many, they might include;

- the perceived mental, and perhaps physical, effort of acting. What Psychologists refer to as the "Law of Least Effort"[29] recognises that, in the words of the Nobel Laureate Daniel Kahneman, "...laziness is built deep into our nature" [23];
- the impact the organisation's goals and incentives might have on the individual's motivation towards or against acting;
- the individual's sense of self-esteem and personal values: for example, if an individual places a high value on the status they believe they hold in the eyes of peers or colleagues, they may embrace the chance to demonstrate their competence and skill. On the other hand, the perception of possible embarrassment if they are not successful could hold them back;
- personal experience; an individual who had experienced criticism from colleagues or managers when they intervened in the past when, it turned out, they didn't need to may be reluctant to put themselves in the position of repeating the experience. On the other hand, if they had been given encouragement for doing what they thought was the right thing even though, as it turned out, it was not necessary, they are likely to be inclined towards intervening;
- training can also play a role in defining what the individual might see as the various costs and benefits associated with deciding whether to intervene.

Supervisory control risk 6: errors in interacting with the automation

The final risk we need to consider in connection with the reliability of an individual in the role of a supervisory controller is that they make some serious error in interacting with the automation. Recall (from Chapter 2) that the definition of the scope of supervisory control adopted for this book includes times when the controller interacts with the automation, such as changing settings or limits or acknowledging alarms, but without taking over manual control of the activity or process under control.

Two different types of interaction errors are possible. The first involve failures effectively to perform tasks that require interacting with the automation's user interface. That could involve any of the three principal human error types – slips (accidentally acting on the wrong object, such as making a data entry error while typing) lapses (forgetting to complete some part of a task sequence) or mistakes (doing the wrong thing, believing it to be right, such as selecting the wrong option from a menu). There is nothing special about these types of errors made by a human supervisory controller than the same errors made by anyone interacting with technology. In both cases they can occur for a variety of reasons. Avoiding them is largely down to the quality of the user interface design (including its resilience to human error) and the competence, experience, and state of the user.

The second type of error interacting with an automated system involves the supervisory controller completing the intended interactive task successfully, but believing, incorrectly, that it has had the intended effect on the automation. That mistaken belief could be due to the individual not understanding how the automation works, or it could be due to not realising what mode or state the automation is in.

An example of the first of those types of automation interaction errors occurred in the events leading up to the shooting down of Korean Airlines flight KAL 007 in 1983.[30] In that incident, the pilot used a physical control (see Figure 6.2 in Chapter 6) to manually change the aircraft's autopilot from Heading model (in which it would follow the selected fixed heading) to flying in Inertial Navigation mode (INS, where it would follow a pre-defined track). The pilot is thought to have believed that simply changing the position of the manual control changed the mode and that, from that moment, the aircraft was following the intended flight path. Though all the pilot's action did was to put the autopilot into a state where it would change to INS mode when software checks found that necessary pre-conditions were met. Because those conditions were never met, the aircraft remained in Heading mode and continued on its fixed heading until it was shot down.

"Mode errors" are among the most common type of errors people make when interacting with technology, whether automated or not. They occur when controls behave in different ways, or instruments mean different things, depending on the mode or state the system is in. There is a great deal of research describing different types of mode errors in a wide range of different systems. An example that is relevant to human supervisory control, also an aviation disaster, occurred during the loss of the Air France Airbus AF447 in 2012. In that incident, after the autopilot suddenly and unexpectedly stopped working due to a loss of airspeed data, the pilots misunderstood which flight mode the aircraft was in. They appear to have believed it was in "normal" mode, when the flight computers make it impossible to fly the aircraft into a stall, when in fact it was in "alternate" mode, when a stall was possible. The pilots flew the aircraft into a stall from which it never recovered. One of the best discussions of mode errors, in human computer interaction with many examples, is contained in the 2nd edition of the classic book by David Woods et al 'Behind Human Error' [24].[31] And in his book "Taming Hal" [25], Asif Degani also describes many examples of mode errors in human interaction with automated systems. He also describes an approach to analysing user interface designs in a way that makes the potential for mode errors very visible and leads to improved designs.

The purpose here is simply to identify this risk as something that those developing, implementing, and regulating the use of highly automated systems need to be aware of. Developing design solutions and other strategies for avoiding or minimising the chances of these types of errors occurring is beyond the scope of this book. Anyone looking for design guidance would do well to start by consulting Mica Endsley and Debra Jones' book "Designing for Situation Awareness" [16].

SUMMARY

This chapter has presented a structure and systematic walk-through of six key risks associated with human supervisory control arising from the comprehensive model described in Chapter 9. We need to conclude however with a reminder of the reality-check given at the start of the chapter. Although the six risks have been discussed and described sequentially and reasonably independently, that is not meant to imply that they will necessarily be independent in the mind of the individual performing the role of a human supervisory controller in the real world.

The perceptual and cognitive processes underlying human performance in real-world situations involves a great deal of parallel activity and interaction and will frequently be far from independent from each other. That complexity and interaction means the risks discussed in this chapter are also unlikely to be independent from each other. To take perhaps the most obvious example, if someone has an unduly high level of trust in the capability and reliability of automation, not only is it likely to mean they don't monitor the automation as often as they should – leading to the potential for not detecting signs of a possible threat - they are also likely to put too much faith in the ability of the automation to deal with the threat. Those responsible for developing, implementing, and regulating systems that provide autonomous control over real-time activities need to be aware of that complexity when assessing the risks associated with the reliability of the human supervisory controllers their systems rely on.

NOTES

1. See for example EU Directives 2013/30/EU (Safety of offshore oil and gas operations); 2016/798 (Railway Safety); 2012/18/EU (Control of major accident hazards), and 2014/87/EURATOM (Nuclear Power). The UK has similar legislation.
2. This may change under recent Executive Orders signed by President Trump.
3. The Nuclear Regulatory Commission in the US currently requires safety analysis reports under 10 CFR Part 50 and 10 CFR Part 52. Both the Bureau of Safety and Environmental Enforcement (BSEE) and the Bureau of Ocean Energy Management (BOEM) require offshore oil and gas operators to identify, analyse and mitigate risks, but do not mandate a full safety case. Similar approaches are adopted by the US chemical industry as well as rail and transportation.
4. At the time of writing, it is not possible to be clear about the situation in the US due to the ongoing review of relevant legislation resulting from President Trump's Executive Order to review all such US regulations.
5. For anyone not familiar with these terms, Wikipedia [4] provides a reasonably concise summary of differences in different countries.
6. I discussed the AHARP principle in my first book [5]. Though I since learned that the concept was first suggested by Professor Erik Hollnagel.
7. As a reminder, Chapter 2 defined the scope of human supervisory control as including; i) monitoring what the automation is doing, ii) making adjustment to system settings or modes, that may include taking over manual control for short periods, but with

the intention of returning to automated control once the adjustment is made, and iii) responding to messages and alerts generated by the system to attract the controller's attention but that do not involve taking over manual control of the activity.

8 Chapter 2 distinguishes between internal and external threats a human supervisory controller needs to be attentive to and provides examples of both types.
9 For examples of best practice in carrying out Safety-critical Task Analysis, see [7] or [8].
10 See [7] for a specific example of a Human Reliability Method tailored to a specific use in one industry.
11 For a review of macrocognition in the context of teamwork, see [9].
12 See for example https://www.youtube.com/watch?v=H8mbpY2bcjE.
13 Issues associated with human supervisory control at the Buncefield incident are summarised and discussed in Chapter 6.
14 These are discussed in Chapter 6.
15 See the discussion on Chapter 7, including Figure 7.2a and b.
16 Drawing on some established psychological theories, and with a degree of speculation, Neville Stanton and Guy Walker provide an interesting exploration of some of the psychological factors that, may have influenced the thinking and decision making of the Class 165 driver leading up to the crash [15].
17 Rather than repeating the detail here, anyone wishing to apply these concepts should read the summary of TSD and the two parameters d' and β in Chapter 8. See also the Wikipedia entry for a more extended and theoretical discussion, as well as more detailed references: https://en.wikipedia.org/wiki/Detection_theory.
18 In Chapter 8 of my previous book [5], I include examples of how a range of decisions associated with the design and implementation of technology can influence d' and β.
19 Level 1 SA refers to detection and perception of information and cues "given off" by the system and objects in the world; Level 2 requires interpreting those cues and information to develop an understanding of what they mean in terms of the current state and performance of the system and objects in the world and how those objects are changing; Level 3 means projecting into the future to predict what those current states and changes might mean in terms of future states, threats, and events. See [16].
20 Think of the crashes of the BOEING 737 MAX incidents summarised in Chapter 6.
21 For a discussion and review of the concept of transparency in automated systems, see [16].
22 Complacency is itself a complex psychological topic. I contributed a chapter, based on a series of studies and workshops conducted for two multinational manufacturing organisations, to a book in which I reviewed the topic and proposed a practical 3-factor model of complacency [18].
23 The Cognitive Reflection Test comprise three questions. The most famous being "If a bat and a ball together cost $1.10, and the bat costs $1 more than the ball, how much does the ball cost? The answer is, of course, five cents, though the very consistent research findings from large populations around the world is that around 60% of people will get at least one of the three questions wrong. I have found these results to be consistent when I've used the test as an exercise with front line operators, engineers, managers and students in innumerable training sessions or lectures in many countries.
24 See for example the section discussing trust in automation in Chapter 7.
25 In 2021 I delivered an invited guest webinar to NASA's Safety and Engineering Centre (the NESC Academy) based on the CIEHF White Paper and focusing on the importance of understanding the concepts of Situation and Context. The webinar included examples of situational and contextual factors based on the investigation into the tragic fire at Grenfell Tower in London in 2017 in which 72 people died and another 70 were injured. A recording of the webinar is available here: https://nescacademy.nasa.gov/video/b05f02755ba64921904e9cf23e392d791d.

26 The events surrounding this crash, and the Human Factors lessons arising from the investigation, are summarised in Chapter 6.
27 This incident is discussed in Chapter 6.
28 The dramatization in the film Sully of the attempts initially to blame Chesley Sullenberger for landing his Airbus A320 on the Hudson river, rather than return to New York's La Guardia airport after both engines were shut down by bird strike immediately after take-off, on 15 January 2009 is a powerful illustration of this tendency.
29 The "Law of Least Effort" was discussed in Chapter 8.
30 This incident is summarised in Chapter 6.
31 A book that should be essential reading for anyone that wants to properly understand the nature of human error.

REFERENCES

1. Regulation (EU) 2023/988 on general product safety, amending Regulation (EU) No 1025/2012 and Directive (EU) 2020/1828, and repealing Directive 2001/95/EC and Directive 87/357/EEC. Accessed 18 April 2025. Available from: https://eur-lex.europa.eu/legal-content/EN/TXT/?uri=celex%3A32023R0988
2. General product safety regulations 2005: Great Britain (Updated 13 December 2024).
3. The Healthcare Foundation Using safety cases in industry and healthcare. 2012. Accessed February 2025. https://www.health.org.uk/sites/default/files/UsingSafetyCasesInIndustryAndHealthcare.pdf
4. Wikipedia entry on ALARP. Available from: https://en.wikipedia.org/wiki/As_low_as_reasonably_practicable#:~:text=As%20low%20as%20reasonably%20practicable%20(ALARP)%2C%20or%20as%20low,as%20far%20as%20reasonably%20practicable
5. McLeod, R.W. Designing for Human Reliability: Human Factors Engineering for the Oil, Gas, and Process Industries. Gulf Professional Publishing. 2015.
6. CIEHF. How to Carry Out Human Factors Assessments of Critical Tasks: Guidance for COMAH Establishments. Chartered Institute of Ergonomics & Human Factors. 2023. Available from: https://ergonomics.org.uk/resource/comah-guidance.html
7. Institute for Energy Technology. The PetroHRA Guideline. 2017. Available from: https://ife.no/en/project/the-petro-hra-project/#:~:text=The%20Petro%2DHRA%20method%20comprises,subsequent%20qualitative%20and%20quantitative%20analyses
8. Energy Institute. Guidance on Human Factors Safety Critical Task Analysis. 2020. Available from: https://www.energyinst.org/technical/publications/topics/human-and-organisational-factors/guidance-on-human-factors-safety-critical-task-analysis
9. NUREG. Building a Psychological Foundation for Human Reliability Analysis. NUREG-2114 INL/EXT-11-23898. 2011.
10. MAIB. Interim Report on the Investigation of the Collision between the Container Ship *Solong* and the Oil/Chemical Tanker *Stena Immaculate*, Resulting in One Fatality, 14 Nautical Miles North-East of Spurn Head at the Entrance to the Humber Estuary, England, on 10 March 2025. Marine Accident Investigation Board. April 2025.
11. CSB. Investigation Digest. The 20th Anniversary of the 2005 fatal BP America refinery explosion in Texas City, TX. Chemical Safety Board. March 2025. Available from: https://www.csb.gov/assets/1/6/csb_bptc_investigation_digest_v3_(004).pdf
12. Moray, N. & Inagaki, T. Laboratory Studies of Trust between Humans and Machines in Automated Systems. *Transactions of the Institute of Measurement and Control*. 1999;21(4/5):203–211.
13. Merritt, S.M., Ako-Brew, A., Bryant, W.J., Staley, A., McKenna, M., Leone, A. & Shirase, L. Automation-Induced Complacency Potential: Development and Validation of a New Scale. *Frontiers in Psychology*. 2019 February. DOI:10.3389/fpsyg.2019.00225

14. Health and Safety Commission. The Ladbroke Grove Rail Enquiry Part 1 Report. HSE Books. 2000.
15. Stanton, N. & Walker, G. Exploring the Psychological Factors Involved in the Ladbroke Grove Rail Accident. *Accident Analysis and Prevention*. 2011;43:1117–1127.
16. Endsley, M. & Jones, D. Designing for Situation Awareness: An approach to User-Centred Design. 3rd Edition. Taylor & Francis. 2025.
17. Fruhen, L.S., Flin, R.H. & McLeod, R.W. Chronic Unease for Safety Managers: A Conceptualisation. *Journal of Risk Research*. 2013;17(8):969–979. DOI:10.1080/13669877.2013.822924
18. McLeod, R.W. The Awareness of Risk, Complacency, and the Normalization of Deviance. Chapter 4 In W.S. Ghanem Al Hashmi (Ed.), Process Safety Management and Human Factors: A Practitioner's Experiential Approach. Butterworth-Heinemann. 2022, p. 35–48.
19. Dekker, S. The Field Guide to Understanding Human Error. Ashgate. 2006.
20. CIEHF. Learning from Adverse Events: A White Paper. Chartered Institute of Ergonomics & Human Factors. 2020. Available from: https://ergonomics.org.uk/resource/learning-from-adverse-events.html
21. Frederick S. Cognitive Reflection and Decision Making. *Journal of Economic Perspectives*. 2005;19(4):25–42. DOI:10.1257/089533005775196732
22. Bainbridge, L. The Ironies of Automation. *Automatica*. 1983;19(6);775–779.
23. Kahneman, D. Thinking, Fast and Slow. Allen Lane. 2010.
24. Woods, D.D., Dekker, S., Cook, R., Johannesen, L, Sarter, N. Behind Human Error. 2nd Edition. Ashgate. 2010.
25. Degani, A. Taming Hal: Designing Interfaces Beyond. Palgrave Macmillan. 2001.

11 Proactive operator monitoring as a Barrier

A major theme running through the previous chapters has been the importance of a human supervisory controller of an automated real-time activity being *proactive* in looking for information about the state both of the automation and of the activity being controlled. The purpose of that proactive monitoring is to try to detect, assess and, if necessary, respond to potential threats to the activity before they develop into problems. The alternative to monitoring proactively is being *reactive*, relying on the system or something else to prompt the human supervisor when there may be a need for them to intervene. Chapter 8 discussed some aspects of proactive operator monitoring (POM) and introduced the concept of the "POM scan".

This chapter explores the potential of POM to be relied on as a control or defence to protect against serious adverse events. The chapter starts by considering just how much reliance is placed on people monitoring proactively. It then explores the characteristics and properties expected of controls relied on to protect against serious adverse events using concepts from the widely used Bowtie Analysis method as the basis: in Bowtie Analysis terms, these controls are known as either "Barriers" or "Key Safeguards".[1] The chapter considers the key issue of whether POM could be capable of meeting the standards of quality, effectiveness and reliability needed for something to be relied on as a Barrier or Key Safeguard. Chapter 12 describes a structured means of evaluating the quality of support provided to the proactive operator monitoring task.

PROACTIVE OPERATOR MONITORING

In the Introduction to the book, POM was defined as the key *task* performed by an individual filling the *role* of a supervisor controller of a process or activity controlled by an automated system. The term "proactive operator monitoring" (POM) refers to the situation where a human self-initiates monitoring of information indicating the state of the operation without being prompted by alarms, or other sources.

POM requires the supervisor actively to seek out and attend to sources of information that can indicate whether the automated activity could be at risk, before it reaches a state where alarms or automated protection systems (if they exist) are triggered, and before the individual is prompted to act by other people or something else. Benefits of proactive monitoring include;

- making operations more robust to failure or loss of reliability of automated systems, or to the occurrence of events for which the automation was not designed to be able to respond without human support;

DOI: 10.1201/9781003679479-14

- making it possible to react earlier and more quickly to potential divergence from the normal or expected state, and;
- increasing the time available to diagnose what is happening, and to plan and implement action to resolve the problem before it develops into a more serious situation.

Proactive operator monitoring can be argued to be among the most important, while simultaneously being one of the least understood, roles that people perform in controlling and managing potentially hazardous operations. The ability of an experienced individual to see, hear, smell, and even feel early indications of potential trouble, together with the cognitive capacities they bring to question, interpret, integrate, recognise, project, and visualise are relied on in many different situations, industrial, commercial, government and domestic, to detect early indications of potential problems. Even in relatively simple situations, such as automated fire-and-gas detection systems, a significant number of events are detected by people before automated detection systems are triggered.

The importance of operator monitoring - both proactive as well as reactive – has been recognised in the process industries for many years. The topic has been the subject of a body of both basic and applied scientific research, as well as significant effort in the design of process control and operator support systems. As far back as the 1960s engineering psychologists such as John Senders [1] were conducting research into the cognitive strategies adopted by control room operators to allocate their limited attentional resources in an optimal way when they were expected to monitor multiple sources of information simultaneously. More recently, a significant amount of work has focused on the design of process control systems and their human machine interfaces to support POM (see for example [2,3]). In industries involving continuous process control operations, equipment suppliers such as Honeywell have developed systems specifically directed at supporting proactive monitoring in control room operators.

While it's importance is most widely recognised in control room operations, POM is also critical in many other areas where there is a need to protect against the possibility of things going seriously wrong. There is however little research or guidance available about how to ensure people are able to perform this critical activity effectively. Consequently, the psychological and organizational conditions that need to be in place to support it are often not recognized or understood. An exception is a study conducted for NASA by Randall Mumaw and colleagues that reviewed how pilots proactively monitor their flight path [4].[2] The importance of people monitoring proactively is also not restricted to automated processes. Throughout industry and commerce people are expected to be proactive in looking for and reacting to signs of possible trouble before they can develop into problems.

Professor Andrew Hopkins noted the reliance on operator monitoring as a key safety defence in oilfield operations. In his 2012 book 'Disastrous Decisions' that examined the role Human and Organisational Factors played in the events leading up to the Deepwater Horizon disaster [5] in the Gulf of Mexico in 2010, Professor Hopkins noted that:

> The design assumption was that the crew would be monitoring the well at all times and that they would quickly recognise when they had lost control of the well.

However;

> For nearly an hour before mud and gas began to spill uncontrollably onto the drill floor there were clear indications of what was about to happen. Had people been monitoring the well, as they were supposed to, they would have recognized these indications and taken preventive action. [5]

Proactive operator monitoring, then, can be argued, at least in some industries, to be among the most important controls relied on to protect against major adverse events. The importance and challenges of POM are however frequently underestimated, and the psychological and organizational conditions that need to be in place to support it properly are rarely recognized.

In my first book [6], I used the 2005 explosion and fire at the Buncefield fuel storage depot as a case study, to examine in detail the role of POM as a Barrier in tank filling operations.[3] My review identified lessons about the impact issues including job design, working arrangements, vigilance and fatigue, control room design, and incentive schemes among others can all have on the ability of operators to monitor proactively.

BOWTIE ANALYSIS (BTA): KEY CONCEPTS AND TERMINOLOGY[4]

Before exploring POM as a safety defence, we need to understand the basic concepts behind Bowtie Analysis (BTA) and especially the concept of "Barriers" as a safety defence. Some of the concepts and terminology used here are slightly different from what is generally considered "best-practice" in conducting BTA in industrial settings. I need to briefly explain why those differences exist.

The content of this chapter draws on experience from four projects which provided insight into the theoretical and practical issues involved in using the BTA method in different domains;

1. Leading a cross-industry team of Human Factors specialists in developing the White Paper 'Human Factors in Barrier Management' published in 2016 by the Chartered Institute of Ergonomics & Human Factors (CIEHF) [7]. The aims of this White Paper were; (i) to bring clarity to areas where there was ambiguity or confusion in the way human performance was generally treated in BTA up to that time and, (ii), to set out recommendations for good practice in developing and managing those elements of barrier systems that either rely on, or can be defeated or degraded by, human performance.
2. Working as a member of a large team drawn from over 30 companies that supported a joint project by the US Centre for Chemical Process Safety (CCPS) and the UK Energy Institute to prepare the book 'Bowties in Risk Management' [8]. Published in 2018, the book was intended as a definitive statement of what constituted best-practice in applying the BTA method for use in process control applications.

3. A series of informal studies conducted on behalf of NHS Education for Scotland (NES) between 2016 and 2020 exploring the potential use of BTA to prospectively identify and evaluate the quality of controls against major risks in health and social care.[5] The principal outcome from these studies was a document intended to provide medical staff and other healthcare professionals who were not specialists in BTA or Human Factors with practical guidance on using BTA in health and social care settings [9,10]. The studies concluded that due to the complexities involved in health and social care, as well as the practicalities and logistics of conducting safety analyses in those settings domain specific and customised guidance was needed.
4. A study for Iannród Éireann (Irish Rail)[6] that used BTA to identify and evaluate the quality of controls relied on to prevent what are known in the rail industry as "overspeed" events [13,15]. Apart from the detailed analysis of the safety issues involved, the study developed several novel graphical objects for representing the results of a BTA, as well as a means of summarising the results of what was a complex and detailed analysis into a single summary graphic that captured the main information senior managers were likely to need.

Figure 11.1 identifies nine key elements used in BTA. A BTA is structured around either three or four core concepts, depending on its' scope,[7] describing a particular issue of concern: Hazards, Top Events, Threats, and Consequences (identified on the figure as items 'A', 'B', 'C' and 'D').

- an individual BTA is based around the risks and controls associated with a specific "Hazard", or "Hazardous Situation" (item 'A' on Figure 11.1). In a vehicle for example, the kinetic energy of the vehicle's mass under motion is the hazard with potential to cause harm. A complex system, such as a

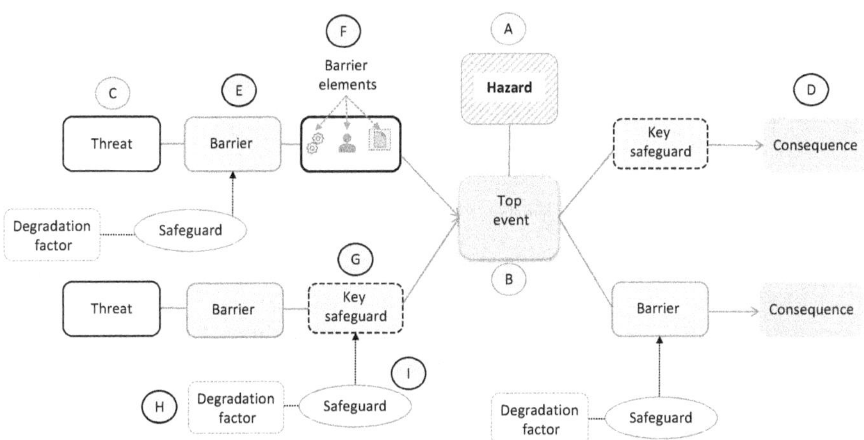

FIGURE 11.1 Summary of key concepts and terminology used in Bowtie Analysis

hospital or an oil refinery will have many hazards, each requiring their own individual analysis;

- each hazard or hazardous situation can be associated with one or more potential "Top Events" (item 'B' on the figure). Each top event represents one of the ways control over the hazard or hazardous situation might be lost, with the potential for harm or loss. In the rail study mentioned above, the top event was a train exceeding its permitted speed. Note that the top event is not itself the harm or loss that may follow;
- there are one or more "Threats" (item 'C' on the figure) associated with the top event. Threats are events or situations that, if they are not controlled, have the potential to lead to the occurrence of the top event. In the rail over-speed study, for example, failure of the train's brakes when approaching a speed restriction was a threat;
- if the top event does occur (meaning the controls on the left of the top event have all failed to block the threat), "Consequences" (item 'D') in the form of the potential harm or loss may follow.

Controls

At the heart of a BTA is the concept of ***Controls***. Controls are all of the things relied on to protect against threats leading to the top event, and from the top event leading to the consequences.[8] In principle, controls can take many different forms: from physical devices and structures such as condoms, dams and blast walls, alarms, and automatic safe-guarding systems, through training and supervision, checklists and cross-checks conducted by other people. The protection expected of controls can be delivered in three main ways;

i. by the use of technology (*Technological* controls);
ii. through efforts to exert control over how people behave or how work is performed (management systems, procedures, etc, collectively referred to as *Organisational* controls);
iii. by relying on competent people, who are fit to work, using their skills and judgement in compliance with work controls and other safe work practices (*Operational* controls).

There are two other important characteristics of controls used in BTA;

- They can be either *active* (they perform some active function), or *passive* (they provide protection through their existence alone);
- Active controls must have the ability to *detect* the threat (or occurrence of the top event), *decide* what to do, and *act* to block the threat.

Delivering the detect-decide-act functionality required of active controls often involves the coordinated activity of two or more *Elements* (item 'F' on Figure 11.1). Individual control elements can themselves be provided by technological, organisational, or operational means. For example, alarms are often proposed as controls.

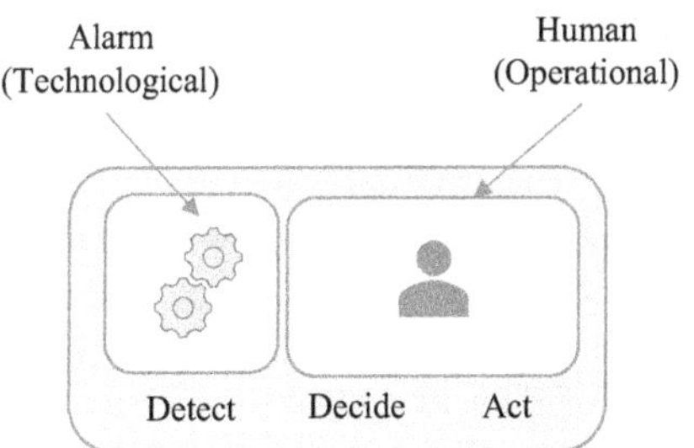

FIGURE 11.2 Illustration of the role of alarms as a control

However, an alarm on its own can do nothing other than alert a human to the occurrence of a situation that may require action. The alarm itself is a technological element that can only contribute to the full functionality needed of an active control when it is combined with the correct and timely response by a human (an organisational or operational element). This is illustrated on Figure 11.2.

Bowtie Analysis has three objectives; first, to identify the controls expected to provide protection against serious adverse events; second, to evaluate the quality of those expected controls to identify those that are sufficiently robust that they can be relied on; and third, to identify circumstances or events that might realistically cause each of those key controls to fail to deliver the expected protection, and to identify what measures need to be in place to prevent such failure.

In BTA, the term "Barrier" is reserved for those (normally few) controls identified as being sufficiently robust that they can be relied on. (On Figure 11.1, barriers are labelled as 'E'). Industries and organisations vary in the criteria used to define exactly what quality of control they are prepared to consider as barriers. Typically, controls are assessed against up to six quality criteria[9]:

i. *Ownership*: Is someone responsible for its implementation and performance?
ii. *Traceability*: Is it directly traceable to the organisation's management system?
iii. *Specific*: Is it specific to the threat?
iv. *Independence*: Is it independent of all other controls expected to protect against the same threat?
v. *Effectiveness*: Is it capable (provided it does what is expected) of preventing the threat from leading to the top event (or the top event leading to the consequences)?
vi. *Auditable*: Is there some sort of record or evidence that can be audited to confirm the control is in place and working as expected.

Except for some passive controls such as physical structures, in one way or another the capabilities of people, and their willingness to intervene when they need to, contributes to most controls. Sometimes that performance takes the form of a direct and active role in performing the control function, such as adjusting set limits or other control parameters, taking manual control over activity, or following a predefined procedure. More often it involves a secondary role through tasks such as supervision, inspection, calibration, or maintenance of technological controls.

An example Bowtie study

To give an example of the application of Bowtie Analysis, between 2016 and 2020, at the initiation of Professor Paul Bowie, who was leading work on Patient Safety on behalf of NHS Education for Scotland (NES), I led a number of informal studies to assess the potential use of BTA as a means of prospectively identifying controls against major risks in healthcare. There was no automation involved in the diagnostic process studied in this analysis, and there was no supervisory control role involved. But the example illustrates how the method is used, and the kind of insights and understanding of risk and how it is controlled a BTA can produce.

The first study[10] involved a workshop to review the controls experienced healthcare professionals believed should be effective in preventing what is referred to as a healthcare "Never Event": something that should never happen in healthcare but does. The top event for the workshop, associated with the hazard presented by carcinogenic drugs, was defined as; "Prescribing systemic oestrogen-only hormone replacement therapy for a patient with an intact uterus". A survey of over 500 General Practitioners had identified that 7% estimated this event to have happened at least once in their practices in the previous year, while 29% believed it was likely to happen again in the next five years.

The workshop initially identified nine controls the professionals attending believed should be in place and effective in ensuring this never event could never happen. When these nine controls were reviewed using the six quality criteria above, only three were considered robust enough to be considered as barriers. Of these three, two relied on several individual elements working successfully together to provide the protection expected. Further analysis identified more than twenty factors capable of defeating or degrading one or more of the three key barriers.

Barriers and Key Safeguards

To summarise, in Bowtie Analysis, the term "Barrier" is reserved specifically to refer to controls that, based on their assessed and assured quality, are relied on to protect against major events. For any organisation, whether an industrial company, a bank, or a government department to declare something as a barrier in the BTA sense is a significant step. Declaring a control as a barrier, and relying on it to protect against a hazard or hazardous situation means at least;

1. The organisation is confident the control can be relied on to block the threat.
2. The organisation is prepared to commit itself to ensuring the barrier is in place and fully functional, and that any degradation in capability can and will be identified and quickly resolved.

Not surprisingly, other than in the most sophisticated and highly engineered systems, there are typically few genuine barriers. Much more often, the things organisations rely on to defend against adverse events are not capable of meeting the barrier quality criteria. Either they are not capable of providing the level of reliable protection required, or they are not implemented, or given the levels of attention and support needed to ensure they will be both effective and reliable when they are needed.

In BTA, the term "Safeguard" is used to refer to all of these other controls expected to contribute to managing the risk associated with a threat. All safeguards are important: they usually include most of the elements in a Safety Management System. But, by their nature, safeguards cannot provide the level of protection against threats required of full barriers. When conducting a BTA, safeguards should not appear as controls on the main path between threats, top events, and consequences. (On Figure 11.1, safeguards are identified as item "I").

Sometimes, however, safeguards are so important that, even although they cannot meet the standards of effectiveness and reliability needed to be considered as full barriers, they nevertheless need to be given the same priority and managed as if they were barriers. These controls are referred to as "Key Safeguards" (shown as item 'G' on Figure 11.1). indeed, sometimes there are no controls capable of meeting the standard required of a full barrier: there are only key safeguards. Unlike standard safeguards, key safeguards can appear on the main threat line on the bowtie diagram.

Degradation factors

There are always situations, events, or states capable of preventing controls, whether barriers, key safeguards, or safeguards, from providing the protection expected of them, whether temporarily or permanently. In BTA, these are referred to as "degradation factors" – things capable of degrading or defeating the performance of a control. A BTA seeks to identify degradation factors that have the potential to defeat or degrade the performance of barriers or key safeguards. Other safeguards should be in place that protect the control from the effects of each degradation factor identified. (On Figure 11.1, degradation factors are identified as item 'H', and their safeguards by item 'I').

Human error is a degradation factor, not a threat

Prior to the publication of the CIEHF White Paper on 'Human Factors in Barrier Management' [7] and the CCPS/EI Guidebook [8], it was widespread practice to model human error in BTA as a threat leading to a top event. Both of those sources however emphasised that this approach was fundamentally mistaken. "Human error" is an event or outcome associated with people performing tasks to try to achieve a goal in a particular situation and context. It is not something that exists in isolation. At the top level of a BTA, "human error" should therefore not be treated as a top event or threat. Rather, it must be seen relative to the performance of a task. In BTA terms, human error should be treated as a degradation factor: something with the potential to defeat or degrade other controls, whether barriers or key safeguards. The six key risks discussed in Chapter 10 are all degradation factors that could degrade or defeat the performance of a human supervisory controller.

BOWTIE ANALYSIS APPLIED TO PROACTIVE OPERATOR MONITORING

To illustrate how proactive operator monitoring can have a place as a control against adverse events, Figure 11.3 shows the left-hand side of a bowtie diagram developed for an autonomous vehicle. The scenario represented on the figure is where the vehicle, under autonomous control, is approaching stationery vehicles in the road ahead. The hazard is the kinetic energy of the vehicle: energy that, if control over the vehicle's speed was lost has the potential to lead to a collision (the top event), with resultant potential for damage, harm, and loss (consequences).[11] The threat is the presence of stationary vehicles in the road ahead. The diagram shows two controls intended to prevent a collision, one technological, the other human;

1. The vehicle's adaptive control system (ADC) comprised of three technological elements: a radar system to detect the stationary vehicles ahead; a software algorithm to recognise the need to reduce the vehicle's speed, and the automated braking system to slow the vehicle down. These three elements form a feedback system capable, if necessary, of bringing the vehicle to a safe stop behind the stationary vehicles ahead.
2. The driver, in the role of a supervisory controller, proactively monitoring the performance of the automation and the state of the road ahead. The driver can detect the threat from the stationery vehicle and decide whether the automation is aware of the threat and able to take action to avoid a collision. And if they decide it is not, the driver can step in to take-over control of the vehicle's speed manually.

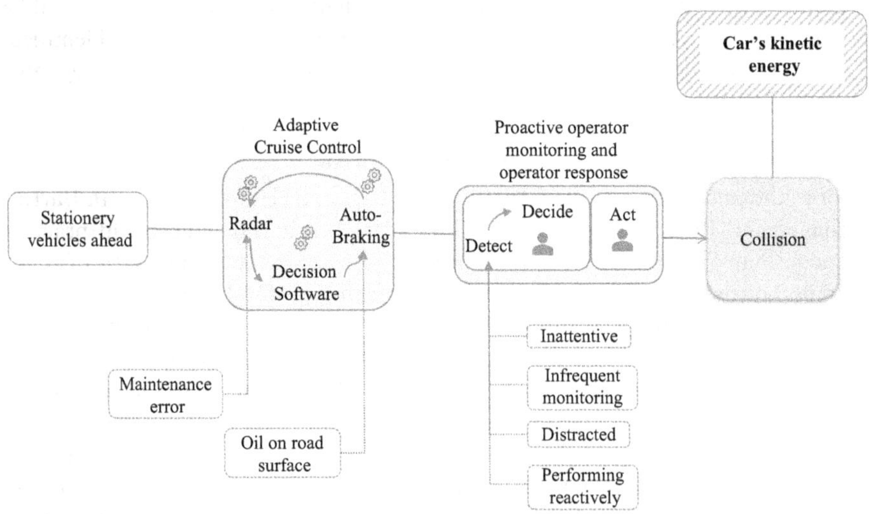

FIGURE 11.3 Example top level Bowtie diagram for Adaptive Cruise Control in a car (left-hand side only)

Figure 11.3 identifies two factors that could defeat performance of the adaptive cruise control system: an error in maintenance of the car's radar system, and oil on the surface of the road that prevents the automatic braking system from stopping the car in time. If either of these were present, the protection afforded by the adaptive cruise control could be defeated. The figure also illustrates four factors that could degrade or defeat the ability of the driver to detect the threat in time to respond: being inattentive; being actively distracted; not checking the road ahead frequently enough; and being reactive and relying on alarms or something else to attract their attention to the threat.[12] The presence of any of these would defeat the ability of the driver to detect and therefore react to the oily road surface in time.

Note that one of the barriers shown on Figure 11.3 comprises two operational barrier elements: "Proactive operator monitoring" and "Operator response". This needs explaining. To be able to think in terms of a defence against serious adverse events, we need to recognise that proactively monitoring can, (a) detect signs of potential threats, (b) decide whether there is a need to intervene, and (c) make adjustments to things like limits and settings. But it cannot include the step of taking over manual control in response to the threat.

From the moment the individual intentionally steps in and takes control, other than for the purpose of making an adjustment to the automation's settings or limits, they are no longer performing the proactive monitoring task. After the decision is made to intervene and take over manual control, many other issues can come into play that are themselves complex, detailed – and indeed, difficult. These include; the need to ensure the supervisory-controller-transitioning-to-manual controller has a sufficiently accurate and up to-date situation awareness both of the state of the activity and of the automation; that they have retained the skills and knowledge to be able to take over manual control effectively; and that they know how to achieve whatever is needed using a system that they may not adequately understand.

Consequently, in order to think in terms of controls that provide all three elements of the detect-decide-act functionality needed for an active barrier, we need to include an operator response that involves taking over manual control. That response is more than POM. Which is why it is identified as a specific barrier element on the figure.

To return to the situation illustrated on Figure 11.3, we are going to assume that the automation would not be able to alert the driver to the presence of the oil on the road at least until the wheels lost traction on the road surface and the vehicle began to skid. Further, that if the human at the wheel was behaving reactively, rather than proactively, by the time they detected the skid (whether through the system generated alert or from their own perception of the vehicles movement) it would be too late for them to take action to present a collision. We are also going to assume that there is some form of indication that could, in principle, make the driver aware of the presence of oil on the road; perhaps sunlight reflecting off the oily surface, a warning sign, or a radio traffic announcement. Provided the driver is attentive, and behaving proactively, they have a chance of being able to detect at least one these indications before the car enters the skid.

The scenario described by Figure 11.3 illustrates the use of Bowtie Analysis to clarify what is relied on to protect against a collision in this hypothetical situation. It's worth briefly considering two other examples of applying BTA to real-world situations. Although neither of these involved automated systems, they both illustrate how Bowtie Analysis makes the reliance on people monitoring proactively in different kinds of operations explicit.

Proactive operator monitoring as a control in oilfield operations

In early 2017 I was engaged by a global company, a leader in providing oilfield services to the international oil and gas industry, to carry out a review of a sample of their corporate bowties for well operations [14]. The review used the CIEHF White Paper [7] and CCPS/EI guidance [8] as baselines. Oilfield operations rely heavily on technological barriers – the most prominent being what is known as the "Blow Out Preventer" – being in place and effective. In field operations, staff of the service company who engaged me work closely with employees of other companies who are expected to monitor the status and availability of those technological barriers. Responsibility for the performance of those barriers usually lies with other companies and is outside the service company's direct control.

The review concluded that most of the oilfield service operations my client company were responsible for relied on at least two of the three full preventive barriers between threats and top events illustrated on Figure 11.4. These are barriers that both; (a) are within the service company's control and, (b) can meet three minimum criteria to be treated as barriers capable of preventing the occurrence of top events (i.e. independence, effectiveness, and auditability). The three barriers were;

i. operations being conducted in compliance with a validated plan;
ii. alarms and operator response (where the operation, equipment and technology involved supports alarms), and;
iii. proactive operator monitoring and response.

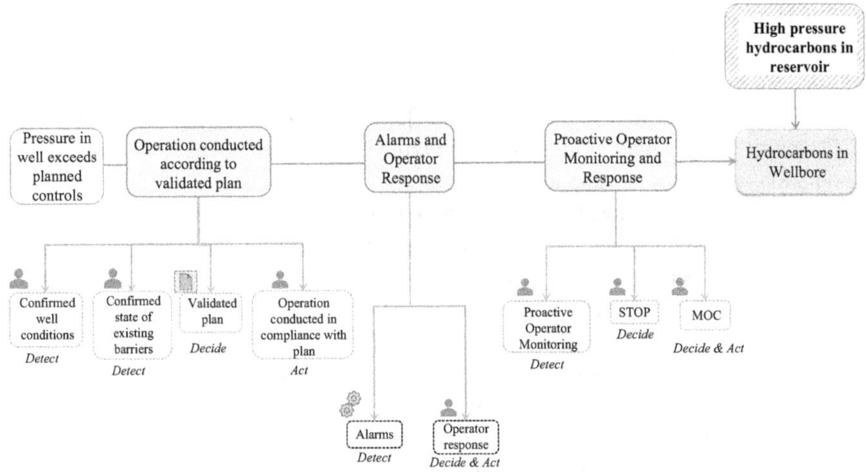

FIGURE 11.4 Left hand side of a top-level bowtie for well service company operations, showing the elements comprising the three main path Barriers. The figure indicates which of the detect-decide-act capabilities are met by each barrier element. MoC = Management of Change. Redrawn from [12]

All three are active barriers. Figure 11.4 illustrates that the required detect-decide-act functionality in each case is achieved through the performance of several individual barrier elements. As the figures illustrates, most of those barrier elements rely on human performance. Proactive operator monitoring, together with the operator's response, was a standalone barrier in its own right.[13]

Proactive operator monitoring as a control against rail overspeed events

Shortly after the oilfield services review, I was engaged as a consultant to lead the study for Iannród Éireann (Irish Rail) referred to earlier. The aim was to use Bowtie Analysis to identify and evaluate the quality of the barriers the company relied on to prevent what are known in the rail industry as "overspeed" events.[14] Overspeed events, which occur when trains exceed their permitted speeds, present a significant threat to the safety of railway operations. Exceeding the permissible speed for a section of track has the potential to result in a range of consequences from passenger discomfort to derailment, multiple fatalities, infrastructure damage and significant financial loss. Unlike driving vehicles on roads, train safety at higher speeds is not predominantly a function of the skill of the driver but of the physics at the interface between the wheel and the rail.

The analysis identified eight different threats, each having the potential to lead to an overspeed event, such as a failure of the train's braking system [13,15]. The detailed analysis, which drew on the experience of 15 highly experienced individuals, including current drivers, identified a wide variety of controls - technological, organisational, and operational - relied on to prevent each of these threats leading to an overspeed event. Figure 11.5 shows the top-level "at-a-glance" graphic summarising the results of the left-hand side of the analysis (i.e. from the threat to the top event). Note that the graphic on Figure 11.5 is unconventional for a Bowtie diagram. The format and symbology was developed to summarise the overall results of the analysis "at-a-glance": detail is kept to lower-level diagrams and the analysis results table. For a user that understands the meaning of the symbology and colour-coding used, the diagram quickly shows;

- the number of threats that have the potential to lead to the top event (eight in this case);
- the number of controls relied on to protect against each threat;
- whether those controls meet the quality standard of "Barriers" or if they are "Key safeguards" (colour coded light green or light blue);
- the number of barriers that rely on technological, organisational or operational (i.e. human) elements;
- the number of degradation factors associated with each control (shown as numbered stars).

For each one of the eight threats identified, at least one of the controls included a reliance on the train driver monitoring what was happening, detecting signs of things potentially going wrong and being prepared to step in and act before the train exceeded its permitted speed.

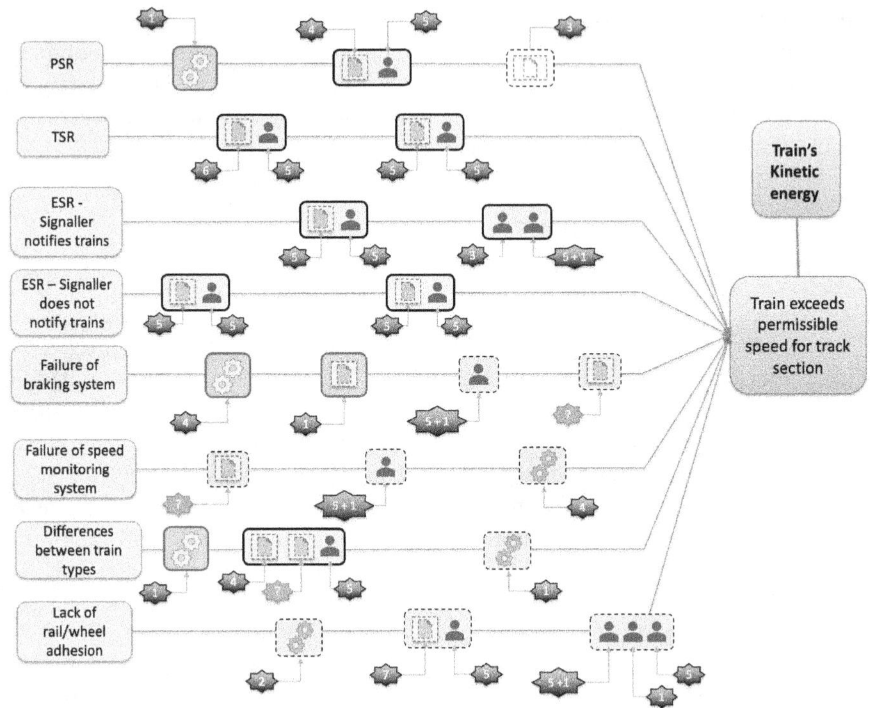

FIGURE 11.5 Example of "at-a-glance" Bowtie graphic developed for analysis of rail overspeed events. Left-hand side of analysis only. From 15

Barriers Must Be Understood to Be Independent

Before moving on to consider whether POM can meet the standards needed to be considered an element of a full barrier in the BTA sense, there is one other issue we need to be aware of. To do that, we need to examine the implications of the Bowtie diagram a little more closely.

The visual representation of a Bowtie Analysis can have a powerful influence on how the users of the diagrams understand and think about how risks are controlled. They can give the impression both;

- that barriers function in the order in which they are laid out on the diagram; and,
- that the barriers located towards the left of each side of the diagram (i.e. between the threat and the top event, or between the top event and consequences) are stronger than those towards the right.

So controls located towards the left of the bowtie might be assumed to both operate earlier and be more effective than those towards the right. If that were true in the case of the scenario of the autonomous vehicle approaching stationery vehicles on the road ahead illustrated on Figure 11.3, Adaptive Cruise Control would be viewed both as the most important control against this threat, as well as operating first. The reliance on proactive operator monitoring would only become necessary if the Adaptive Cruise Control system had failed. That assumption is wrong and can be seriously misleading. The power of the graphical representation has the potential to lead people who lack an understanding of the theoretical basis of Bowtie Analysis to misunderstand both the nature, and the expected functioning of the controls relied on to prevent against major events.

This issue actually goes deeper than simply the representation of Bowtie diagrams. As Professor Andrew Hopkins has pointed out [5], the problem is inherent in the visual representation of the Swiss Cheese Model of accident causation made famous by Professor Jim Reason. Professor Hopkins argued that immediately prior to the blow-out and explosion on the night of April 20th, 2010, the crew of the Deepwater Horizon were working under the dangerous assumption that, even if one of the controls they were relying on to prevent a blow-out (in this case, the cement job that was meant to secure the well) failed, others would work successfully;

> ...many people in their own minds were relying on subsequent defences to function properly should the cement job in fact fail. [5]

Every control relied on as a barrier is expected not only to be independent of the other barriers, but to be capable, on their own, of blocking the threat: they all need to be working simultaneously. Not only do barriers need to be independent, but everyone with a responsibility for managing critical activities needs to understand the controls and to recognise the importance of their independence. Most critically, they must not assume that other controls either have or will work as a reason for not ensuring that every one of them performs to the expected standard. Using the analogy of falling dominoes, Professor Hopkins demonstrated how, in the case of Deepwater Horizon, underlying human and organizational factors led to the defeat of every one of the controls that had been relied on not only to prevent a blow-out, but also to prevent the escalation that occurred.

Professor Hopkins' argument is that the controls that were expected to prevent exactly the type of incident that occurred failed precisely because operators did not treat them as being independent: they did not give the attention needed to individual controls because they assumed – explicitly or implicitly – that other controls either had or would work. They therefore failed to ensure that each control did its job properly;

> '...it was the whole strategy of defence-in-depth that failed' and that '...the reasons it failed are likely to operate in many other situations where reliance is placed on this strategy'. [5]

Figure 11.6 uses the same details as were shown on Figure 11.3 for the Adaptive Cruise Control example to illustrate a way of avoiding the potential for a Bowtie diagram to create this misleading impression. The diagram makes clear that there is no implication of any barrier operating before or being more important than any other: they all need to be active, all of the time, if the relied-on protection is to be achieved. If a human supervisory controller is to be relied on as a control against adverse events in highly automated systems, they must, whatever other barriers they believe to be

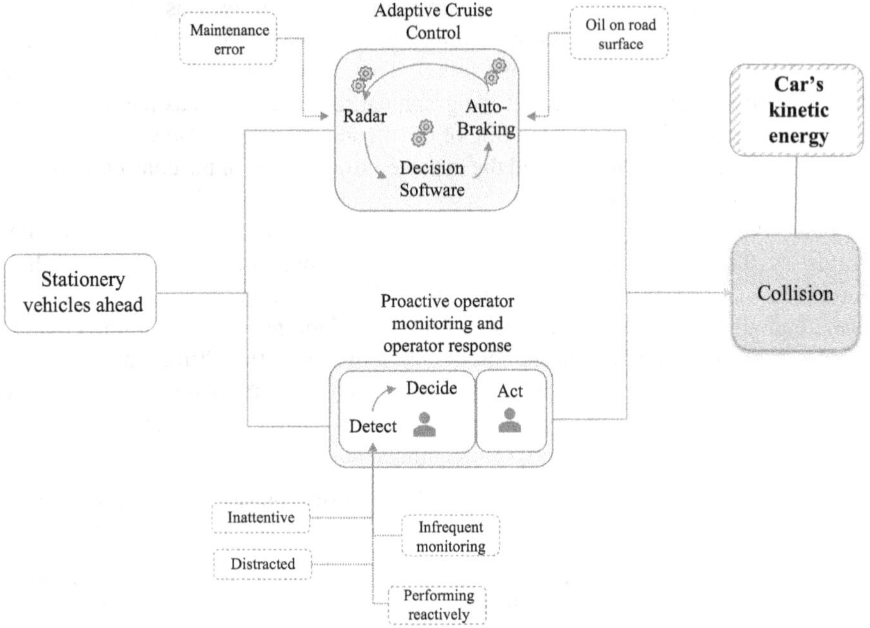

FIGURE 11.6 Suggested alternative means of representing the result of a Bowtie Analysis avoiding the implication that controls operate sequentially, and those to the left are stronger and of higher priority

in place, be proactive in monitoring for threats sufficiently often to be able to detect them before they become a problem. In the autonomous vehicle scenario illustrated on Figure 11.3, the driver cannot simply rely on the ADC system and assume they will be able to be able to step in if it fails. They must carry on monitoring proactively simultaneously with the automation.

This is a real challenge: how to design user interfaces to highly automated and autonomous real-time control systems and preparing and supporting the individuals who will need to act as supervisory controllers of those systems in a way that allows them to be effective as proactive monitors.

CAN POM BE A BARRIER ELEMENT?

So far, this chapter has argued that, even it is rarely explicitly recognised, proactive operator monitoring is widely relied on as a control against major adverse events. It has also argued that little thought is generally given to understanding what is needed to provide confidence that an individual expected to monitor proactively will be able to do so reliably and effectively. The chapter introduced a framework (using Bowtie Analysis) that can be useful in thinking about the quality needed for controls to protect against major events. And it has argued that there are cases where POM can be, and frequently needs to be, relied on as an element of key safeguards.

This final section gets to the heart of the matter. It considers what would be needed to ensure POM could satisfy the quality criteria needed for controls to be considered as an element (along with the action provided by the operator response) of a full barrier in the strict, Bowtie Analysis, sense. The focus is on situations where individuals are required to act as human supervisory controllers of automated real-time systems and who, therefore, are expected to perform the task of POM reliably and to a high standard.

We need to start by recognising that, in terms of the ability to implement, influence and assure controls against major events, there are at least two different situations – or use cases - in which someone might be relied on to be effective monitoring proactively;

1. Where the automation is implemented and used in the service of an organisation of some kind, and where the individual(s) relied on as supervisory controllers are expected to fill the role as part of their employment or other contractual obligations. In this situation, the people involved have no choice whether to fill the role of a supervisory controller: it forms part of their work responsibilities. These individuals are non-discretionary human supervisory controllers.
2. Where the decision to use a highly automated product or system is the personal choice of the individuals who will need to fill the role of its supervisory controller. In this case, use of the automation does not form part of any responsibilities to an employer or anyone else. Though responsibility for any adverse consequences will lie, partly at least, with the individual. This situation is most obviously associated with consumer products such as autonomous vehicles. Though as ever more aspects of our everyday lives become automated, the number of situations where this discretionary supervisory control role is going to exist is only going to grow.[15]

The essential difference between these two use cases is that, in the first, there is an authority that can exert some degree of control over the circumstances in which the automated system is used. And therefore the conditions (the situation as well as aspects of the context) under which the human supervisory controller performs the POM task.

Table 11.1 contains reflections and suggestions on the kind of activities that might be carried out, and what other kind of controls might be put in place to provide assurance over the ability of the human supervisory controller to perform the POM task. The table treats the case of the discretionary and the non-discretionary user separately. The content of the table includes consideration of the period when individuals are expected to transition from being a manual control of the process or activity, to being a human supervisory controller. The table has customised the general barrier quality criteria described earlier in this chapter and made them specific to the case where proactive operator monitoring is being considered as a potential barrier element. The table considers the kind of things that might be done at three different stages in the life of a new highly automated product or system to give confidence that POM could be relied on as a control;

- at the start of the product or system's life cycle when the concept is being developed and is going through design and development;
- during the process of verifying that the new product or system is safe and suitable to be released and obtaining certification or other approval for it to be used;
- when the new product or system is being implemented and introduced into use for the first time. In this last stage, a distinction is made between the cases of the discretionary and the non-discretionary users.

Table 11.1 is no more than my own reflections and suggestions, based on my experience working as a Human Factors specialist on the research, design, development and analysis of complex industrial and defence systems over more than four decades. Hopefully, the table will stimulate others to think about these issues and come up with other activities or controls that will help improve the quality of support to POM as a control against major adverse events in future highly automated systems.

Despite the limited personal perspective, several conclusions can be drawn from Table 11.1. These conclusions are made against a background, as has been argued earlier in this chapter, that, however difficult it is to ensure the effectiveness and reliability of any kind of complex human performance as a safety defence, the fact is that many organisations that operate complex systems - industrial, military, commercial, financial as well as consumer products – already do rely on people proactively monitoring for signs of trouble nearly all of the time.

The first conclusion is that, while some of it may not be normal practice in the development or use of modern products and systems, there are a lot of things that could, in principle, be done that would strengthen the role of POM as a defence against serious adverse events. The key is to recognise, from the earliest stages of thinking about the concept for the new product or system, the importance of the role of human supervisory control, and the reliance on the individuals filling that role monitoring proactively. Ensuring good practice is followed in applying Human Factors principles in the design of automated systems is the least that could be expected (such as, for example [16,17]).

Statements, typically prepared early in the development lifecycle, such as the automation policy and the design and use concept need to recognise that there is going to be a reliance on people performing the role of a human supervisory controller, and therefore monitoring proactively. Those statements also need to identify when there is going to be a reliance on people changing from a role where they were previously acting as a manual controller and transitioning to becoming a supervisory controller. And they need to recognise that there may be a need to provide support to those users during that transition phase, at least until the user has gained sufficient experience and understanding of the system, it's features, capabilities, and limits.

Table 11.1 makes clear that there is a need for those responsible for developing and using highly automated systems to be explicit, from as early as possible, about where the balance of ability, accountability, control and responsibility are going to lie,[16] and, especially, how, and when these allocations are going to change and how the human supervisory controller will be aware of both the current allocation and the changes.

TABLE 11.1
Reflections on activities that could help the task of Proactive Operator Monitoring (POM) meet the quality criteria needed to be considered an element of a full barrier against major adverse events, with a focus on transition to use of the automation (In the table, "HSC" means human supervisory control)

	What could be done to ensure POM meets the barrier quality criteria, especially during transition to use of the automation?			
Barrier quality criteria	**During design and development**	**During verification and certification**	**During introduction and initial use**	
			Discretionary users (Consumer products)	**Non-discretionary users (industrial/military/commercial systems)**
Ownership: Does the individual expected to transition to the role of supervisory controller understand the importance of monitoring proactively?	The system's Automation Policy and other material defining how the system is to be used should recognise the need to transition to or introduce human supervisory control and should acknowledge the need to support the transition phase. The policy should be clear about the balance of allocation of accountability, authority, control and responsibility between the human and automated elements and how the human allocations are to be supported during initial implementation.			
	Design concept should recognise that the system will introduce the need for HSC, and the conditions when it will replace existing manual control. The concept should identify any features or functions that need to be incorporated in the design to support POM, including during transition to automated use.	The product or system Safety Case (or other demonstration that the product or system is fit to be certified for use) should explicitly recognise the introduction of or transition to HSC and emphasise the importance of POM. The safety case should identify the characteristics of the individuals likely to fill the role of a supervisory controller and whether they will need to transition from manual controller of the process or activity being automated. It should identify what training or support is expected to be in place to ensure the individuals will understand the importance of POM, especially during the transition to use of the new product or system.	Manufacturer's sales and marketing material should explicitly recognise the reliance on HSC and POM. User manuals, training and support materials should emphasise that HSC's need to monitor proactively. User Support material should be clear about the balance of	Material controlling or guiding the use of the system should recognise the need for HSC and the importance of POM, e.g. Safety Management System; Operations philosophy and Manning policy; Local Work Instructions and Procedures; Job Descriptions, etc. Operations managers and work supervisors need to understand the importance of POM and ensure the individuals assigned the HSC role

			allocation of accountability, authority, control and responsibility between the human and automated elements.	also understand the role and the POM task. They should be available to provide additional support during transition when the automation is first introduced. Incentive schemes and other means of motivating staff should not create priorities or incentives that could lead those in the HSC role to give a low priority to performing the POM task.
Traceability: Is the importance of POM recognised and documented in an authoritative source controlling use of the automated system?	Not applicable. (This criterion refers to the point when the automation is in use).	The product or system Safety Case (or similar) should explicitly recognise the need for HSC and should emphasise the importance of POM. It should identify how those responsible for use of the system will know POM is expected to ensure the system can be operated safely and reliably.	User manuals and training material, including online instruction, should clearly emphasise the importance of POM.	Documents controlling how the system is to be used should recognise the need for HSC and the importance of POM, e.g. Safety Management System; Operations philosophy and Manning policy; Local Work Instructions and Procedures; Job Descriptions.

(Continued)

TABLE 11.1
(Continued)

Barrier quality criteria	What could be done to ensure POM meets the barrier quality criteria, especially during transition to use of the automation?			
	During design and development	During verification and certification	During introduction and initial use	
			Discretionary users (Consumer products)	Non-discretionary users (industrial/military/commercial systems)
Specificity: Is it possible to be specific about the kind of information that would indicate the presence of a threat as well as the kind of events or situations that would require them to intervene?	Information and Task analysis, or research supporting system development, should identify the kind of threats POM should be looking for, as well as the characteristics of information and sources that could indicate potential threats. Wherever risk and safety analyses identify a reliance on POM, it should be specific about characteristics of the information that need to be monitored as well as where that information might exist.	The product or system Safety Case (or similar) should demonstrate that sufficient analysis has been completed to ensure information and sources indicating key threats the HSC is relied on to detect and respond to have been identified. It should be possible to demonstrate how the results of the analyses have influenced the design of the user interface, user training or support materials.	User manuals and training material, including online instruction, should include details of the information and sources that POM should be looking for.	User manuals and training material, including online instruction should include details of the information and sources that POM should be looking for. Competence assessments should include assessment of individuals assigned to the HSC role to perform POM effectively.

	Operational experience as well as detailed knowledge of the activity or process being controlled should be used to advise on information and sources relied on to monitor for external threats.			
Independence: Does the individual expected to perform POM also have a role as part of another control to protect against the same threat?	Safety and risk analysis should be clear about the kind of people likely to fill the role of HSC including any other simultaneous responsibilities. It should identify whether the individuals filling the HSC role will also play a role in other key controls expected to protect the product or system.	If the product or system Safety Case includes reliance on HSC and POM as a control against major events, it should include an explanation of how independence is achieved for the POM task.	Unlikely to be achievable for discretionary single user products.	Job Design and Job Description for the individuals expected to perform the HSC role should ensure no conflict between performance of the POM task and responsibility for other safety-related tasks.

(Continued)

TABLE 11.1
(Continued)

Barrier quality criteria	What could be done to ensure POM meets the barrier quality criteria, especially during transition to use of the automation?			
	During design and development	During verification and certification	During introduction and initial use	
			Discretionary users (Consumer products)	Non-discretionary users (industrial/military/commercial systems)
Effectiveness: Providing the operator is proactive and monitors for threats sufficiently often, can they be relied on to detect and respond to all significant threats to the safe performance of the controlled activity?	The system's Automation Policy should be clear about the allocation of accountability, authority, control and responsibility. Design of the user interface, as well as user support materials should be clear about the characteristics of the individuals expected to fill the role of HSC and be expected to perform the POM task. The system design concept should recognise where existing manual controllers are going to have to transition to become HSCs and should recognise the potential need for additional support during the transition	The product or system Safety Case (or similar material) should be able to demonstrate what consideration has been given in design and development to supporting the POM task. It should specifically demonstrate that any special needs of people transitioning to the HSC role have been identified, and how the system supports them during the period. It should be able to cite evidence from user trials or other sources that demonstrate the support provided for POM. The safety case should be clear about the limits or constraints to POM assurance: i.e. examples of situations or threats where POM cannot be assured. The safety case should be clear about the assumptions made to support a case for POM effectiveness (such as Job design to avoid task conflicts, avoidance of conflicting priorities or incentives).	User manuals and training material, including online instruction can help ensure effective POM. Clear legal responsibilities on the user and insurance conditions, for example, can help encourage POM behaviour. Due to limited opportunities to exert control over the user's behaviour while using a product, POM effectiveness will be difficult to ensure for discretionary single user products.	Material controlling operations should emphasise and seek to ensure the effectiveness of POM. For example; Safety Management System; Operations philosophy and Manning policy; Local Work Instructions and Procedures; Job Descriptions. Operations managers and work supervisors need to understand the importance of POM and ensure the individuals assigned the HSC role are competent to and in a fit state to monitor proactively. They need to ensure other task and responsibilities do not conflict with allocating sufficient time for POM. They should be available to provide additional support during

Design of the user interface should provide effective support for POM, especially during transition when the user may have little understanding of the system's features, limits, and capabilities. The interface design concept could include features intended to support user engagement with the controlled activity while under automated/ autonomous control. The system should not have features, or behave in ways, that interfere with or distract from the ability of the individual in the HSC role to monitor proactively.	transition when the automation is first introduced. Incentive schemes and other means of motivating staff should not create priorities or incentives that could lead those in the HSC role to give a low priority to performing the POM task.

(Continued)

TABLE 11.1
(Continued)

Barrier quality criteria	What could be done to ensure POM meets the barrier quality criteria, especially during transition to use of the automation?			
	During design and development	During verification and certification	During introduction and initial use	
			Discretionary users (Consumer products)	Non-discretionary users (industrial/military/commercial systems)
Assurance: Are there any measures or indicators that would confirm whether the assigned individual is carrying out POM sufficiently frequently to be effective?	Design concept and system Automation Policy could potentially include features to monitor state of alertness and activity of the individual in the HSC role. Given the largely perceptual/ cognitive nature of POM identifying traceable evidence of its' performance is difficult.	N/A	Difficult to ensure for discretionary single user products. In principle, features could be designed into product or systems to monitor and report on the user's state of alertness and behaviour. Though there may be legal constraints or problems of customer acceptance.	In principle, features could be designed into products or systems to monitor and report on the user's state of alertness and behaviour. Though there may be legal constraints or problems of user acceptance. Records could potentially be kept of POM indicators (such as user interactions or possibly missed threats). Given the largely perceptual/ cognitive nature of POM identifying traceable evidence of its' performance is difficult.

The table also suggests that the other key document that can have a major impact on the effectiveness of POM in a new system is the product or system Safety Case - or whatever other demonstration is needed to gain approval or certification to sell or use the system. That document should serve as a repository of information demonstrating both that the issues of transitioning to human supervisory control have been recognised and understood, and that effective and sufficient measures have been put in place to ensure POM is properly supported.

The Safety Case also provides the opportunity to demonstrate that the developers and those who put the automated control system to work have addressed wider issues. Not least the design of the user interface to the new product or system, as well, in the case of non-discretionary users, as the way jobs are designed and the operational controls such as supervision, work practices and procedures put in place to control how work is carried out. As well as documenting how the organisation deploying the automation will ensure that that things like motivation and incentive schemes (whether explicit or implicit) don't encourage human supervisors to adopt priorities that conflict with their supervisory role.

The key question is whether POM (with the addition of the element of "Operator Response", though that would need to be evaluated as a barrier element in its own right) could realistically be relied on as an element of a full barrier in the sense it is used in Bowtie Analysis. That means a control element that can be relied on, even if every other control fails, to detect threats to the safe and/or effective performance of an automated real-time activity and to make good decisions about the need to intervene to prevent the threat disrupting the activity. For some, perhaps many organisations, such as in the process control industries and others that have historically been based on heavily engineered control systems, taking the step of explicitly stating a reliance on human performance in any form as an element of a full barrier against major incidents can be a step too far. Though even if they cannot state the reliance on human performance as a defence explicitly, it is clear from investigations into numerous industrial accidents that many organisations do so implicitly. Perhaps the most powerful example of this is Professor Andrew Hopkins' conclusions referred to earlier in this chapter about the issues that led to the explosion, fire, and subsequent sinking of the Deepwater Horizon drilling platform in 2010 – an incident that led to the largest environmental disaster in US history, and that came close to bringing one of the world's largest companies to its knees.

Unfortunately – and frustratingly - so far as the scope of this book is concerned, the answer to the question whether the task of proactive operator monitoring can be considered as an element of a full barrier has to be that it depends. It depends on how much effort and attention has gone into supporting the POM task throughout design and development as well as during the initial implementation and ongoing support of the product or system.

At the very least, even if POM cannot be treated as an element of a full barrier, it is certainly frequently relied on as a key safeguard. It is something that, even if it cannot fully satisfy the criteria for being treated as a barrier element in its' own right, nonetheless is relied on. As such, it needs to be treated as if it was a barrier element. That means giving it serious attention throughout the design, development and certification of a new product or system that will introduce automation to a real-time process or activity. And it means ensuring that the automation is implemented and supported in a way that makes people as effective and reliable in the POM as can realistically be achieved.

NOTES

1 Although this chapter uses concepts from the Bowtie Analysis method, similar arguments could be developed if other types of safety analyses were used, such as Safety Integrity Level (SIL) Analysis or Layers of Protection Analysis (LOPA). Difficulties will be encountered however if there is a need to try to quantify the likelihood of human failure in the proactive monitoring task. Trying to quantify the probability of human error with any degree of credibility remains challenging even for relatively simple human tasks, never mind something as inherently psychologically complex as proactive monitoring.
2 There is a description of this study in Chapter 8.
3 Lessons for human supervisory control from the Buncefield incident are summarised in Chapter 6.
4 Some academics have dismissed Bowtie Analysis as being fundamentally unsuitable for analysing safety controls in complex systems on the grounds that it assumes there is a linear cause-and-effect relationship between threats and consequences. This is based on some early users of the method treating a bowtie as an inverted cause-effect diagram. It is however a misunderstanding of the conceptual basis and modern use of Bowtie Analysis. For more detailed refutations of this argument, see [7,8,10].
5 See for example [11].
6 See [13].
7 Sometimes, a BTA may decide to focus exclusively on either the left-hand side (from threats to the occurrence of the top event), or the right-hand side (from the top event to the consequences) of the risk picture.
8 In industry, government and the media as well as everyday life, a range of terms are used to describe the same concept, including "checks and balances", "assurances"," protections", etc.
9 These six quality criteria are as defined in the CIEHF White Paper [7]. The CCPS guidebook [8] relies on criteria iv, v and vi.
10 Reported in [11].
11 Note that I have chosen to use the collision as the top event here, and the resulting loss, harm and damage as consequences on the grounds that other controls are possible that could prevent, or mitigate, the collision from leading to the consequences. An alternative approach would be possible to treat the top event as "loss of control over the car's speed".
12 All of these were discussed in Chapter 10 as aspects of the six key risks to effective human supervisory control.
13 Examples of several oilfield service tasks identified as relying on proactive operator monitoring are reported in [12]. See also Chapter 12.
14 A summary of this study is reported in [13].
15 The recent growth in things like automated lawn mowers and domestic vacuum cleaners are examples of other types of automated consumer products that involve automated control of real-time activities. Clearly, with products like these, both the demands on the human supervisory controller as well as the potential consequences if things go wrong are trivial compared with the case of automated vehicles. Though automation of other types of real-time consumer activities that are both more demanding on the human supervisor as well as of having more significant potential consequences is only going to grow.
16 See Chapter 2 for a discussion of the importance of these issues.

REFERENCES

1. Senders, J. The Human Operator as a Monitor and Controller of Multidegree of Freedom Systems. *IEEE Transactions in Human Factors in Electronics.* 1964;HFE-5:2–5. Reproduced in Moray N. Ergonomics Major Writings Volume 3: Psychological mechanisms and Models in Ergonomics. Taylor & Francis. 2004.
2. Smallman, H.S. & Cook, M.B. Proactive Supervisory Decision Support from Trend-Based Monitoring of Autonomous and Automated Systems: A Tale of Two Domains. *Lecture Notes in Computer Science.* 2013 July;8022:320–329. DOI:10.1007/978-3-642-39420-1_34
3. Cook, M.B., Smallman, H.S. & Reith, C.A. Increasing the Effective Span of Control: Advanced Graphics for Proactive Trend-Based Monitoring. *IIE Transactions on Occupational Ergonomics and Human Factors.* 2014;2:1370151.
4. Mumaw, R, Billmanm, D. & Feary, M. Analysis of Pilot Monitoring Skills and a Review of Training Effectiveness. NASA/TM-220210000047. 2020 December.
5. Hopkins, A. Disastrous Decisions. CCH Australia Ltd. 2012.
6. McLeod, R.W. Designing for Human Reliability: Human Factors Engineering for the Oil, Gas and Process Industries. Gulf Professional Publishing. 2015.
7. CIEHF. Human Factors in Barrier Management. Chartered Institute for Ergonomics & Human Factors. 2016. Available from: https://ergonomics.org.uk/resource/human-factors-in-barrier-management.html
8. CCPS/Energy Institute. Bowties for Risk Management: A Concept Book for Process Safety. Wiley. 2018.
9. NES. Barrier Management for Health and Social Care Safety: A Comprehensive Guide to Conducting Bowtie Analysis (BTA). NHS Education for Scotland. 2021.
10. McLeod, R.W. & Bowie, P. Guidance on Customizing Bowtie Analysis for Use in Healthcare. Contemporary Ergonomics. 2020.
11. McLeod, R.W. & Bowie, P. Bowtie Analysis as a Prospective Risk Assessment Technique in Primary Healthcare. Policy and Practice in Health and Safety. 2018 May;177–193. DOI:10.1080/14773996.2018.1466460
12. McLeod, R.W., Novia, M. & Nonno, L. Proactive Operator Monitoring as a Safety Barrier for a Global Service Company. SPE-190690-MS. SPE International Conference on health, Safety, Security, Environment and Social Responsibility. Abu Dhabi, UAE. 2018.
13. Balfe, N., Byrne, K. & McLeod, R.W. Applying Barrier Analysis to Prevent Overspeed Events in Rail Operations. In R. Charles & D. Golightly (Eds.), Ergonomics & Human Factors. 2021. Available from: https://publications.ergonomics.org.uk/uploads/05_29.-edited.pdf
14. McLeod, R.W., Novia, M, Nonno, L, Acton, S. & Easton, N. A Method for Assessing the Quality of Proactive Operator Monitoring (POM) as a Safety Barrier in Service Company Operations. SPE International Conference on Health, Safety, Security, Environment and Social Responsibility, Abu Dhabi, UAE. SPE-190670-MS. 2018.
15. McLeod, R.W. A Bowtie Analysis of Overspeed Events. Ron McLeod Ltd. 2020.
16. Endsley, M. & Jones, D. Designing for Situation Awareness: An Approach to User-Centred Design. 3rd Edition. Taylor & Francis. 2025.
17. Lee, J.D. & Seppelt, B.D. Human Factors in Automation Design. 2009 January. DOI: 10.1007/978-3-540-78831-7_25

12 The Proactive Operator Monitoring Assessment Tool for supervisory control (POMATsc)

To conclude Part III of the book this chapter suggests a structured approach (i.e. a "tool") that can be used to estimate the quality of implementation and support expected to be provided to a human supervisory controller who is expected to be effective and reliable proactively monitoring an automated real-time system. The assessment is based on subjective ratings and, inevitably, must make assumptions. The extent of subjectivity involved is reduced, however, by trying to ensure the individuals making the ratings draw on evidence gained through consideration of a broad range of relevant issues. The method draws on the discussion of issues and risks that can impact on the effectiveness of human supervisory control and the POM task throughout Parts II and III of the book.

The approach suggested here is based on the Proactive Operator Monitoring Assessment Tool (POMAT) tool reported in [1]. POMAT was developed following a detailed analysis of safety controls relied on by a global oilfield services company.[1] A review of the company's incident database and safety analyses had concluded that, even for operations performed largely manually, POM played a significant role in the company's defences against their personnel causing or contributing to incidents in oilfield operations. Having recognised the key role that POM played, a study was conducted to assess the feasibility and potential value of being able to predictively assess the quality and reliability of monitoring tasks carried out by their staff in the field. The aim was to develop a structured approach that could be used by experienced staff in an office environment; for many reasons it was not feasible to carry out these assessments at the worksite. The POMAT tool was the result of that work. The tool is described, together with the results of its use on eight separate monitoring tasks involved in two different oilfield service operations,[2] in [1].

The original POMAT tool was developed for application to manual tasks that were neither automated nor involved real-time control. The POMATsc tool presented here has been amended from the original tool both to capture knowledge and experience gained since the original study, as well as to focus on assessing proactive operator monitoring of automated real-time systems.

"SIMPLE LITTLE TOOLS"

Before introducing the POMATsc tool, I need to say a word about its status and role. Human Factors professionals place a lot of value on structured tools and techniques to

DOI: 10.1201/9781003679479-15

perform many different forms of analysis. Often, these provide the evidence, insight or knowledge to support design decisions or make judgements about the expected safety, usability or utility of different design solutions or proposed working practices. Some of these tools, such as NASA's Task Loading Index (TLX) for measuring mental workload [2], or the Situation Awareness Rating Technique (SART) [4] are the result of years of scientific research and development. Similarly, tools intended to analyse the risk of human error in safety-critical systems have been the subject of extensive research and development involving the collaboration of experts from around the world over many years.

Tools and technique such as these need to be applied in a rigorous and precise manner. Their use is facilitated by individuals with specific training and experience both in using the tool and in its theoretical basis. And they are usually subject to a high degree of quality assurance to ensure the results are valid and repeatable. Without that assurance, it would be difficult to have confidence that the numbers produced mean what it is claimed they do. Where such tools are used to support decisions around the safety of industrial processes, or safety-critical operations, the effort and rigour involved is justified by the seriousness of the decisions the results support.

The academic community seems to place a particularly heavy emphasis and trust in these formalised tools and methods. They are not, however, close to being the full picture of how tools and techniques are used to support Human Factors analyses in industry. There are many tools used by Human Factors professionals in industrial and commercial settings that produce great value, not by trying to measure something, but by the role they play in helping engineers, designers, customers, equipment users and other stakeholders think through complex human performance issues in a way that would not be possible without the use of some sort of structured approach.

The other side of the picture from these formalised and scientifically validated tools is what the late Harrie Rensinke[3] used to call "simple little tools". What he meant, was structured approaches to analysing Human Factors issues that can help engineering project teams recognise and understand the kind of Human Factors issues and risks involved in the projects they were conducting. While they are structured and require a competent person to facilitate their use, these "simple little tools" are not supported by years of scientific research and have not been subject to academic validation through peer-reviewed research publications. Rather, they are developed in response to a need to address specific challenges applying Human Factors in an industrial context. "Simple little tools" that can be applied quickly and efficiently and can be quickly tailored or customised to the specific needs of a situation. To be effective, "simple little tools" need to have high face validity: those providing the information and making use of the results need to clearly see the relevance of what is being done.

Over the course of my more than 40-year career, there have been many occasions when I, sometimes with colleagues, have needed to develop my own "simple little tools" to help understand a problem, or to help non-Human Factors professionals systematically think through the issues they are facing. Over the course of time, some of these have grown well beyond their original role as tools intended to be used by myself or others under my supervision and have become much more widely used.[4]

To be clear, there is no suggestion that the POMATsc tool presented here is a scientifically validated tool. Or that it has been experimentally demonstrated to produce results that can be relied on as an objective indication of how well the POM tasks is likely to be performed in any situation. It is not and does not pretend to be. Rather, it should be seen in the context of the many other "simple little tools" Human Factors professionals as well as the companies that employ or contract them have found to be useful and effective. That effectiveness usually lies in delivering insight and understanding of the issues and risks associated with human performance in complex systems. Frequently, the use of such simple tools leads to improved design solutions and working practices.

The POMATsc tool should be viewed as a framework that provokes deeper and broader thinking, discussion, and review of what is needed to support effective POM in supervising automated real-time activities than is typically the case. Having gone to the effort of thinking deeply and broadly about the issues, the tool provides a simple approach to capturing the results of the assessment numerically. The validity of the ratings produced in terms of their ability to predict actual performance with the eventual system can of course be challenged. Though generating numerical ratings, even if they do tend towards "wet finger" rather than the "hard science" end of the validity spectrum, has several benefits;

- If the assessment produces a particularly low numerical rating, this alone can draw attention to potentially significant weakness in the likelihood of a system providing effective support to the POM task;
- it allows relative comparison of how well different POM tasks are likely to be supported, or how well POM appears to be supported for different kinds of threats. Those relative comparisons can highlight potential weak spots in a system;
- it makes it possible to make judgements about the likely impact of different design solutions or different means of supporting the user in the role of supervisory controller.

There will, perhaps inevitably, always be a focus on the numbers produced by a POMATsc analysis. The real value however – as is the case with many of the different types of analyses performed by Human Factors and Safety professionals - usually lies in the questioning, insight and shared understanding that comes from the process of helping a team of experienced people to think both broadly and deeply and to share their opinions about issues associated with a new product or system.

OVERVIEW

A POMATsc assessment can be used in at least two ways;

1. To support the assessment and verification of a new design intended to automate all or part of a real-time activity by estimating how well the design and implementation plans appear to support people who will be relied on to monitor proactively.

2. As part of an assessment of the risks of human failure (or the reliability of human performance) associated with an existing product or system that automates a real-time activity.

Indeed, a POMATsc assessment could potentially contribute to a product or system's safety case, by demonstrating both the recognition that POM may have a crucial role to play in the system's defences, as well as the depth of consideration that has been given to supporting POM tasks during design and implementation of the system.

POMATsc is intended to be used in a workshop format in an office environment. There is no inherent reason not to use it at the worksite if that is logistically and practically possible. The workshop should be attended by – or have contributions from – enough stakeholders having knowledge and experience both of the product or system as well as of the activity or process being controlled and the environment it is likely to exist in. That might mean, for example;

- individuals involved in the design of the automation;
- individuals involved in the design of the user interface;
- someone responsible for ensuring the product or system is safe for use;
- representatives of the likely users of the product or system who have experience of the activity or process that will be controlled;
- representatives of the customers or organisation that will operate the new system in support of its activities.

The workshop should be facilitated by someone who understands the background and intended purpose of the tool and has the necessary technical competence. That may, though not necessarily, mean a Human Factors professional with experience of human supervisory control, or someone who has gained competence in Human Factors skills.[5] As a minimum, it means someone who understands the background to the tool, and who has the skills to facilitate and encourage all participants to contribute to the workshop, without dominating the discussion with their own opinions.

A POMATsc assessment is conducted in three stages:

Stage 1. Threat identification: identifies the key threats the human supervisory controller will be expected to detect and respond to.

Stage 2. Situating: ensure the individuals who will make ratings about the quality of support provide to POM tasks have thought about a broad scope of relevant issues before they are asked to express opinions and make judgements (in stage 3).

Stage 3. Assessment: reviews and discusses the quality of POM support against four POM quality dimensions. Captures subjective assessments of the quality of support expected to be provided for each dimension and produces a numerical PMQsc score.

Stage 1: Threat Identification

The aim of Stage 1 is to identify and understand as clearly as possible the key threats the system's human supervisory controllers will be expected to detect and respond to without relying on prompts from the system or elsewhere: that is, by monitoring proactively.

Stage 1 involves engaging with as wide a group of stakeholders, representing as many diverse views on the likely use of the system as is necessary to generate sufficient understanding of the kind of threats the system's human supervisory controllers are going to be relied on to detect. That might mean perhaps the person leading the design of the user interface to the automation and a representative of the intended user population. Or it might mean a range of different technical specialists from within the system design team, as well as representatives of sub-contractors supplying technical components, customers, insurers, and others. The extent of engagement will depend on things like the novelty of the system and the hazards and risks associated with it, as well as the capital investment and effort involved in developing it, and the range of potential customers and users of the new product or system. Of course, commercial sensitivity and perhaps security concerns associated with a new product or system may constrain access to stakeholders.

Whatever range of stakeholders it is decided to engage with, and however it is organised, Stage 1 involves inviting those stakeholders to complete the threat identification table (Table 12.1). Stakeholders complete the table with up to five examples of threats they expect the human supervisor to detect and respond to that would otherwise interfere with the performance of the automation. The table includes details to allow threats to be prioritised such that a manageable number can be taken forward for further analysis in stages 2 and 3.

Instructions that could be sent to participating stakeholders to help them complete the table are included below.[6]

The GIGO principle

The principle of "Garbage-In-Garbage-Out" (GIGO) applies to many processes that aim to transform some form of input material or information into something else: (leaving aside processes whose purpose is to clean or improve the quality of a source material) if the quality of source material is poor, the quality of the resulting output is also going to be poor. The GIGO principle seemed very important to me as a student. I vividly remember as a postgraduate research assistant, my colleagues and I being passionate about the quality of our music systems and trying to outdo each other by investing in better record players (these were the days before CD players), amplifiers and speakers.[7] Many Saturdays were spent visiting music stores and reading specialist HiFi magazines (they were also the days before the internet) to identify the best equipment we could afford with the limited finances our positions as university research assistants allowed. And the growing excitement, after buying a new needle for the record player and carefully fiddling to install it, of selecting the perfect album to be amazed at the vastly improved sound quality the new needle produced.

The GIGO principle applies to the process of conducting a POMATsc analysis as much as it does to the quality of sound produced by a music system or anything else. If the quality of information generated in Stage 1 about the nature and characteristics of the threats a supervisory controller is relied on to detect is poor, then the quality of the resulting analysis and numerical score will also be poor, perhaps meaningless. There is an onus on those responsible for instructing and organising a POMATsc

TABLE 12.1
Identification of threats (including example data from a vehicle with autonomous driving capability)[8]

Threat	Internal or External?	Function	In what circumstances might it occur	Possible frequency		Implications	Detection Difficulty	Notes
				Of the circumstances	Of the threat			
Sensors not able to detect lane markings	External	Lane Tracing Assist	Whenever the lane boundary markings are significantly degraded.	Often	Certain	Important	Not difficult	The graphic on the driver interface will show that lane boundaries are not being detected, but the LTA function will remain active.

analysis to ensure Stage 1 of the process generates the best quality information about threats it can. Similarly, anyone using or reviewing the resulting assessment and scores – such as those using the results to make changes to the product or system design, or especially regulators or those tasked with certifying a product or system for use - should seek re-assurance that the threats the analysis is based on were properly considered and are valid and realistic.

Instructions for completing the threat identification table

The term "threat" means anything that could lead to the automation being unable to provide effective control over the activity or process it is designed to control without the intervention of a human. Threats can be specific events (such as the failure of a sensor or other technical component), or situations (such as several things happening in the world around the same time).

Using Table 12.1, record examples of what you believe are up to five main threats a user will be expected to detect to support the use of the system. Think about both Internal and External threats;

- ***Internal threats*** are things that form part of the design of the system that have the potential to prevent the automation from doing what it is designed to do should they not work correctly. Examples include technical components, such as sensors or activators going out of calibration, errors in software algorithms, or data becoming corrupted.
- ***External threats*** are events, conditions or objects that exist in the external world and do not form part of the design of the automation but that can represent risk to the ability of the automation to do what it is designed to do. Examples include conditions outside the automation's design envelope or a loss of information or data the automation relies on from external sources.

For each of the threats;

a. Indicate whether you would expect the threat to affect performance of the entire system or be associated with a specific function.
b. Indicate whether it is Internal or External to the system design;
c. Give an example of the kind of circumstances in which the threat could occur;
d. Make an estimate of both how often those circumstances might exist, and, if they do occur, how likely it is that the threat will also occur;
   - Frequency: Very often; Often; Not often; Rarely.
   - Likelihood: Certain; Very likely; Quite likely; Unlikely.
e. If the threat does occur and the user does not detect or react to it properly, estimate how critical you think it could be to the ability of the automation to continue to work as designed;
   - Critical: "I don't think the automation will be able to do what it was designed to do. It would stop working. I would expect the user to be alerted and must take over".

- Not critical: "I don't think the automation will be able to do everything it was designed to do but it would not stop working. I would expect the user to be alerted and be able to decide whether to step in".
- Important: I think the automation should be able to continue, but its abilities would be degraded. I would expect the system to show an indication of its reduced ability, but the user would not need to be actively alerted. The user could decide whether to step in.

f. Give an indication of how difficult you think it might be for someone who is alert and paying attention to detect the threat;
   - Extremely difficult; Difficult; Not difficult; Easy.

Completing stage 1

Once the stakeholders have returned the completed threat identification and assessment tables, the results need to be collated. The POMATsc organiser may need to discuss suggested threats with individuals with the technical competence and knowledge of the system design and automation concept as well as relevant operational experience to advise on the credibility and importance of suggested threats.

If many potential threats have been suggested, and found to be credible, the organiser will need to take a view, probably in consultation with those responsible for the development of the system, of which and how many threats should be taken forward for detailed assessment and rating. It may be possible to group several of them together based on their nature or the cues indicating their existence and treat them as a generic type rather than individual threats.

The organiser should prepare a short summary of each selected threat, or type of threat, with sufficient detail to allow those contributing to Stages 2 and 3 to understand their nature and characteristics.

Stage 2: situating

Stage 2 involves organising a small team of relevant stakeholders to attend a workshop. The team should include individuals who, between them, have a good understanding of the automation and the system being developed, the user interface design concept, how the system is likely to be used, as well as safety issues associated with the system's use. Everyone attending the workshop should be able to understand the threat(s) selected for assessment.

The workshop itself forms Stage 3 of the POMATsc process. The purpose of Stage 2 is to ensure those contributing to the workshop have given sufficient thought to the nature of the threats, as well as the situation and the context in which the supervisory controller is expected to be able to detect them, to be able to make an informed contribution to the workshop.

The rationale for this situating stage is based on observation and experience that workshops conducted to assess safety in large complex systems are frequently conducted in a kind of vacuum that misses the reality of the situation and context people will find themselves in when working with technological products and systems under real-world pressures and constraints. This is the difference between what Professor Erik Hollnagel has described as "work-as-imagined" and "work-as-done" [5]. The purpose of the situating questions is partly to try to bring an awareness of "work-as-done" into the minds of participants at the time they make POMAT judgements in Stage 3.

In Stage 2, participants are sent copies of the situating questions contained in Table 12.2, as well as summaries of each of the threats identified as forming the basis of the POMATsc assessment (from Stage 1). Situating questions cover a range of issues that can interfere with the ability of people to monitor proactively. The questions are not definitive and are not intended to be comprehensive: if other questions are considered more important or more relevant to a specific product or system, they should be used. Participants are not expected to try to answer all the questions. Rather, they are intended as prompts to encourage thinking broadly about the nature of the monitoring tasks users of the new product or system will be relied on to perform. Participants should consider them in advance of attending the workshop in stage 3. They should come to the workshop informed and prepared to contribute to the discussion of the situating questions.

Stage 3: Assessment

The final stage in a POMATsc assessment involves a workshop attended by those individuals who took part in Stage 2. Prior to the workshop, the facilitator needs to decide whether to carry out a single assessment taking account of all the threats identified, or whether to conduct individual assessments for individual threats. Clearly, more time and effort will be needed to conduct individual assessments. But if the identified threats are very different in their nature, characteristics, and/or importance, there may be little value in carrying out a single generic assessment.

During the workshop, the facilitator leads a discussion around the monitoring task(s) faced by the human supervisory controller in proactively detecting the threat(s) selected for analysis. The discussion should be prompted by the situating questions that seem most relevant to the threat(s) or to the product or system.

Following the discussion, the workshop team use the POMATsc tool to decide on the extent to which they agree or disagree with twelve statements describing the nature of the task of proactively monitoring for each of the threats being assessed. Statements cover the dimensions of Ownership, Threat Detectability, Context and Resilience. The dimensions are summarised below and the detailed statements to be assessed are shown on Table 12.3

Specifically, the question asked for each dimension is:

To what extent do you agree or disagree with each of the statements?

Possible responses are;

- Definitely agree;
- Probable agree;
- Not sure;
- Probably disagree;
- Definitely disagree.

TABLE 12.2
Examples of situating questions to be considered by workshop participants (In the table, "HSC" means human supervisory controller; "POM" means proactive operator monitoring)

Topic	Example situating questions
Users	• Is it possible to describe the kind of people likely to fill the role of HSC of the automation in terms of characteristics, such as their job or profession? • Will the automation be used in a working context, or could it be used by members of the public? • Will the users be able to choose whether they want to use the automation, or will they have to use it as part of their job or other responsibilities? • Can users be expected to have specialist professional or technical qualifications or skills? • Are they likely to have passed through recruitment or selection tests to enable them to be in the position to act as a user of the automation? • Can the users be assumed to have completed specific training to teach them how to use the automation? Would they have to have passed some form of test to be a user? • Are there reasons to be confident that the individual expected to fill the HSC role will know that they are responsible for POM? • Are users likely to be using the automation in the presence of a supervisor, manager, or colleagues? Or are they more likely to be on their own?
Threats	• Will users filling the HSC role understand that POM means they cannot rely on reacting to alarms or other controls to protect against adverse events? • What is the balance between threats that are internal to the automation (such as limitations in the systems design envelope, or failure of technical components) as opposed to external threats (events or situations in the outside world that the automation may not be able to deal with)? • Is there any inherent prioritisation of the threats? For example, could some develop especially quickly? Do some impose more severe limits on the time the user will have to detect, assess, and respond? Are some associated with more severe consequences than others? • Are there specific events or periods of time when any of the threats are more likely to occur? Or are they equally likely to occur at any time? • Are there times or situations that would lead the user to anticipate the occurrence of particular threats, and to be especially vigilant in monitoring for them? • How quickly could things go wrong that would be detectable by the monitoring task?

(Continued)

TABLE 12.2
(Continued)

Sources	• Are the signals or information indicating the presence of threats distributed across numerous different sources? Or is all the information needed for POM available by paying attention to one or a few sources? • Are the sources of information about possible threats mainly visual? Or could a user detect signs of trouble through other senses – hearing, feeling, movement or smell? • Will the sources of threat information always be available to the user? Or could some be hidden of inaccessible at some times? • How much of the information users need to detect threats is presented directly through the automation user interface and how much needs to be detected from other sources? • To what extent is the quality of information about threats conveyed via other sources under the control of the system design? • Will the user always be confident in the information conveyed by each source? Could some sources be known to be unreliable?
Signals	• Is it clear what kinds of information would tell the user about the presence of each threat? For example, would it be possible, in principle, to create a simulation scenario to train the user to detect and identify each threat? • Is the information indicating the presence of a threat indicated directly? Or does the user have to infer the presence of a threat by interpreting or seeing relationships between different pieces of information? • How inherently salient is the information conveying signs of threats? I.e. to what extent can it be relied on to attract the users' attention without relying on any alarms or other alerting system? • What format is the information the user is expected to detect likely to be in? (e.g., text or graphics or symbology on visual displays, movement of objects in the external world, sounds, etc)? • Will it always be easy for the user to know immediately that information or signals they have detected means something could potentially go wrong? Or could there be uncertainty in the user's mind about whether something was a threat? Would they need to keep monitoring for a period to decide? • Is it likely that a user could misinterpret, confuse, or misunderstand the meaning or implications of information about a potential threat? • Is there anything that might cause the operator to doubt whether the information is accurate, or to misbelieve it?

Context	• What is the longest time someone could be expected to perform the role of supervisory controller of the system, being proactive in monitoring for signs of trouble? • Can the user be expected to know what the automation is doing most of the time? Could the automation do things that would be difficult to explain to a user? • Is it likely that the user could have simultaneous competing tasks or priorities that could interfere with their ability to pay attention to the monitoring? • Is it possible that the user could be expecting or anticipating a different threat at the time? • Is there anything that could incentivise or motivate an operator not to pay enough attention to monitoring proactively? • Is the user likely to be mentally engaged with what the system is doing? Or could lack of things to do cause them to be bored distracted or focusing their attention somewhere else? • Does the system have a lot of different modes? I.e. Could the way information is presented to the user, or the way the user interacts with the system, change depending on what mode the system is in?
A2CR	• Is the user likely to be clear about what they are responsible for? • Could their responsibilities change depending on whether they or the automation is in control? • Are they likely to be clear what authority to act they have? • Could there be any reason the user is reluctant to act when they believe they have detected a threat? Such as concerns over accountability or authority, if they will be blamed if they intervene when they didn't need to, or if they doubt their abilities?
Resilience	• Are there any threats that, if the user did not proactively detect the threat and respond, would be likely to lead directly to a serious event or system failure? • If the user did not detect or respond to signs of a threat through POM, is there anything else that would draw their attention to it before an adverse event occurred?

Ownership: Ownership considers whether users can be assumed to understand and accept their responsibility and the importance of them being proactive in monitoring for signs of possible threats. It also considers the extent to which users might be aware of the kind of events or conditions that could be a threat to the ability of the automation to continue doing what is expected of it successfully. Ownership comprises two criteria:

a. POM Responsibility;
b. Threat Awareness,

Threat Detectability: Threat detectability is concerned with the characteristics of the sources and the kind of information users will expected to use to detect the threat(s) being assessed The dimension takes account of aspects such as; the number of different sources that need to be monitored, their location and accessibility, the nature and format of the information, whether its presentation is compatible with the way the user is likely to think about it, and whether they are likely to trust the information if a potential problem is detected. Threat detectability is assessed against six criteria:

c. Sources;
d. Salience;
e. Format;
f. Cognitive compatibility;
g. Anticipation;
h. Trust.

Context: The Context dimension considers issues such as whether users could have simultaneous competing demands for their time and attention; whether there may be a resistance to intervene or act if they think they may have detected signs of potential threat but are not certain; and whether, if they do detect signs of threats, they will have the necessary time to decide what needs to be done and to act. Context is assessed using three criteria:

i. Competing demands;
j. Reluctance to act;
k. Time to act.

Resilience[9]: Resilience is concerned with whether there is likely to be sufficient redundancy and/or flexibility associated with use of the automation such that failure to detect early signs of threats through proactive operator monitoring is likely to be detected elsewhere (whether by technology or other people) with sufficient time to intervene before the automation fails. Resilience comprises one criterion:

l. Opportunities for recovery.

Table 12.3 summarises the dimensions and criteria and shows the ratings scales used to make assessments of the extent to which the workshop participants agree with

TABLE 12.3
POMATsc assessment statements: To what extent do you agree or disagree with each of the statements?

Ratings: Definitely agree; Probable agree; Not sure; Probably disagree; Definitely disagree

Dimension	Criteria	Statements	Rating
Ownership	(a) POM Responsibility	Users can be relied on to understand their responsibility to monitor proactively.	
	(b) Threat awareness	Users can be relied on to understand the nature of the threats they are expected to detect and respond to.	
Threat detectability	(c) Sources	Information about potential threats is available from no more than two sources. The sources will always be easily accessible from the user's position. Source could never be obscured, hidden or otherwise not accessible.	
	(d) Salience	Signs of potential threats will be easy to detect. They are likely to attract the user's attention even if they are doing something else.	
	(e) Format	The format of information indicating signs of potential threats will be easy for the user to understand. *(For example, the format would not be clear if the information could be difficult to see or read; if the user needed to understand the meaning of colour, symbology, or acronyms; or if it used technical terms the user might not be familiar with, etc.).*	
	(f) Cognitive compatibility	The information indicating potential threats is likely to be compatible with the way the user will think about and reason with it. *(Information would have low cognitive compatibility if, for example; the user had to draw on information from memory to understand what something meant; they needed to put two or more pieces of information together; they needed to mentally convert data from one format to another; or they needed to mentally compare a number or value against a target to be able to understand the implications).*	
	(g) Anticipation	The user will usually be able to anticipate the presence of a potential threat and to be vigilant in looking for signs of it.	
	(h) Trust	Users can always be expected to trust the sources of information they use to look for signs of potential threats. Sources will not have a history of being unreliable.	

(Continued)

TABLE 12.3
(Continued)

Dimension	Criteria	Statements	Rating
Context	(i) Competing demands	It is unlikely that users could be distracted while performing the monitoring task. *(They will not have competing demands for their time or attention and will have no incentives causing them not to attend to the system as often as they should).*	
	(j) Reluctance to act	There is no reason a user might be reluctant to intervene if they thought they had detected signs of a potential threat. *(For example, there could be a reluctance to act if the cues the HSC had noticed had a lot of uncertainty associated with them, or varied a lot, and the costs or consequences of making a mistake were believed to be high. Or if they thought they might be blamed if they intervened when they didn't need to).*	
	(k) Time to act	Users will always have sufficient time to take the necessary action if they are attentive to monitoring for signs of potential threats.	
Resilience	(l) Opportunities for recovery	If the user did not notice signs of a threat, something else – other people or other systems - would detect it in sufficient time to intervene to prevent the performance of the automation being affected.	

each statement. (Note that throughout the table, the term "user" refers to the individual filling the role of the human supervisory controller who is expected to be proactive in monitoring for signs of potential threats).

THE PROACTIVE MONITORING QUALITY SCORE (PMQsc SCORE)

The nominal ratings of the twelve statements in Table 12.3, can be converted into an overall score for the assessed quality of the task of monitoring for each threat (referred to as the "PMQsc score"). The PMQsc score can be seen as an indicator of the expected robustness, or strength, of POM as an element in the controls against each threat.

Calculating the PMQsc score is done by assigning the numerical values shown on Table 12.4 to the nominal ratings agreed for the eleven statements in the Ownership, Signals and Context dimensions and calculating the sum of the assigned values. The maximum possible score if the workshop participants "definitely agreed" with all eleven statements (statements a to k on Table 12.3) would be +22. The PMQsc score

TABLE 12.4
PMQ scores assigned to the subjective ratings

Rating	Score
Definitely agree	+2
Probably agree	+1
Not sure	0
Probably disagree	−1
Definitely disagree	−2

for the task of proactively monitoring for each of the assessed threats can then be expressed as a percentage of the maximum possible score;

$$PMQ \ = \ Sum \ (ratings \ a-k) \ / \ 22 \ * \ 100.$$

The PMQsc calculation does not include the Resilience dimension. The resilience of the system to catch failures to detect the presence of a threat is assumed to be independent of the monitoring task itself. The combination of the PMQsc score and the Resilience rating, taken together, gives an indication of the need for action: a relatively low score may be acceptable in a highly resilient situation, whereas it would not be acceptable in a situation with low resilience.

Centring the scale on zero has two implications in terms of the interpretation of the meaning of the resulting PMQsc score;

i. The sign of the score indicates whether or not, overall, the assessment team feel the design and implementation arrangements support the POM task. A negative sign indicates that, overall, the team disagree that POM is supported, while a positive sign indicates an overall judgement that POM is supported at least to some extent;
ii. The numerical size of the score indicates how strongly the team believes POM is or is not supported. A large negative score indicates a strong feeling that the task is not well supported, while a large positive score indicates the team strongly feel POM is well supported.

Weighting the dimensions

Note that all eleven of the ratings used in calculating the PMQsc score make the same contribution to the PMQsc score: they are all considered as being equally important. It could easily be argued that some of the criteria are more important than others in ensuring a high quality of POM performance. For example;

- if the individual filling the role of a human supervisory controller did not understand the importance of their being proactive in monitoring for signs of potential threats, and simply relied on reacting to alarms, that might have a bigger impact on the chances of not detecting signs of a threat than some of the other criteria. So it could be argued that criteria (a) on Table 12.3 should be treated as, perhaps, twice as important as the others: it should perhaps be given a weighting of 2, while the other ten criteria were given a weighting of 1. A workshop rating "Definitely disagree" with statement (a) would therefore contribute -4 points to the overall PMQsc score, rather than –2;

- alternately, if users were always able to anticipate situations where threats were likely and could therefore always be expected to be especially vigilant looking for signs of threats at those times, that might be thought as having a major impact on ensuring effective POM performance. In that case, criteria (g) might perhaps be given a rating of 5. So a workshop rating of "Definitely agree" with statement (g) would contribute +10 points to the overall PMQsc score, rather than +2. A big difference in the resulting PMQsc score.[10]

However, in the absence of any evidence of the relative importance of each of the criteria, and without any compelling grounds for making such relative judgements, for the purposes here all eleven criteria are scored equally with a weighted value of 1.

IN CONCLUSION

APPLYING POMATSC TO TRANSITIONING TO OWNING A SELF-DRIVING CAR

By way of illustrating the use of the POMATsc method, as well as completing this chapter, and rounding off the book, I applied the POMATsc approach to my experience as the owner of my Lexus NX450h+.[11] Although this involved using the method retrospectively, as a tool for personal reflection, rather than as a means of prospectively collecting the views of a range of stakeholders with relevant experience and knowledge, it made an interesting exercise.

Table 12.5 shows the result of my reflections applying the POMATsc method to my own experience as the supervisory controller of my new car. As I described in detail in Part I of the book, the initial days, weeks and even few months after I took ownership of the car was a period when I had to transition from being someone who had more than 40 years experience driving cars manually, to someone who, when I chose to engage the combination of driver support features that allowed the vehicle to perform autonomously and drive itself, was now required to adopt the role of the vehicle's supervisory controller.

To create the table, I reflected both on what I could remember of my experience in the first days and weeks using the car's autonomous features, prompted by what I had written in Chapters 3–5 of the book (most of which was written while the car was still new to me, and contemporaneously with the experiences described). The table shows my assessment of POMATsc ratings both in those initial days, and how I view the experience now – when I am what was described in Chapter 5 as being a "highly engaged supervisor" of the car when I choose to put it into autonomous mode.

There is no need here to go into the details behind each of the ratings shown in Table 12.5. But the PMQsc score of –14% table clearly captures the difficulties I experienced filling the role of supervisory controller in the early days: knowing when

TABLE 12.5
Example POMATsc ratings and PMQsc scores for my experience transitioning to owning a self-driving car

Dimension		Criteria	POMATsc rating	
			First day	**"Highly engaged supervisor"**
Ownership	a	POM Responsibility	Definitely agree	Definitely agree
	b	Threat awareness	Definitely disagree	Definitely agree
Threat detectability	c	Sources	Definitely disagree	Probably agree
	d	Salience	Definitely disagree	Probably disagree
	e	Format	Definitely disagree	Probably agree
	f	Cognitive Compatibility	Definitely disagree	Probably disagree
	g	Anticipation	Definitely disagree	Probably agree
Context	h	Trust	Not sure	Definitely agree
	i	Competing demands	Definitely agree	Probably agree
	j	Resistance to act	Definitely agree	Definitely agree
	k	Time to act	Definitely agree	Definitely agree
		PMQsc Score	**–14%**	**+64%**

I could trust the car's abilities and when I felt the need to take over control; what the information on the drivers' user interface meant and where to look. The table also shows how much easier the task became once I had gained experience and confidence in the car's abilities and features.

In conclusion, and to reiterate what was said at the start of the chapter, the POMATsc tool, and the PMQsc scores it produces, should be seen in the context of a large number of "simple little tools" that many Human Factors professionals as well as the companies that employ or contract them have found to be effective. The tool, and the PMQsc scores it produces, should be viewed as a framework that provides structure to provoke deeper and broader thinking, discussion, review and insight into what is needed to provide effective and reliable support to a human supervisory controller performing the POM task.

NOTES

1 This study was briefly outlined in Chapter 11.
2 For those with a knowledge of oilfield service operations, the tasks evaluated were Coiled tubing and Tubular Cutting and Milling.
3 Harrie was Shell International's first Global Discipline Lead for Human Factors Engineering. I am, as always, grateful to Harrie for his willingness to share his deep insight into applied Human Factors and his emphasis on the importance of "simple little tools" to help engineers, managers and others understand and apply the principles of Human Factors Engineering in their work.
4 An example is the HFE Screening Tool that is included in the International Oil and Gas Producer's Association (IOGP) publication 454. This tool was developed as a "simple little tool" by a small team, principally myself, Harrie Rensinke and Johan Hendrikse and was intended for internal use within Shell's capital projects. Its inclusion in the

original IOGP report in 2011 stimulated it's use by many other oil and gas companies and contractors around the world. The tool was retained in the revised and updated 2nd Edition of IOGP 454 published in 2020 [3].

5 For example, the Chartered Institute of Ergonomics & Human Factors (CIEHF) offers professional development courses for individuals who are not HF professionals, leading to accreditation as a Technical member of the Institute.

6 All the materials needed to carry out the three stages of a POMATsc assessment, including a spreadsheet to calculate PMQsc scores in Stage 3, can be downloaded from the book's website at [www.routledge.com/9781041150053].

7 I still own and regularly use the NAD 3020 amplifier and BLQ speakers I was able to buy in 1985 with an unexpected windfall courtesy of a visit to our laboratory by Dr Ken Boff from the Harry G. Armstrong Aerospace Medical Research Laboratory and his invitation to write a number of entries on the topic of human response to vibration for the Engineering Data Compendium on Human Perception and Performance they were in the process of preparing.

8 Note that this assessment, and the comment in the Notes, are based on my experience with my Lexus NX450+h as described in Part I of the book.

9 The resilience dimension is a simplification of a much more complex concept. See for example [6].

10 Although, of course, the denominator would also increase accordingly.

11 The three chapters in Part I of the book went into some detail in discussing and exploring the difficulties I experienced in the days and weeks following my purchase of a Lexus NX 450h+ which, in some circumstances can drive itself autonomously. To understand the basis of the ratings in table 12.5 it will be necessary to have read Part I of the book.

REFERENCES

1. McLeod, R.W., Novia, M, Nonno, L, Acton, S. & Easton, N. A Method for Assessing the Quality of Proactive Operator Monitoring (POM) as a Safety Barrier in Service Company Operations. SPE International Conference on Health, Safety, Security, Environment and Social Responsibility, Abu Dhabi, UAE. SPE-190670-MS. 2018.
2. Hart, S.G. & Staveland, L.E. Development of NASA-TLX (Task Load Index): Results of Empirical and Theoretical Research. *Advances in Psychology*. 1988;52:139–183.
3. IOGP 454. Human Factors Engineering in Projects. 2nd Edition. Human Factors Engineering in Projects | IOGP Publications library International Oil and Gas Producers Association. 2020 June.
4. Taylor, R.M. Situational Awareness Rating Technique (SART): The Development of a Toll for Aircrew Systems Design. In AGARD Conference Proceedings No 478, Situational Awareness in Aerospace Operations. Aerospace Medical Panel Symposium, Copenhagen, 2nd–6th October 1989.
5. Hollnagel, E. Can We Ever Imagine How Work is Done? *Hindsight*. 2017 Summer (25). Available from: https://www.eurocontrol.int/sites/default/files/publication/files/hindsight25.pdf
6. Hollnagel, E., Woods, D.D. & Leveson, N. Resilience Engineering. Ashgate. 2006.

Index

Note: **Bold** page numbers refer to tables, *italic* page numbers refer to figures and page numbers followed by "n" refer to end notes.

U

W

For Product Safety Concerns and Information please contact our EU representative GPSR@taylorandfrancis.com
Taylor & Francis Verlag GmbH, Kaufingerstraße 24, 80331 München, Germany

www.ingramcontent.com/pod-product-compliance
Lightning Source LLC
Chambersburg PA
CBHW060814220326
41598CB00022B/2613

* 9 7 8 1 0 4 1 1 5 0 0 5 3 *